FUZZY SETS AND SYSTEMS

Theory and Applications

This is Volume 144 in
MATHEMATICS IN SCIENCE AND ENGINEERING
A Series of Monographs and Textbooks
Edited by RICHARD BELLMAN, *University of Southern California*

The complete listing of books in this series is available from the Publisher
upon request.

FUZZY SETS AND SYSTEMS
Theory and Applications

Didier Dubois

Department of Automatics (DERA)
ONERA-CERT
Toulouse, France

Henri Prade

CNRS, Languages and Computer Systems (LSI)
Université Paul Sabatier
Toulouse, France

 1980

ACADEMIC PRESS

A Subsidiary of Harcourt Brace Jovanovich, Publishers

New York   London   Toronto   Sydney   San Francisco

ACADEMIC PRESS, INC.
111 Fifth Avenue, New York, New York 10003

United Kingdom Edition published by
ACADEMIC PRESS, INC. (LONDON) LTD.
24/28 Oval Road, London NW1 7DX

Library of Congress Cataloging in Publication Data

Dubois, Didier J
    Fuzzy sets and systems.

    (Mathematics in science and engineering ; v.   )
    Includes bibliographies and indexes.
    1. Set theory. 2. System analysis. I. Prade,
Henri M., joint author. II. Title. III. Series.
QA248.D74   511'.3   79−25992
ISBN 0−12−222750−6

The more, the fuzzier

Contents

FOREWORD

When I first met Henri Prade and Didier Dubois, I was impressed at once by their unusual breadth of knowledge about all facets of the theory of fuzzy sets and their youthful enthusiasm for a theory that challenges the traditional reliance on two-valued logic and classical set theory as a basis for scientific inquiry.

Later on, when they told me about their plans for writing an up-to-date research monograph on fuzzy sets and systems, I was rather skeptical that it could be done although the earlier five-volume work of Professor Arnold Kaufmann had covered the basic ground both comprehensively and with great authority.

The publication of this volume shows that my skepticism was unwarranted. Dubois and Prade have produced a comprehensible research monograph that covers almost all of the important developments in the theory of fuzzy sets and in their applications that have taken place during the past several years—developments that include their own significant contributions to fuzzy arithmetic and the analysis of fuzzy relations.

In presenting the work of others, Dubois and Prade have contributed many useful insights and supplied a number of examples which aid materially in understanding of the subject matter. Inevitably, there are some instances where one could take issue with their choice of topics, their interpretations, and their conclusions. But what is remarkable is that they have been able to cover so much ground—within the compass of a single volume—in a field that is undergoing rapid growth and spans a wide variety of applications ranging from industrial process control to medical diagnosis and group decision processes.

Like other theories that have broken away from tradition, the theory of fuzzy sets has been and will continue to be controversial for some time to come. The present volume may or may not convince the skeptics of the utility of fuzzy sets. But it will certainly be of great value to those who are interested in acquainting themselves with the basic aspects of the theory and in exploring its potentialities as a methodology for dealing with phenomena that are too complex or too ill-defined to be susceptible to analysis by conventional means.

LOTFI A. ZADEH

PREFACE

Since Lotfi A. Zadeh published his now classic paper almost fifteen years ago, fuzzy set theory has received more and more attention from researchers in a wide range of scientific areas, especially in the past few years. This theory is attractive because it is based on a very intuitive, although somewhat subtle, idea capable of generating many intellectually appealing results that provide new insights to old, often-debated questions. Opinion is still divided about the importance of fuzzy set theory. Some people have argued that many contributions were simply exercises in generalization. However, several significant and original developments have recently been proposed, which should convince those who are still reluctant. Anyway, fuzziness is not a matter of aesthetics; neither is it an ingredient to make up arid formal constructions; it is an unavoidable feature of most humanistic systems and it must be dealt with as such.

This book is intended to be a rather exhaustive research monograph on fuzzy set theory and its applications. The work is based on a large compilation of the literature* in English, French, and German. Approximately 550 publications or communications[†] are referred to; it is hoped that they are representative of about a thousand papers existing in the world. Whenever possible we have tried to cite published easy-to-find versions of works rather than rare research memoranda. Of course, some original contributions may have been missed; this is unavoidable in such a fast growing field of research.

It is not intended here to embed fuzzy set theory in a pure mathematics framework. Sophisticated formalisms, such as that of category theory, do not seem suitable in working with concepts at an early stage of their development. No high-level mathematical tool will be used in the exposition.

We do not propose that this work be used as a textbook, but only as a research compendium. As such, topics are developed unequally according

---

*Throughout this book NF stands for references to the nonfuzzy literature.
[†] Appearing between 1965 and 1978.

to our own state of knowledge and fields of interest. Hence some chapters are only modest surveys of existing works, while others may appear more original and detailed. More specifically, there are very few tutorial numerical examples and no exercises; however, some hints or ideas at their early stage of development can be found, which we hope will be of some use for further research.

This book is a structured synthesis in an attempt to unify existing works. Such an attempt is made necessary because several research directions have been investigated, often independently.

In spite of the relative lack of mathematical ambition within the work, some may find the material rather hard to read because it covers a wide range of topics within a comparatively small number of pages. Thus, this monograph is aimed at readers at the graduate level, involved in research dealing with human-centered systems.

This synthesis is organized in five parts, respectively devoted to (1) a short informal discussion on the nature of fuzziness; different kinds of uncertainty are pointed out; (2) a structured exposition in five chapters of the mathematics of fuzzy sets; (3) a description of fuzzy models and formal structures: logic, systems, languages and algorithms, and theoretical operations research; (4) a survey of system-oriented applied topics dealing with fuzzy situation; (5) a brief review of results in existing fields of applications.

DIDIER DUBOIS
HENRI PRADE

Purdue University
West Lafayette, Indiana
September 1978

ACKNOWLEDGMENTS

This book was written while the authors were supported by scholarships from the Institut de Recherche d'Informatique et d'Automatique 78150 Le Chesnay, France.

The authors wish to thank Professor Lotfi A. Zadeh for his helpful discussions and constant encouragement; Professor King-Sun Fu for welcoming them in his laboratory and for his helpful suggestions; Professor A. Kaufmann who urged them to write this book; I. G. M. Pélegrin, Professor J. F. Le Maitre, Professor J. Delmas, and Dr. G. Giralt for their early encouragement; Professor Thomas O. Binford for accepting one of them on his research team and for his broadmindedness; and Gérard Bel, Pat MacVicar-Whelan, and Lucia Vaina for their friendly and helpful discussions.

Part I

INTRODUCTION

Fuzziness is not a priori an obvious concept and demands some explanation. "Fuzziness" is what Black (NF 1937) calls "vagueness"[†] when he distinguishes it from "generality" and from "ambiguity." Generalizing refers to the application of a symbol to a multiplicity of objects in the field of reference, ambiguity to the association of a finite number of alternative meanings having the same phonetic form. But, the fuzziness of a symbol lies in the lack of well-defined boundaries of the set of objects to which this symbol applies.

More specifically, let X be a field of reference, also called a universe of discourse or universe for short, covering a definite range of objects. Consider a subset $\tilde{A}$ where transition between membership and nonmembership is gradual rather than abrupt. This "fuzzy subset" obviously has no well-defined boundaries. Fuzzy classes of objects are often encountered in real life. For instance, $\tilde{A}$ may be the set of tall men in a community X. Usually, there are members of X who are definitely tall, others who are definitely not tall, but there exist also borderline cases. Traditionally, the grade of membership 1 is assigned to the objects that completely belong to $\tilde{A}$—here the men who are definitely tall; conversely the objects that do not belong to $\tilde{A}$ at all are assigned a membership value 0. Quite naturally, the grades of membership of the borderline cases lie between 0 and 1. The

---

[†] However, it must be noticed that Zadeh (1977a) [Reference from IV.2] has used the word "vagueness" to designate the kind of uncertainty which is *both* due to fuzziness and ambiguity.

more an element or object x belongs to $\tilde{A}$, the closer to 1 is its grade of membership $\mu_{\tilde{A}}(x)$. The use of a numerical scale such as the interval $[0, 1]$ allows a convenient representation of the gradation in membership. Precise membership values do not exist by themselves, they are tendency indices that are subjectively assigned by an individual or a group. Moreover, they are context-dependent. The grades of membership reflect an "ordering" of the objects in the universe, induced by the predicate associated with $\tilde{A}$; this "ordering," when it exists, is more important than the membership values themselves. The membership assessment of objects can sometimes be made easier by the use of a similarity measure with respect to an ideal element. Note that a membership value $\mu_{\tilde{A}}(x)$ can be interpreted as the degree of compatibility of the predicate associated with $\tilde{A}$ and the object x. For concepts such as "tallness," related to a physical measurement scale, the assignment of membership values will often be less controversial than for more complex and subjective concepts such as "beauty."

The above approach, developed by Zadeh (1964), provides a tool for modeling human-centered systems (Zadeh, 1973). As a matter of fact, fuzziness seems to pervade most human perception and thinking processes. Parikh (1977) has pointed out that no nontrivial first-order-logic-like observational predicate (i.e., one pertaining to perception) can be defined on an observationally connected space;[†] the only possible observational predicates on such a space are not classical predicates but "vague" ones. Moreover, according to Zadeh (1973), one of the most important facets of human thinking is the ability to summarize information "into labels of fuzzy sets which bear an approximate relation to the primary data." Linguistic descriptions, which are usually summary descriptions of complex situations, are fuzzy in essence.

It must be noticed that fuzziness differs from imprecision. In tolerance analysis imprecision refers to lack of knowledge about the value of a parameter and is thus expressed as a crisp tolerance interval. This interval is the set of possible values of the parameters. Fuzziness occurs when the interval has no sharp boundaries, i.e., is a fuzzy set $\tilde{A}$. Then, $\mu_{\tilde{A}}(x)$ is interpreted as the degree of possibility (Zadeh, 1978) that x is the value of the parameter fuzzily restricted by $\tilde{A}$.

The word *fuzziness* has also been used by Sugeno (1977) in a radically different context. Consider an arbitrary object x of the universe X; to each nonfuzzy subset A of X is assigned a value $g_x(A) \in [0, 1]$ expressing the

---

[†] Let $\alpha > 0$. A metric space is α-connected if it cannot be split into two disjoint nonempty ordinary subsets A and B such that $\forall x \in A$, $\forall y \in B$, $d(x, y) \geqslant \alpha$, where d is a distance. A metric space is observationally connected if it is α-connected for some α smaller than the perception threshold.

"grade of fuzziness" of the statement "x belongs to A." In fact this grade of fuzziness must be understood as a grade of *certainty*: according to the mathematical definition of g, $g_x(A)$ can be interpreted as the probability, the degree of subjective belief, the possibility, that x belongs to A. Generally, g is assumed increasing in the sense of set inclusion, but not necessarily additive as in the probabilistic case. The situation modeled by Sugeno is more a matter of guessing whether $x \in A$ rather than a problem of vagueness in the sense of Zadeh. The existence of two different points of view on "fuzziness" has been pointed out by MacVicar-Whelan (1977) and Skala (Reference from III.1). The monotonicity assumption for g seems to be more consistent with human guessing than does the additivity assumption. Moreover, grades of certainty can be assigned to fuzzy subsets $\tilde{A}$ of X owing to the notion of a fuzzy integral (see II.5.A.b). For instance, seeing a piece of Indian pottery in a shop, we may try to guess whether it is genuine or counterfeit; obviously, genuineness is not a fuzzy concept. x is the Indian pottery; A is the crisp set of genuine Indian artifacts; and $g_x(A)$ expresses, for instance, a subjective belief that the pottery is indeed genuine. The situation is slightly more complicated when we try to guess whether the pottery is old: actually, the set $\tilde{A}$ of old Indian pottery is fuzzy because "old" is a vague predicate.

It will be shown in III.1 that the logic underlying fuzzy set theory is multivalent. Multivalent logic can be viewed as a calculus either on the level of credibility of propositions or on the truth values of propositions involving fuzzy predicates. In most multivalent logics there is no longer an excluded-middle law; this situation may be interpreted as either the absence of decisive belief in one of the sides of an alternative or the overlapping of antonymous fuzzy concepts (e.g., "short" and "tall").

Contrasting with multivalent logics, a fuzzy logic has been recently introduced by Bellman and Zadeh (Reference from III.1). "Fuzzy logic differs from conventional logical systems in that it aims at providing a model for approximate rather than precise reasoning." In fuzzy logic what matters is not necessarily the calculation of the absolute (pointwise) truth values of propositions; on the contrary, a fuzzy proposition induces a possibility distribution over a universe of discourse. Truth becomes a relative notion, and "true" is a fuzzy predicate in the same sense as, for instance, "tall."

As an example, consider the proposition "John is a tall man." It can be understood in several ways. First, if the universe is a set of men including John and the set of tall men is a known fuzzy set $\tilde{A}$, then the truth-value of the proposition "John is a tall man" is $\mu_{\tilde{A}}(\text{John})$. Another situation consists in guessing whether John, about whom only indirect information is available, is a tall man; the degree of certainty of the proposition is expressed by $g_{\text{John}}(\tilde{A})$. In contrast, in fuzzy logic we take the proposition "John is a tall man" as assumed, and we are interested in determining the information it conveys. "Tall" is then in a universe of heights a known fuzzy set that fuzzily restricts John's height. In other words, "John is a tall man" translates into a possibility distribution $\pi = \mu_{\text{tall}}$. Then $\mu_{\text{tall}}(h)$ gives a value to the possibility that John's height is equal to h. The possibility that John's height lies in the interval $[a, b]$ is easily calculated as

$$g_{\text{John}}([a, b]) = \sup_{a \leqslant h \leqslant b} \mu_{\text{tall}}(h),$$

as explained in II.5.B. It can also be verified, using a fuzzy integral, that $g_{\text{John}}(\text{tall}) = 1$, when "tall" is normalized (see II.1.A). This is consistent with taking the proposition "John is a tall man" as assumed.

One of the appealing features of fuzzy logic is its ability to deal with approximate causal inferences. Given an inference scheme "if P, then Q" involving fuzzy propositions, it is possible from a proposition P' that matches only approximately P, to deduce a proposition Q' approximately similar to Q, through a logical interpolation called "generalized modus ponens." Such an inference is impossible in ordinary logical systems.

APPENDIX: SOME HISTORICAL AND BIBLIOGRAPHICAL REMARKS

Fuzzy set theory was initiated by Zadeh in the early 1960s (1964; see also Bellman *et al.*, 1964). However, the term *ensemble flou* (a posteriori the French counterpart of *fuzzy set*) was coined by Menger (1951) in 1951. Menger explicitly used a "max-product" transitive fuzzy relation (see II.3.B.c.β), but with a probabilistic interpretation. On a semantic level Zadeh's theory is more closely related to Black's work on vagueness (Black, NF 1937), where "consistency profiles" (the ancestors of fuzzy membership functions) "characterize vague symbols."

Since 1965, fuzzy set theory has been considerably developed by Zadeh himself and some 300 researchers. This theory has begun to be applied in a wide range of scientific areas.

There have already been two monographs on fuzzy set theory published: a tutorial treatise in several volumes by Kaufmann (1973, 1975a, b, 1977; and others in preparation) and a mathematically oriented concise book by Negoita and Ralescu (1975). There are also two collections of papers edited by Zadeh *et al.* (1975) and Gupta *et al.* (1977).

Apart from Zadeh's excellent papers, other introductory articles are those of Gusev and Smirnova (1973), Ponsard (Reference from II.1), Ragade and Gupta (Reference from II.1), and Kandel and Byatt (1978). Rationales and discussions can also be found in Chang (1972), Ponsard

(1975), Sinaceur (1978), Gale (1975), Watanabe (1969, 1975), and Aizerman (1977).

Several bibliographies on fuzzy sets are available in the literature, namely, those of De Kerf (1975), Kandel and Davis (1976), Gaines and Kohout (1977), and Kaufmann (1979).

REFERENCES

Aizerman, M. A. (1977). Some unsolved problems in the theory of automatic control and fuzzy proofs. *IEEE Trans. Autom. Control* **22**, 116–118.

Bellman, R. E., Kalaba, R., and Zadeh, L. A. (1964). "Abstraction and Pattern Classification," RAND Memo, RM-4307-PR. (Reference also in IV.6, 1966.)

Chang, S. S. L. (1972), Fuzzy mathematics, man and his environment. *IEEE Trans. Syst., Man Cybern.* **2**, 92–93.

De Kerf, J. (1975). A bibliography on fuzzy sets. *J. Comput. Appl. Math.* **1**, 205–212.

Gaines, B. R., and Kohout, L. J. (1977). The fuzzy decade: A bibliography of fuzzy systems and closely related topics. *Int. J. Man-Mach. Stud.* **9** 1–69. (Also in Gupta *et al.*, 1977, pp. 403–490).

Gale, S. (1975). Boundaries, tolerance spaces and criteria for conflict resolution. *J. Peace Sci.* **1**, No. 2, 95–115.

Gupta, M. M., and Mamdani, E. H. (1976). Second IFAC round table on fuzzy automata and decision processes. *Automatica* **12**, 291–296.

Gupta, M. M., Saridis, G. N., and Gaines, B. R., eds. (1977). "Fuzzy Automata and Decision Processes." North-Holland Publ., Amsterdam.

Gusev, L. A., and Smirnova, I. M. (1973). Fuzzy sets: Theory and applications (a survey). *Autom. Remote Control (USSR)* No. 5, 66–85.

Kandel, A., and Byatt, W. J. (1978). Fuzzy sets, fuzzy algebra and fuzzy statistics. *Proc. IEEE* **68**, 1619–1639. (Reference from II.5.)

Kandel, A., and Davis, H. A. (1976). "The First Fuzzy Decade. (A Bibliography on Fuzzy Sets and Their Applications)," CSR-140. Comput. Sci. Dep., New Mexico Inst. Min. Technol., Socorro.

Kaufmann, A. (1973), "Introduction à la Théorie des Sous-Ensembles Flous. Vol. 1: Eléments Théoriques de Base." Masson, Paris.

Kaufmann, A. (1975a). "Introduction à la Théorie des Sous Ensembles Flous. Vol. 2: Applications à la Linguistique, à la Logique et à la Sémantique." Masson, Paris.

Kaufmann, A. (1975b). "Introduction à la Théorie des Sous-Ensembles Flous. Vol. 3: Applications à la Classification et à la Reconnaissance des Formes, aux Automates et aux Systèmes, au Choix des Critères." Masson, Paris.

Kaufmann, A. (1975c). "Introduction to the Theory of Fuzzy Subsets. Vol. 1: Fundamental Theoretical Elements." Academic Press, New York. (Engl. trans. of Kaufmann, 1973.)

Kaufmann, A. (1977). "Introduction à la Théorie des Sous-Ensembles Flous. Vol. 4: Compléments et Nouvelles Applications." Masson, Paris.

Kaufmann, A. (1980). Bibliography on fuzzy sets and their applications. *BUSEFAL* No. 1–3 (LSI Lab, Univ. Paul Sabatier, Toulouse, France).

Kaufmann, A., Dubois, T., and Cools, M. (1975). "Exercices avec Solutions sur la Théorie des Sous-Ensembles Flous." Masson, Paris.

MacVicar-Whelan, P. J. (1977). Fuzzy and multivalued logic. *Int. Symp. Multivalued Logic, 7th, N.C.* pp. 98–102. (Reference from IV.1.)

Menger, K. (1951). Ensembles flous et fonctions aléatoires. *C. R. Acad. Sci.* **232**, 2001–2003.

Negoita, C. V., and Ralescu, D. A. (1975). "Applicatior.s of Fuzzy Sets to Systems Analysis," ISR.11. Birkhaeuser, Basel.

Parikh, R. (1977). "The Problem of Vague Predicates," Res. Rep. No. 1-77. Lab. Comput. Sci., MIT, Cambridge, Massachusetts.

Ponsard, C. (1975). L'imprécision et son traitement en analyse economique. *Rev. Econ. Polit.*, No. 1, 17–37. (Reference from V, 1975a.)

Prévot, M. (1978). "Sous-Ensembles Flous—Une Approche Theorique," I.M.E. No. 14. Editions Sirey, Paris.

Sinaceur, H. (1978). "Logique et mathématique du flou," *Critique* No. 372, pp. 512–525.

Sugeno, M. (1977). Fuzzy measures and fuzzy integrals: A survey. In Gupta *et al.*, 1977, pp. 89–102. (Reference from II.5.)

Watanabe, S. (1969). Modified concepts of logic, probability and information based on generalized continuous characteristic function. *Inf. Control* **15**, 1–21.

Watanabe, S. (1975). Creative learning and propensity automata. *IEEE Trans. Syst., Man Cybern.* **5**, 603–609.

Zadeh, L. A. (1964). "Fuzzy Sets," Memo. ERL, No. 64-44. Univ. of California, Berkeley. (Reference also in II.1, 1965.)

Zadeh, L. A. (1973). Outline of a new approach to the analysis of complex systems and decision processes. *IEEE Trans. Syst. Man Cybern.* **3**, 28–44. (Reference from III.3.)

Zadeh, L. A. (1978). Fuzzy sets as a basis for a theory of possibility. *Int. J. Fuzzy Sets Syst.* **1**, No. 1, 3–28. (Reference from II.5.)

Zadeh, L. A., Fu, K. S., Tanaka, K., and Shimura, M., eds. (1975). "Fuzzy Sets and Their Applications to Cognitive and Decision Processes." Academic Press, New York.

Part **II**

MATHEMATICAL TOOLS

This part is devoted to an extensive presentation of the mathematical notions that have been introduced in the framework of fuzzy set theory.

Chapter 1 provides the basic definitions of various kinds of fuzzy sets, set-theoretic operations, and properties. Lastly, measures of fuzziness are described.

Chapter 2 introduces a very general principle of fuzzy set theory: the so-called extension principle. It allows one to "fuzzify" any domain of mathematics based on set theory. This principle is then applied to algebraic operations and is used to define set-theoretic operations for higher order fuzzy sets.

Chapter 3 develops the extensive theory of fuzzy relations.

Chapter 4 is a survey of different kinds of fuzzy functions. The extremum over a fuzzy domain and integration and differentiation of fuzzy functions of a real variable are emphasized. Fuzzy topology is also outlined. Categories of fuzzy objects are sketched.

Chapter 5 presents Sugeno's theory of fuzzy measures. In this chapter the link between such topics as probabilities, possibilities, and belief functions is pointed out.

Chapter *1*

FUZZY SETS

This chapter deals with naïve set theory when membership is no longer an all-or-nothing notion. There is no unique way to build such a theory. But, all the alternative approaches presented here include ordinary set theory as a particular case. However, Zadeh's fuzzy set theory may appear to be the most intuitive among them, although such concepts as inclusion or set equality may seem too strict in this particular framework—many relaxed versions exist as will be shown. Usually the structures embedded in fuzzy set theories are less rich than the Boolean lattice of classical set theory. Moreover, there is also some arbitrariness in the choice of the valuation set for the elements: the real interval [0, 1] is the most commonly used, but other choices are possible and even worth considering: these are summarized under the label "*L*-fuzzy sets." Fuzzy structured sets, such as fuzzy groups and convex fuzzy sets, are also presented. Lastly, a survey of scalar measures of fuzziness is provided.

A. DEFINITIONS

Let X be a classical set of objects, called the *universe*, whose generic elements are denoted x. Membership in a classical subset A of X is often viewed as a characteristic function μ_A from X to $\{0,1\}$ such that

$$\mu_A(x) = \begin{cases} 1 & \text{iff} & x \in A, \\ 0 & \text{iff} & x \notin A. \end{cases}$$

(N.B.: "iff" is short for "if and only if.")
$\{0,1\}$ is called a *valuation set*.

If the valuation set is allowed to be the real interval $[0,1]$, A is called a *fuzzy set* (Zadeh, 1965). $\mu_A(x)$ is the grade of membership of x in A. The closer the value of $\mu_A(x)$ is to 1, the more x belongs to A. Clearly, A is a subset of X that has no sharp boundary.

A is completely characterized by the set of pairs

$$A = \{(x, \mu_A(x)), x \in X\}. \tag{1}$$

A more convenient notation was proposed by Zadeh (Reference from II.2, 1972). When X is a finite set $\{x_1, \ldots, x_n\}$, a fuzzy set on X is expressed as

$$A = \mu_A(x_1)/x_1 + \cdots + \mu_A(x_n)/x_n = \sum_{i=1}^{n} \mu_A(x_i)/x_i. \tag{2}$$

When X is not finite, we write

$$A = \int_X \mu_A(x)/x. \tag{3}$$

Two fuzzy sets A and B are said to be *equal* (denoted $A = B$) iff

$$\forall x \in X, \quad \mu_A(x) = \mu_B(x).$$

Remarks 1   A fuzzy set is actually a generalized *subset* of a classical set, as pointed out by Kaufmann. However, we keep the term "fuzzy set" for the sake of convenience.

2   What we call a universe is never fuzzy.

The *support* of a fuzzy set A is the ordinary subset of X:

$$\text{supp}\, A = \{x \in X, \mu_A(x) > 0\}.$$

The elements of x such that $\mu_A(x) = \frac{1}{2}$ are the *crossover points* of A. The *height* of A is $\text{hgt}(A) = \sup_{x \in X} \mu_A(x)$, i.e., the least upper bound of $\mu_A(x)$. A is said to be *normalized* iff $\exists x \in X$, $\mu_A(x) = 1$; this definition implies $\text{hgt}(A) = 1$. The *empty set* $\emptyset$ is defined as $\forall x \in X$, $\mu_\emptyset(x) = 0$; of course, $\forall x, \mu_X(x) = 1$.

N.B.:   Elements with null membership can be omitted in Eq. (2). Using this convention, (2) can be extended to represent finite support fuzzy sets.

Examples 1 $X = \mathbb{N} = \{\text{positive integers}\}$. Let

$$A = 0.1/7 + 0.5/8 + 0.8/9 + 1.0/10 + 0.8/11 + 0.5/12 + 0.1/13.$$

A is a fuzzy set of integers approximately equal to 10.

2 $X = \mathbb{R} = \{\text{real numbers}\}$. Let

$$\mu_A(x) = \frac{1}{1 + \left[\frac{1}{5}(x - 10)\right]^2}, \quad \text{i.e.,} \quad A = \int_{\mathbb{R}} \frac{1}{1 + \left[\frac{1}{5}(x - 10)\right]^2} \Big/ x.$$

A is a fuzzy set of real numbers clustered around 10.

B. SET-THEORETIC OPERATIONS

a. Union and Intersection of Fuzzy Sets

The classical union ($\cup$) and intersection ($\cap$) of ordinary subsets of X can be extended by the following formulas, proposed by Zadeh (1965):

$$\forall x \in X, \quad \mu_{A \cup B}(x) = \max(\mu_A(x), \mu_B(x)), \tag{4}$$

$$\forall x \in X, \quad \mu_{A \cap B}(x) = \min(\mu_A(x), \mu_B(x)), \tag{5}$$

where $\mu_{A \cup B}$ and $\mu_{A \cap B}$ are respectively the membership functions of $A \cup B$ and $A \cap B$.

These formulas give the usual union and intersection when the valuation set is reduced to $\{0, 1\}$. Obviously, there are other extensions of $\cup$ and $\cap$ coinciding with the binary operators.

A justification of the choice of max and min was given by Bellman and Giertz (1973): max and min are the only operators f and g that meet the following requirements:

(i) The membership value of x in a compound fuzzy set depends on the membership value of x in the elementary fuzzy sets that form it, but not on anything else:

$$\forall x \in X, \quad \mu_{A \cup B}(x) = f(\mu_A(x), \mu_B(x,))$$

$$\mu_{A \cap B}(x) = g(\mu_A(x), \mu_B(x)).$$

(ii) f and g are commutative, associative, and mutually distributive operators.

(iii) f and g are continuous and nondecreasing with respect to each of
their arguments. Intuitively, the membership of x in $A \cup B$ or
$A \cap B$ cannot decrease when the membership of x in A or B
increases. A small increase of $\mu_A(x)$ or $\mu_B(x)$ cannot induce a strong
increase of $\mu_{A \cup B}(x)$ or $\mu_{A \cap B}(x)$.

(iv) $f(u, u)$ and $g(u, u)$ are strictly increasing. If $\mu_A(x_1) = \mu_B(x_1) > \mu_A(x_2)$
$= \mu_B(x_2)$, then the membership of x_1 in $A \cup B$ or $A \cap B$ is certainly
strictly greater than that of x_2.

(v)   Membership in $A \cap B$ requires more, and membership in $A \cup B$
less, than the membership in one of A or B:

$$\forall x \in X, \quad \mu_{A \cap B}(x) \leqslant \min(\mu_A(x), \mu_B(x)),$$

$$\mu_{A \cup B}(x) \geqslant \max(\mu_A(x), \mu_B(x)).$$

(vi)   Complete membership in A and in B implies complete membership
in $A \cap B$. Complete lack of membership in A and in B implies
complete lack of membership in $A \cup B$:

$$g(1, 1) = 1, \qquad f(0, 0) = 0.$$

The above assumptions are consistent and sufficient to ensure the
uniqueness of the choice of union and intersection operators.

Fung and Fu (1975) also found max and min to be the only possible
operators. They use a slightly different set of assumptions. They kept (i)
and added the following:

(ii') f and g are commutative, associative, and idempotent.

(iii') f and g are nondecreasing.

(vii) f and g can be recursively extended to $m \geqslant 3$ arguments.

(viii) $\forall x \in X, f(1, \mu_A(x)) = 1, g(0, \mu_A(x)) = 0$.

The interpretation of these axioms was given in the framework of group
decision-making with a slightly more general valuation set (see IV.3.C).

b.   Complement of a Fuzzy Set

The *complement* $\overline{A}$ of A is defined by the membership function (Zadeh,
1965)

$$\forall x \in X, \quad \mu_{\overline{A}}(x) = 1 - \mu_A(x). \tag{6}$$

The justification of (6) is more difficult than that of (4) and (5). Natural
conditions to impose on a complementation function h were proposed by

Bellman and Giertz (1973):

(i) $\mu_{\overline{A}}(x)$ depends only on $\mu_A(x)$: $\mu_{\overline{A}}(x) = h(\mu_A(x))$.

(ii) $h(0) = 1$ and $h(1) = 0$, to recover the usual complementation when A is an ordinary subset.

(iii) h is continuous and strictly monotonically decreasing, since membership in $\overline{A}$ should become smaller when membership in A increases.

(iv) h is involutive: $h(h(\mu_{\overline{A}}(x))) = \mu_A(x)$.

The above assumptions do not determine h uniquely, not even if we require in addition $h(\frac{1}{2}) = \frac{1}{2}$. However, $h(u) = 1 - u$ if we introduce the following fifth requirement (Gaines, Reference from III.1, 1976b):

(v) $\forall x_1 \in X$, $\forall x_2 \in X$, if $\mu_A(x_1) + \mu_A(x_2) = 1$, then $\mu_{\overline{A}}(x_1) + \mu_{\overline{A}}(x_2) = 1$.

Instead of (v), Bellman and Giertz have proposed the following very strong condition:

(vi) $\forall x_1 \in X$, $\forall x_2 \in X$, $\mu_A(x_1) - \mu_A(x_2) = \mu_{\overline{A}}(x_2) - \mu_{\overline{A}}(x_1)$, which means that a certain change in the membership value in A should have the same effect on the membership in $\overline{A}$.

(i), (ii), and (vi) entail $h(u) = 1 - u$.

However, there may be situations where (v) or (vi) may appear to be not really necessary assumptions. Sugeno (Reference from II.5, 1977) defines the λ-*complement* $\overline{A}^{\lambda}$ of A

$$\mu_{\overline{A}^{\lambda}}(x) = (1 - \mu_A(x))/(1 + \lambda\mu_A(x)), \qquad \lambda \in \,]-1, +\infty) \tag{7}$$

λ-complementation satisfies (i), (ii), (iii), and (iv).

Lowen (1978) has developed a more general approach to the complementation of a fuzzy set in the framework of category theory.

When A is an ordinary subset of X, the pair $(A, \overline{A})$ is a partition of X provided that $A \neq \varnothing$ and $A \neq X$. When A is a fuzzy set $(\neq \varnothing, \neq X)$, the pair $(A, \overline{A})$ is called a *fuzzy partition*; more generally an m-tuple $(A_1, \ldots, A_m)$ of fuzzy sets $(\forall i, A_i \neq \varnothing$ and $A_i \neq X)$ such that

$$\forall x \in X, \quad \sum_{i=1}^{m} \mu_{A_i}(x) = 1 \qquad (orthogonality) \tag{8}$$

is still called a fuzzy partition of X (Ruspini, Reference from IV.6, 1969).

c.　Extended Venn Diagram

Venn diagrams in the sense of ordinary subsets no longer exist for fuzzy sets. Nevertheless, Zadeh (1965) and Kaufmann (1975) use the graph of μ_A as a representation in order to visualize set-theoretic operators, as in Fig. 1.

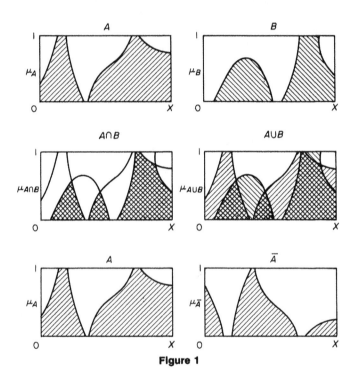

Figure 1

d.　Structure of the Set of Fuzzy Subsets of X

Let $\mathcal{P}(X)$ be the set of ordinary subsets of X. $\mathcal{P}(X)$ is a Boolean lattice for $\cup$ and $\cap$.

Let us recall some elementary definitions from lattice theory. A set L equipped with a partial ordering (reflexive and transitive relation $\leqslant$) is a *lattice* iff

$$\forall a \in L, \quad \forall b \in L, \quad \begin{cases} \exists! c \in L, & c = \inf(a,b), \\ \exists! d \in L, & d = \sup(a,b). \end{cases}$$

inf and sup mean respectively greatest lower bound and least upper bound. $\exists!$ is short for there exists one and only one.

L is *complemented* iff

$$\exists 0 \in X, \quad \exists 1 \in X, \quad \forall a \in L, \quad \exists \bar{a} \in L,$$
$$\inf(a,\bar{a}) = 0 \quad \text{and} \quad \sup(a,\bar{a}) = 1$$
$$\text{and} \quad \bar{a} \neq 0 \text{ if } a \neq 1, \quad \bar{a} \neq 1 \text{ if } a \neq 0.$$

0 and 1 are respectively the least and the greatest element of L. ($\forall a \in L$, $\inf(a,0) = 0$, $\sup(a,1) = 1$). A lattice with a 0 and a 1 is a complete lattice. L is distributive iff sup and inf are mutually distributive.

A complemented distributive lattice is said to be Boolean. In a Boolean lattice the complement $\bar{a}$ of a is unique.

The structure of $\mathcal{P}(X)$ may be viewed as induced from that of $\{0,1\}$, which is a trivial case of a Boolean lattice.

Let $\tilde{\mathcal{P}}(X)$ be the set of fuzzy subsets of X. Its structure can be induced from that of the real interval $[0, 1]$. $[0, 1]$ is a pseudocomplemented distributive lattice where max and min play the role of sup and inf, respectively. The pseudocomplementation is complementation to 1. It is not a genuine complementation. $\tilde{\mathcal{P}}(X)$, considered as the set of mappings from X to $[0,1]$, is thus also a pseudocomplemented distributive lattice. More particularly, we have the following properties for $\cup$, $\cap$, and $^-$:

(a) Commutativity: $A \cup B = B \cup A; A \cap B = B \cap A$.
(b) Associativity: $A \cup (B \cup C) = (A \cup B) \cup C, A \cap (B \cap C) = (A \cap B) \cap C$.
(c) Idempotency: $A \cup A = A, A \cap A = A$.
(d) Distributivity: $A \cup (B \cap C) = (A \cup B) \cap (A \cup C), A \cap (B \cup C) = (A \cap B) \cup (A \cap C)$.
(e) $A \cap \emptyset = \emptyset, A \cup X = X$.
(f) Identity: $A \cup \emptyset = A, A \cap X = A$.
(g) Absorption: $A \cup (A \cap B) = A, A \cap (A \cup B) = A$.
(h) De Morgan's laws: $\overline{(A \cap B)} = \bar{A} \cup \bar{B}, \overline{(A \cup B)} = \bar{A} \cap \bar{B}$.
(i) Involution: $\bar{\bar{A}} = A$.
(j) Equivalence formula: $(\bar{A} \cup B) \cap (A \cup \bar{B}) = (\bar{A} \cap \bar{B}) \cup (A \cap B)$.
(k) Symmetrical difference formula:
$(\bar{A} \cap B) \cup (A \cap \bar{B}) = (\bar{A} \cup \bar{B}) \cap (A \cup B)$.

N.B.: λ-complementation is also involutive and satisfies De Morgan's laws.

The only law of ordinary fuzzy set theory that is no longer true is the excluded-middle law:

$$A \cap \bar{A} \neq \emptyset, \quad A \cup \bar{A} \neq X.$$

The same holds for the λ-complementation.

Since the fuzzy set A has no definite boundary and neither has $\bar{A}$, it may seem natural that A and $\bar{A}$ overlap. However, the overlap is always limited, since

$$\forall A, \quad \forall x, \quad \min(\mu_A(x), \mu_{\bar{A}}(x)) \leqslant \tfrac{1}{2}.$$

For the same reason, $A \cup \bar{A}$ do not exactly cover X; however, $\forall A, \forall x$, $\max(\mu_A(x), \mu_{\bar{A}}(x)) \geqslant \tfrac{1}{2}$.

For example, if X is a set of colored objects, and A is the subset of red ones, $\mu_A(x)$ measures the degree of redness of x. A pink object has a membership value close to $\tfrac{1}{2}$, and thus belongs nearly equally to A and $\bar{A}$.

N.B.: A Zermelo–Fraenkel-like axiomatization, formulated in ordinary first-order logic with equality, was first investigated by Netto (1968), and completely developed by Chapin (1974). In this approach fuzzy sets are built ab initio, without viewing them as a superstructure of a predetermined theory of ordinary sets. The only primitive relation used in the theory is a ternary relation, interpreted as a membership relation. There are 14 axioms, some of which have a strongest version. However, as pointed out by Goguen (1974), the difficulty with such a theory is in showing that its only model is in fact the universe of fuzzy sets. Goguen, to cope with this flaw, sets forth axioms for fuzzy sets in the framework of category theory.

e. Alternative Operators on $\tilde{\mathcal{P}}(X)$

Other operators can be defined for union and intersection. First, there are the following probabilisticlike operators:

Intersection,

$$\forall x \in X, \quad \mu_{A \cdot B}(x) = \mu_A(x) \cdot \mu_B(x) \qquad (product); \qquad (9)$$

Union,

$$\forall x \in X, \quad \mu_{A \hat{+} B}(x) = \mu_A(x) + \mu_B(x) - \mu_A(x) \cdot \mu_B(x)$$

$$(probabilistic\ sum). \qquad (10)$$

Under these operators and the usual pseudocomplementation, $\tilde{\mathcal{P}}(X)$ is only a pseudocomplemented nondistributive structure. More particularly, $\hat{+}$ and $\cdot$ satisfy only commutativity, associativity, identity, De Morgan's laws, and $A \cdot \varnothing = \varnothing$, $A \hat{+} X = X$. Such operators reflect a trade-off between A and B, and are said to be *interactive*, as opposed to min and max. Using these latter operators, a modification of A (or B) does not necessarily imply an alteration of $A \cap B$ or $A \cup B$. $\cap$ and $\cup$ are said to be *noninteractive*.

Second, let $\cup$ be the *bounded sum* operator (according to Zadeh, also called *bold union* by Giles (1976)),

$$\forall x \in X, \quad \mu_{A \cup B}(x) = \min(1, \mu_A(x) + \mu_B(x)); \tag{11}$$

and let $\cap$ be the associated operator called *bold intersection*,

$$\forall x \in X, \quad \mu_{A \cap B}(x) = \max(0, \mu_A(x) + \mu_B(x) - 1). \tag{12}$$

With $\cup$, $\cap$, and the usual pseudocomplementation, $\tilde{\mathcal{P}}(X)$ is a complemented nondistributive structure. More particularly, idempotency, distributivity, and absorption are no longer valid, but commutativity, associativity, identity, De Morgan's laws, $A \cap \varnothing = \varnothing$, $A \cup X = X$, and even excluded-middle laws are satisfied. In this set theory $\overline{A}$ is the real complement of A (see Giles, 1976).

A fuzzy partition in the sense of Eq. (8) is an ordinary partition in the sense of $\cup$ and $\cap$:

$$\left(\forall x \in X, \quad \sum_{i=1}^{m} \mu_{A_i}(x) = 1 \right) \quad \text{implies} \quad \begin{cases} A_1 \cup A_2 \cup \cdots \cup A_m = X, \\ \forall i \neq j, \quad A_i \cap A_j = \varnothing. \end{cases}$$

The converse is false for $m > 2$. A partition in the sense of $\cup$ and $\cap$ is more general than a fuzzy partition.

The existence of the excluded-middle law is consistent with a situation in which an experiment is made whose result can be modeled as a fuzzy set A: $A \cap \overline{A} = \varnothing$ means that a given event cannot happen at the same time as the complementary one. Nevertheless, a complete interpretation of the operators $\cup$ and $\cap$ has not yet been provided.

Lastly, let us notice that the following properties hold. Writing

$$A \cup A \cup \cdots \cup A \quad (m \text{ times}) = \cup^m A$$

and

$$A \cap A \cap \cdots \cap A \quad (m \text{ times}) = \cap^m A,$$

$$\forall x, \quad \mu_{\cup^m A}(x) = \min(1, m\mu_A(x)), \quad \mu_{\cap^m A}(x) = \max(0, m\mu_A(x) - m + 1)$$

so that

$$\lim_{m \to \infty} \mu_{\cup^m A}(x) = 1 \quad \text{iff} \quad \mu_A(x) \neq 0,$$

$$\lim_{m \to \infty} \mu_{\cap^m A}(x) = 0 \quad \text{iff} \quad \mu_A(x) \neq 1.$$

More details on the above operators ($\cup$ and $\cap$) and the three lattice structures ($\tilde{\mathcal{P}}(X), \cup, \cap$), ($\tilde{\mathcal{P}}(X), \hat{+}, \cdot$), ($\tilde{\mathcal{P}}(X), \cup, \cap$) are provided in Section E.

N.B.: The aforementioned intersection operators $\min(a, b)$, $a \cdot b$, $\max(0, a + b - 1)$, are known to be *triangular norms*: A triangular norm T is a 2-place function from $[0, 1] \times [0, 1]$ to $[0, 1]$ that satisfies the following conditions (Schweizer & Sklar, NF 1963):

(i) $T(0,0) = 0$; $T(a,1) = T(1,a) = a$;
(ii) $T(a,b) \leqslant T(c,d)$ whenever $a \leqslant c$, $b \leqslant d$;
(iii) $T(a,b) = T(b,a)$;
(iv) $T(T(a,b),c) = T(a,T(b,c))$.

Moreover, every triangular norm satisfies the inequality

$$T_w(a,b) \leqslant T(a,b) \leqslant \min(a,b)$$

where

$$T_w(a,b) = \begin{cases} a & \text{if } b = 1 \\ b & \text{if } a = 1 \\ 0 & \text{otherwise} \end{cases}$$

The crucial importance of $\min(a,b)$, $a \cdot b$, $\max(0, a + b - 1)$ and $T_w(a,b)$ is emphasized from a mathematical point of view in Ling (NF 1965) among others.

f. More Operators

Some other operators are often used in the literature:
Bounded difference $| - |$ (Zadeh, Reference from II.3, 1975a)

$$\forall x \in X, \quad \mu_{A|-|B}(x) = \max(0, \mu_A(x) - \mu_B(x)). \tag{13}$$

$A| - |B$ is the fuzzy set of elements that belong to A more than to B. It extends the classical $A - B$.

Symmetrical differences: In the framework of fuzzy set theory there may be different ways to define a symmetrical difference. First, the fuzzy set $A \triangledown B$ of elements that belong more to A than to B or conversely is defined as

$$\forall x \in X, \quad \mu_{A \triangledown B}(x) = |\mu_A(x) - \mu_B(x)|. \tag{14}$$

$\triangledown$ is not associative.

Secondly, the fuzzy set $A \triangle B$ of the elements that approximately belong to A and not to B or conversely to B and not to A is defined as

$$\forall x \in X, \quad \mu_{A \triangle B}(x)$$
$$= \max\left[\min(\mu_A(x), 1 - \mu_B(x)), \min(1 - \mu_A(x), \mu_B(x))\right]. \tag{15}$$

It can be shown that $\triangle$ is associative; moreover,

$$A \triangle B \triangle C = (\bar{A} \cap \bar{B} \cap C) \cup (\bar{A} \cap B \cap \bar{C}) \cup (A \cap \bar{B} \cap \bar{C}) \cup (A \cap B \cap C).$$

mth power of a fuzzy set: A^m is defined as (Zadeh, Reference from II.2,

1972)

$$\mu_{A^m}(x) = \left[\mu_A(x) \right]^m \qquad \forall x \in X, \quad \forall m \in \mathbb{R}^+. \tag{16}$$

This operator will be used later to model linguistic hedges (see IV.2.B.b).

Let us notice that the mth power and the probabilistic sum of m identical fuzzy sets have the same behavior as $\cap^m$ and $\cup^m$, respectively.

Convex linear sum of min and max. A combination of fuzzy sets A and B that is intermediary between $A \cap B$ and $A \cup B$ is $A \|_\lambda B$ such that

$$\forall \lambda \in [0, 1], \quad \forall x \in X, \quad \mu_{A \|_\lambda B}(x)$$
$$= \lambda \min(\mu_A(x), \mu_B(x)) + (1 - \lambda) \max(\mu_A(x), \mu_B(x)).$$

$\|_\lambda$ is commutative and idempotent, but not associative. It is distributive on $\cap$ and $\cup$, but not on $\|_{1-\lambda}$ except when $\lambda \in \{0, \frac{1}{2}, 1\}$. Moreover, $\overline{(A \|_\lambda B)} = \bar{A} \|_{1-\lambda} \bar{B} \; \forall A, B \in \tilde{\mathcal{P}}(X)$.

Other formulas for intersection were suggested by Zimmermann (Reference from IV.1) after experimental studies: the arithmetic mean and geometric mean of membership values. (See also Rödder, Reference from IV.1.) The former does not yield an intersection for classical sets.

C. α-Cuts

When we want to exhibit an element $x \in X$ that typically belongs to a fuzzy set A, we may demand its membership value to be greater than some threshold $\alpha \in]0, 1]$. The ordinary set of such elements is the *α-cut* A_α of A, $A_\alpha = \{x \in X, \mu_A(x) \geqslant \alpha\}$. One also defines the *strong α-cut* $A_{\bar{\alpha}} = \{x \in X, \mu_A(x) > \alpha\}$.

The membership function of a fuzzy set A can be expressed in terms of the characteristic functions of its α-cuts according to the formula

$$\mu_A(x) = \sup_{\alpha \in]0, 1]} \min(\alpha, \mu_{A_\alpha}(x)),$$

where

$$\mu_{A_\alpha}(x) = \begin{cases} 1 & \text{iff} \quad x \in A_\alpha, \\ 0 & \text{otherwise.} \end{cases}$$

It is easily checked that the following properties hold:

$$(A \cup B)_\alpha = A_\alpha \cup B_\alpha, \qquad (A \cap B)_\alpha = A_\alpha \cap B_\alpha.$$

However, $(\bar{A})_{\bar{\alpha}} = \overline{(A_{1-\alpha})} \neq \overline{(A_\alpha)}$ if $\alpha \neq \frac{1}{2}$ ($\alpha \neq 1$). This result stems from the fact that generally there are elements that belong neither to A_α nor to $(\bar{A})_\alpha$ $(A_\alpha \cup (\bar{A})_\alpha \neq X)$.

Radecki (1977) has defined *level fuzzy sets* of a fuzzy set A as the fuzzy sets $\tilde{A}_\alpha$, $\alpha \in]0, 1[$, such that

$$\tilde{A}_\alpha = \{(x, \mu_A(x)), x \in A_\alpha\}.$$

The rationale behind this definition is the fact that in practical applications it is sufficient to consider fuzzy sets defined in only one part of their support—the most significant part—in order to save computing time and computer memory storage. Radecki has developed an algebra of level fuzzy sets. However, $(\widetilde{\overline{A}})_\alpha$, the approximation of $\overline{A}$, cannot be obtained from $\tilde{A}_\alpha$ $((\widetilde{\overline{A}_\alpha}) \neq (\widetilde{\overline{A}})_\alpha)$, which creates some difficulties.

N.B.:   In the literature, α-cuts are also called *level sets*.

D.   CARDINALITY OF A FUZZY SET

a.   Scalar Cardinality

When X is a finite set, the *cardinality* $|A|$ of a fuzzy set A on X is defined as

$$|A| = \sum_{x \in X} \mu_A(x).$$

$|A|$ is sometimes called the *power* of A (see De Luca and Termini, 1972b). $\|A\| = |A|/|X|$ is the *relative cardinality*. It can be interpreted as the proportion of elements of X that are in A.

When X is not finite, $|A|$ does not always exist. However, if A has a finite support, $|A| = \sum_{x \in \text{supp} A} \mu_A(x)$. Otherwise, if X is a measurable set and P is a measure on X $(\int_X dP(x) = 1)$, $\|A\|$ can be the weighted sum $\int_X \mu_A(x) dP(x)$. The introduction of the weight function P looks like a "fuzzification" of the universe X. This can be done more directly by choosing a fuzzy set $\tilde{X}$ on X as the most significant part of the universe. $\tilde{X}$ is assumed to have finite support or finite power. The relative cardinality of A will then be $\|A \cap \tilde{X}\|$.

b.   Fuzzy Cardinality of a Fuzzy Set

Strictly speaking the cardinality of a fuzzy set should be a "fuzzy number." When A has finite support, its fuzzy cardinality is (Zadeh, Reference from III.1, 1977a)

$$|A|_f = \sum \alpha / |A_\alpha| = \{(n, \alpha), n \in \mathbb{N} \quad \text{and} \quad \alpha = \sup\{\lambda, |A_\lambda| = n\}\},$$

where A_α denotes the α-cut of A.

E. INCLUSIONS AND EQUALITIES OF FUZZY SETS

a. Inclusion in the Sense of Zadeh   (1965)

A is said to be included in $B(A \subseteq B)$ iff

$$\forall x \in X, \quad \mu_A(x) \leqslant \mu_B(x). \tag{17}$$

When the inequality is strict, the inclusion is said to be strict and is denoted $A \subset B$. $\subseteq$ and $\subset$ are transitive. $\subseteq$ is an order relation on $\tilde{\mathscr{P}}(X)$; however, it is not a linear ordering. Obviously, $A = B$ iff $A \subseteq B$ and $B \subseteq A$.

b. Examples. Comparison of Operators

It is easy to check that

$$\forall A, B \in \tilde{\mathscr{P}}(X), \quad \begin{cases} A \cap B \subseteq A \cdot B \subseteq A \cap B, \\ A \cup B \subseteq A \hat{+} B \subseteq A \cup B. \end{cases}$$

See Fig. 2, where $\mu_A(x) = a$, $\mu_B(x) = u$.

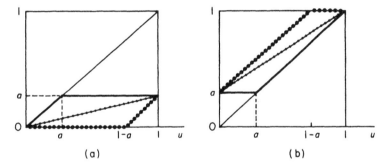

(a)   (b)

Figure 2   (a) ●●●● $A \cap B : \max(0, a + u - 1)$ ⁃⁃⁃⁃ $A \cdot B : a \cdot u$ ⸺ $A \cap B : \min(a, u)$. (b) ⸺ $A \cup B : \max(a, u)$ ⁃⁃⁃⁃ $A \hat{+} B : a + u - au$ ●●●● $A \cup B : \min(1, a + u)$.

It is patent from Fig. 2 that the probabilistic operators $(+, \cdot)$ are a median between $(\cup, \cap)$ and $(\cup, \cap)$. The respective algebraic structures support this evidence. Moreover, the operator $\cap$ is sensitive to only significant overlapping of membership functions.

Convex combination of fuzzy sets (Zadeh, 1965):   Let A, B, and Λ be arbitrary fuzzy sets on X. The *convex combination* of A, B, and Λ is denoted by $(A, B; \Lambda)$. It is such that

$$\forall x \in X, \quad \mu_{(A, B; \Lambda)}(x) = \mu_\Lambda(x) \mu_A(x) + (1 - \mu_\Lambda(x)) \mu_B(x).$$

A basic property of the convex combination is

$$\forall \Lambda, \quad A \cap B \subseteq (A, B; \Lambda) \subseteq A \cup B.$$

Conversely, $\forall C$ such that $A \cap B \subseteq C \subseteq A \cup B$, $\exists \Lambda \in \tilde{\mathcal{P}}(X)$, $C = (A, B; \Lambda)$. The membership function of Λ is given by

$$\mu_\Lambda(x) = (\mu_C(x) - \mu_B(x))/(\mu_A(x) - \mu_B(x)).$$

c. Other Inclusions and Equalities

Zadeh's definitions of inclusion and equality may appear very strict, especially because precise membership values are by essence out of reach.

α. Weak Inclusion and Equality

A first way to relax fuzzy-set inclusion is given by the definitions:
x α-belongs to A iff $x \in A_\alpha$;
A is *weakly included* in *B*, denoted $A \prec_\alpha B$, as soon as all the elements of *X* α-belong to $\bar{A}$ or to *B*; mathematically,

$$A \prec_\alpha B \quad \text{iff} \quad x \in (\bar{A} \cup B)_\alpha \qquad \forall x \in X, \tag{18}$$

which is equivalent to

$$\forall x \in X, \quad \max(1 - \mu_A(x), \mu_B(x)) \geq \alpha.$$

Practically, $A \prec_\alpha B$ is not true as soon as

$$\exists x \in X, \quad \mu_A(x) > 1 - \alpha \quad \text{and} \quad \mu_B(x) < \alpha.$$

As such $\prec_\alpha$ is transitive only for $\alpha > \frac{1}{2}$. Transitivity for $\alpha = \frac{1}{2}$ can be recovered by slightly modifying the above condition and stating

$$A \prec_{\frac{1}{2}} B \quad \text{iff} \quad \forall x \in X, \quad \mu_A(x) \leq \frac{1}{2} \quad \text{or} \quad \mu_B(x) > \frac{1}{2}. \tag{19}$$

We may want to impose the condition that Zadeh's inclusion ($\subseteq$) be a particular case of $\prec_\alpha$, i.e.,

$$\text{if} \quad A \subseteq B, \quad \text{then} \quad A \prec_\alpha B.$$

This holds only for $\alpha \leq \frac{1}{2}$. Hence, the only transitive $\prec_\alpha$ consistent with $\subseteq$ is $\prec_{\frac{1}{2}}$ (abbreviated $\prec$), provided that we adopt the above slight modification.[†]

N.B.: If $\alpha > \frac{1}{2}$, Zadeh's inclusion does not imply $\prec_\alpha$ because the elements $x \in X$ such that $1 - \alpha < \mu_A(x) \leq \mu_B(x) < \alpha$ never belong to $(\bar{A} \cup B)_\alpha$ (see (18)).

The set equality $\succ\!\!\prec$ associated with $\prec$ is defined as $A \succ\!\!\prec B$ iff $A \prec B$ and $B \prec A$, i.e.,

$$A \succ\!\!\prec B \quad \text{iff} \quad \forall x \in X,$$
$$\min\left[\max(1 - \mu_A(x), \mu_B(x)), \max(\mu_A(x), 1 - \mu_B(x))\right] \geq \frac{1}{2}.$$

[†] After modification, $\prec_{\frac{1}{2}}$ is still consistent with $\subseteq$.

which is equivalent to

$$\forall x \in X, \quad \max\left[\min(\mu_A(x), \mu_B(x)), \min(1 - \mu_A(x), 1 - \mu_B(x))\right] \geq \tfrac{1}{2}.$$

The *weak equality* $A \bowtie B$ is thus interpreted as follows. Both membership values $\mu_A(x)$ and $\mu_B(x)$ are either greater than or equal to $\tfrac{1}{2}$ or both smaller than or equal to $\tfrac{1}{2}$. This weak equality is not transitive. Lack of transitivity does not contradict our intuition concerning weak inclusion or equality. However, to recover the transitivity of $\bowtie$, we could use (19) to define equality.

Lastly, $\bowtie$ is related to the symmetrical difference $\triangle$ through

$$A \bowtie B \qquad \text{iff} \qquad \forall x \in X, \quad \mu_{A \triangle B}(x) < \tfrac{1}{2}.$$

Similarly, the other symmetrical difference $\triangledown$ is related to Zadeh's set equality ($=$):

$$A = B \qquad \text{iff} \qquad A \triangledown B = \emptyset.$$

β. ϵ-Inclusions and ϵ-Equalities

Another way of defining less strong equalities or inclusions is to use some scalar measures S of similarity or "inclusion grades" I between two fuzzy sets A and B. A threshold ϵ is chosen such that

$$A \subset_\epsilon B \quad \text{iff} \quad I(A,B) \geq \epsilon, \qquad A =_\epsilon B \quad \text{iff} \quad S(A,B) \geq \epsilon.$$

$\subset_\epsilon$ and $=_\epsilon$ denote respectively ϵ-inclusion and ϵ-equality. According to the definitions of I and S, $\subset_1$ and $=_1$ may coincide with $\subseteq$ and $=$, respectively. We must state at least the following conditions. If $A \subseteq B$, then $A \subset_1 B$; if $A = B$, then $A =_1 B$. Moreover, S must be symmetrical.

Inclusion grades and similarity measures are very numerous in the literature. An informal presentation of such indices follows; X is supposed finite.

Inclusion grades
Based on intersection and cardinality:

$$I_1(A,B) = \|A \cap B\| / \|A\|$$

(Sanchez, Reference from II.3, 1977c). When $A \subseteq B$, $I_1(A,B) = 1$.
Based on inclusion and cardinality:

$$I_2(A,B) = \left\| \overline{(A_{|-|}B)} \right\| = \|\overline{A} \cup B\| \qquad \text{(Zadeh's inclusion)};$$

(Goguen, Reference from III.1) when $A \subseteq B$, $I_2(A,B) = 1$.

$$I_3(A,B) = \|\overline{A} \cup B\| \text{ (weak inclusion)};$$

when $A \prec B$, $I_3(A,B) \geq \tfrac{1}{2}$.

Based on inclusion:

$$I_4(A, B) = \inf_{x \in X} \mu_{\overline{A| - |B}}(x) = \inf_{x \in X} \mu_{\overline{A} \cup B}(x);$$

when $A \subseteq B$, $I_4(A, B) = 1$.

$$I_5(A, B) = \inf_{x \in x} \mu_{\overline{A} \cup B}(x);$$

when $A \prec B$, $I_5(A, B) \geqslant \frac{1}{2}$.

Similarity measures
Based on intersection, union and cardinality:

$$S_1(A, B) = \|A \cap B\| / \|A \cup B\|;$$

when $A = B$, then $S_1(A, B) = 1$.
Based on equality and cardinality:

$$S_2(A, B) = 1 - \|A \bigtriangledown B\| = \|\overline{A \bigtriangledown B}\|;$$

$A = B$ iff $S_2(A, B) = 1$.

$$S_3(A, B) = 1 - \|A \bigtriangleup B\| = \|\overline{A \bigtriangleup B}\|;$$

if $A \succ\!\!\prec B$, then $S_3(A, B) \geqslant 1/2$.

N.B.: $1 - S_2(A, B)$ is the *relative Hamming distance* between A and B (Kaufmann, 1975). Kacprzyk (Reference from V) employed a slightly different version of this distance, i.e., $\sum_{x \in X} |\mu_A(x) - \mu_B(x)|^2$.

Based on equality:

$$S_4(A, B) = 1 - \sup_{x \in X} \mu_{A \bigtriangledown B}(x) = \inf_{x \in X} \mu_{\overline{A \bigtriangledown B}}(x);$$

$A = B$ iff $S_4(A, B) = 1$.

$$S_5(A, B) = 1 - \sup_{x \in X} \mu_{A \bigtriangleup B}(x) = \inf_{x \in X} \mu_{\overline{A \bigtriangleup B}}(x);$$

$A \succ\!\!\prec B$ iff $S_5(A, B) \geqslant 1/2$.

N.B.: $1 - S_4(A, B)$ is a distance between A and B which was used by Nowakowska (Reference from IV.1) and Wenstøp (Reference from IV.2, 1976a).

It is interesting to notice that

$$S_i(A, B) \leqslant \min(I_i(A, B), I_i(B, A)) = S_i'(A, B) \qquad \text{for } i = 1, 2, 3;$$

$$S_i(A, B) = \min(I_i(A, B), I_i(B, A)) \qquad \text{for } i = 4, 5.$$

Consistency-like indices:
Consistency (Zadeh):

$$C(A, B) = \sup_{x \in X} \mu_{A \cap B}(x)$$

$C(A, B) = 0$ means that A and B are separated. Indeed, $1 - C(A, B)$ is often used as a *separation index* between fuzzy sets. $C(A, B) = 1$ means that it is possible to exhibit an element $x \in X$ (finite) which totally belongs to A and B.

Other indices:

Note that $C(A, B) = 1 - I_5(A, \overline{B})$.

Similarly,

$$1 - I_4(A, \overline{B}) = \sup_{x \in X} \mu_{A \cap B}(x);$$

$$1 - I_2(A, \overline{B}) = \|A \cap B\|;$$

$$1 - I_3(A, \overline{B}) = \|A \cap B\|.$$

hgt$(A \cap B)$ behaves as a consistency. When $\|A \cap B\| = 0$, A and B are separated; but if $\|A \cap B\| = 1$, then $A = B = X$. The same holds for $\|A \cap B\|$.

γ. *Remark: Representation of a Fuzzy Set Using a Universe of Fuzzy Sets*

Willaeys and Malvache (1976) employed consistency to describe a fuzzy set A in terms of a given finite family $R_1, \ldots, R_p$ of fuzzy sets. A is characterized by hgt$(A \cap R_i)$, $i = 1, p$. They proved that the information that was lost in the representation was the "least significant." This representation was adopted in order to save computer memory storage. To achieve such a representation, it is clear that indices other than consistency may be tried.

N.B.:   In this way any element x of X may be viewed as a fuzzy set $\tilde{x}$ on $\{ R_i, i = 1, p \} : \tilde{x} = \sum_{i=1}^{p} \mu_{R_i}(x) / R_i$.

F.   CONVEX FUZZY SETS AND FUZZY STRUCTURED SETS

a.   Convex Fuzzy Sets

The notion of convexity can be generalized to fuzzy sets of a universe X, which we shall assume to be a real Euclidean N-dimensional space (Zadeh, 1965).

A fuzzy set A is *convex* iff its α-cuts are convex. An equivalent definition of convexity is: A is convex iff

$$\forall x_1 \in X, \quad \forall x_2 \in X, \quad \forall \lambda \in [0, 1],$$
$$\mu_A(\lambda x_1 + (1 - \lambda)x_2) \geqslant \min(\mu_A(x_1), \mu_A(x_2)). \tag{20}$$

Note that this definition does not imply that μ_A is a convex function of x

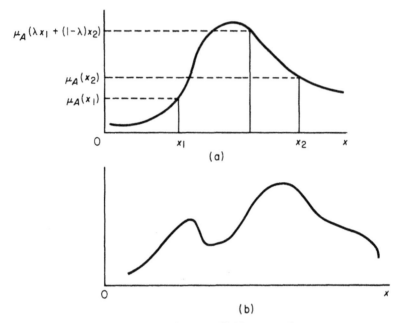

Figure 3 (a) Convex fuzzy set. (b) Nonconvex fuzzy set.

(see Fig. 3). If A and B are convex, so is $A \cap B$. An element x of X can also be written $(x^1, x^2, \ldots, x^N)$ since X has N dimensions. The *projection* (*shadow*) of A (Zadeh, 1965) on the hyperplane $H = \{x, x^i = 0\}$ is defined to be a fuzzy set $P_H(A)$ such that

$$\mu_{P_H(A)}(x^1, \ldots x^{i-1}, x^{i+1}, \ldots, x^N) = \sup_{x^i} \mu_A(x^1, \ldots, x^N).$$

When A is a convex fuzzy set, so is $P_H(A)$. Moreover, if A and B are convex and if $\forall H, P_H(A) = P_H(B)$, then $A = B$.

N.B.: *Definition*: A *fuzzy number* is a convex normalized fuzzy set A of the real line $\mathbb{R}$ such that

(a) $\exists! x_0 \in \mathbb{R}, \mu_A(x_0) = 1$ (x_0 is called the *mean value* of A);
(b) μ_A is piecewise continuous.

N.B.: Gitman and Levine (Reference from IV.6) defined *symmetric* and *unimodal* fuzzy sets as follows: Let X be equipped with a metric d, and let Γ_{x_i} be the $\mu_A(x_i)$-cut of a fuzzy set A. A is said to be unimodal iff Γ_{x_i} is connected $\forall x_i$. If A is convex, A is unimodal. Let x_0 be the unique element of X such that $\mu_A(x_0) = \sup_x \mu_A(x)$, and $\Gamma_{x_{id}} = \{x, d(x_0, x) \leqslant d(x_0, x_i)\}$. A is symmetric iff $\Gamma_{x_i} = \Gamma_{x_{id}}, \forall x_i \in X$.

b. Fuzzy Structured Sets

Fuzzy sets can be equipped with algebraic structures. Let $*$ be a composition law on X. A fuzzy set A is *closed* under $*$ iff (Rosenfeld, 1971)

$$\forall x_1 \in X, \quad \forall x_2 \in X_1 \quad \mu_A(x_1 * x_2) \geqslant \min(\mu_A(x_1), \mu_A(x_2)) \tag{21}$$

If $(X, *)$ is a group, a fuzzy subgroup A of X satisfies the above inequality and the equality $\mu_A(x^{-1}) = \mu_A(x)$, where $x^{-1}x = e$ and e is the identity.

If X is a real Euclidean space and $x_1 * x_2 = \lambda x_1 + (1 - \lambda)x_2$, $\lambda \in [0, 1]$, we see that a convex fuzzy set is a particular case of a fuzzy structured set.

Other fuzzy structured sets, such as fuzzy ideals (Rosenfeld, 1971) or fuzzy modules (Negoita and Ralescu, 1975b), have already been defined.

G. *L*-FUZZY SETS

a. Definitions

Let L be a set. An *L-fuzzy set* A is associated with a function μ_A from the universe X to L (Goguen, 1967). If L has a given structure, such as lattice or group structure, $\mathscr{P}_L(X)$, the set of L-fuzzy sets on X, will have this structure too. Several structures are worth considering.

First, let L be a lattice. The intersection and the union of L-fuzzy sets can be induced in the following way:

$$\forall x \in X, \quad \mu_{A \cap B}(x) = \inf(\mu_A(x), \mu_B(x)), \tag{22}$$

$$\forall x \in X, \quad \mu_{A \cup B}(x) = \sup(\mu_A(x), \mu_B(x)), \tag{23}$$

where inf and sup denote respectively the greatest lower bound and the least upper bound. Note that membership values of L-fuzzy sets cannot always be compared unless L is linearly ordered. Moreover, distributivity and complementation require a richer structure to be defined.

A *Brouwerian lattice* is a lattice L such that $\forall a \in L$, $\forall b \in L$, $\{x \in L, \inf(a, x) \leqslant b\}$ has a least upper bound, denoted $a \, \alpha \, b$. $a \, \alpha \, b$ is a relative pseudocomplement of a with respect to b. For example, a linearly ordered set having a greatest element $\mathbb{1}$ is a Brouwerian lattice.

A dual Brouwerian lattice is a lattice L such that $\forall a \in L$, $\forall b \in L$, $\{x \in L, \sup(a, x) \geqslant b\}$ has a greatest lower bound, denoted $a \, \epsilon \, b$. For instance, $[0, 1]$ is a complete Brouwerian and dual Brouwerian lattice:

$$a \, \alpha \, b = \begin{cases} 1 & \text{if } a \leqslant b \\ b & \text{if } b < a \end{cases}; \qquad a \, \epsilon \, b = \begin{cases} b & \text{if } a < b \\ 0 & \text{if } b \leqslant a. \end{cases}$$

The following theorem relates distributivity and Brouwerian lattices (Birkhoff, NF 1948):

A complete lattice is Brouwerian iff inf is totally distributive over sup, i.e.,

$$\forall I \subseteq L, \quad \forall a \in L, \quad \inf\left(a, \sup_{b \in I} b\right) = \sup_{b \in I} \inf(a, b).$$

A complete lattice is dual Brouwerian iff sup is totally distributive over inf. Thus, if L has such properties, $\cap$ and $\cup$ are mutually distributive (De Luca and Termini, 1972a). Moreover, a complete lattice L that is both Brouwerian and dual Brouwerian is Boolean iff $\forall a \in L$, $a \, \alpha \, 0 = a \, \epsilon \, 1$. This property does not hold in $L = [0, 1]$.

In a Boolean lattice, $a \, \alpha \, b = \sup(\bar{a}, b)$, where $\bar{a}$ is the complement of a. Brown (1971) studied L-fuzzy sets when L is a Boolean lattice. The complement of an L-fuzzy set A is then the $\bar{A}$ such that $\mu_{\bar{A}}(x)$ is the complement of $\mu_A(x) \, \forall x$. Brown also gives some results about the convexity and the connectivity of L-fuzzy sets.

Negoita and Ralescu (1975b) considered other kinds of structure for L, for instance, semigroup and semiring structures.

b. Interpretation

There may occur some situations for which valuation sets different from $[0, 1]$ are worth considering (De Luca and Termini, 1974).

For instance, if m ordinary fuzzy sets A_i $(i = 1, m)$ in X correspond to m properties, it is possible to associate with each $x \in X$ the vector of membership values $[\mu_{A_i}(x)]$ that represent the degree with which x satisfies the properties. A function from X to the set $L = [0, 1]^m$ has been built. L is a complete lattice that is not a linear ordering.

Now assume that each element x of X is described by means of only one property among $A_1, \ldots, A_m$, supposedly the most significant one for x. The property that best describes an element $x' \neq x$ may be different from that which describes x. We obtain in this way a partition of X into m classes. Obviously, it is meaningless to compare membership values of elements in different classes. Thus, the valuation set is here a collection of m disjoint linear orderings.

c. Flou Sets

An *m-flou set* is an m-tuple $A = (E_1, \ldots, E_m)$ of ordinary subsets of X such that

$$E_1 \subseteq \cdots \subseteq E_m. \tag{24}$$

Operators on flou sets are defined, with $A = (E_1, \ldots, E_m)$, $B = (F_1, \ldots, F_m)$, as follows:

union:

$$A \cup B = (E_1 \cup F_1, \ldots, E_m \cup F_m);$$

intersection:

$$A \cap B = (E_1 \cap F_1, \ldots, E_m \cap F_m);$$

complementation:

$$\overline{A} = (\overline{E_m}, \ldots, \overline{E_1});$$

inclusion:

$$A \subseteq B \quad \text{iff} \quad E_i \subseteq F_i, \qquad i = 1, m.$$

It is easy to check that the set $\mathcal{F}_m(X)$ of m-flou sets is a pseudocomplemented distributive lattice. $\mathcal{F}_m(X)$ has the same structural properties as $\tilde{\mathcal{P}}(X)$ (see B.d). Generally, $A \cup \overline{A} \neq (X, \ldots, X)$ and $A \cap \overline{A} \neq (\emptyset, \ldots, \emptyset)$.

The concept of flou set was introduced by Gentilhomme (1968). For $m = 2$, an m-flou set may be interpreted as follows: E_1 is the set of the "central" elements in A, and $E_2 - E_1$, the set of "peripheral" ones. The elements of E_1 are considered to belong more to A than the elements of $E_2 - E_1$. m-flou sets are particular cases of L-fuzzy sets where L is the finite linearly ordered set of $m + 1$ elements $(\alpha_0, \alpha_1, \ldots, \alpha_m)$ with $\alpha_i = i/m$; there is a structural isomorphism f between the set $\mathcal{P}_{Lm}(X)$ of these L-fuzzy sets and $\mathcal{F}_m(X)$,

$$\mathcal{P}_{Lm}(X) \to \mathcal{F}_m(X)$$
$$\tilde{A} \mapsto f(\tilde{A}) = (A_{\alpha_m}, \ldots, A_{\alpha_1}) = A$$

where A_{α_i} is the α_i-cut of $\tilde{A}$. For instance,

$$f(\overline{\tilde{A}}) = ((\overline{A})_{\alpha_m}, \ldots, (\overline{A})_{\alpha_1}) = (\overline{(A_{\alpha_1})}, \ldots, \overline{(A_{\alpha_m})}) = \overline{A},$$

since $(\overline{A})_{\alpha_i} = (\overline{A})_{\overline{\alpha_{i-1}}} = \overline{(A_{1-\alpha_{i-1}})}$ and $1 - \alpha_i = \alpha_{m-i}$.

N.B.: $A_{\overline{\alpha_i}}$ denotes the strong α_i-cut of A. Q.E.D.

We also have $f(\tilde{A} \cap \tilde{B}) = f(\tilde{A}) \cap f(\tilde{B})$; $f(\tilde{A} \cup \tilde{B}) = f(\tilde{A}) \cup f(\tilde{B})$.

More general kinds of flou sets are studied by Negoita and Ralescu (1975b).

Since there is in fact no sharp boundary between the sets of central and peripheral elements, we may define more general flou sets as m-tuples of ordinary fuzzy sets that satisfy (24), i.e., as fuzzy m-flou sets.

d. Type m Fuzzy Sets

Type m fuzzy sets are defined recursively as follows:

a type 1 fuzzy set is an ordinary fuzzy set in X;

a type m fuzzy set $(m > 1)$ in X is an L-fuzzy set whose membership values are type $m - 1$ fuzzy sets on $[0, 1]$.

Let $\tilde{\mathcal{P}}_m(X)$ be the set of type m fuzzy sets in X. $\tilde{\mathcal{P}}_1(X) = \tilde{\mathcal{P}}(X)$. This notion was introduced by Zadeh (Reference from IV.2, 1971).

Union, intersection, and complementation of type m fuzzy sets can also be recursively defined by induction from the structure of the valuation set. Let us denote these operators by $\cup_m$, $\cap_m$, $^{-m}$, for instance,

$$\cup_1 = \cup; \qquad \mu_{A \cup_m B}(x) = \mu_A(x) \cup_{m-1} \mu_B(x), \quad m > 1.$$

Type 2 fuzzy sets are the most easily interpreted and thus seem to be the most useful. Mizumoto and Tanaka (1976) were the first to study them. Type 2 fuzzy sets are fuzzy sets whose grades of membership are themselves fuzzy. They are intuitively appealing because grades of membership can never be obtained precisely in practical situations.

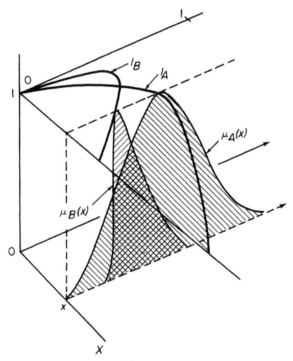

Figure 4

In Fig. 4 a representation of two fuzzy sets of type 2 is given, where $\mu_A(x)$ is assumed to be a fuzzy number; $\forall x$, l_A is the set of the maxima of $\mu_A(x)$ when x ranges over X.

Although $\cup_2$, $\cap_2$, $^{-2}$ are canonical operators, it can easily be shown that they are inconsistent with our intuitions concerning union, intersection, and complementation of type 2 fuzzy sets, and even with the corresponding operators in the original fuzzy set theory itself. To prove this, let $\mu_A(x)$ and $\mu_B(x)$ be fuzzy numbers whose mean values are respectively $l_A(x)$ and $l_B(x)$. Then $\mu_A(x)$ and $\mu_B(x)$ intuitively mean "approximately $l_A(x)$" and "approximately $l_B(x)$." We wish the membership value of x in $A \cup_2 B$ to be "approximately $\max(l_A(x), l_B(x))$," i.e., a fuzzy number whose mean value is $\max(l_A(x), l_B(x))$. However, using the above canonical definition of $\cup_2$, we get $\mu_A(x) \cup \mu_B(x)$, which is generally nonconvex and hence not a fuzzy number (see Fig. 4). As a matter of fact, we obtain a set of two elements that are approximately $l_A(x)$, and $l_B(x)$. Other set-theoretic operators are thus needed for type 2 fuzzy sets. These operators will be provided in the next chapter thanks to an "extension principle."

Special kinds of type 2 fuzzy sets that can be found in the literature include:

Classical sets of type 2 (Zadeh, 1975):   The membership function of a classical set of type 2 is a mapping from X to the set $\mathcal{P}(\{0, 1\})$ of classical subsets of $\{0, 1\}$, $\mathcal{P}(\{0, 1\}) = \{\varnothing, \{0\}, \{1\}, \{0, 1\}\}$. A possible interpretation of the four membership values is:

$$\mu_A(x) = \varnothing: \quad x \in A \ (\text{as } x \notin A) \text{ is undefined or absurd;}$$

$$\left. \begin{array}{ll} \mu_A(x) = \{0\}: & x \notin A \\ \mu_A(x) = \{1\}: & x \in A \end{array} \right\} \quad \text{"}\in\text{" has its ordinary meaning here.}$$

$$\mu_A(x) = \{0, 1\}: \quad \text{We do not know if } x \in A \text{ or if } x \notin A.$$

Φ-*fuzzy sets* (Sambuc, 1975): Φ-fuzzy sets are mappings from X to the set of the closed intervals in $[0, 1]$, i.e., interval-valued fuzzy sets. (See also Grattan-Guiness, 1975; Jahn, 1975.)

Many-valued quantities (Grattan-Guiness, 1975):   These are mappings from $X = \mathbb{R}$ to $\mathcal{P}([0, 1])$.

e. Probabilistic Sets   (Hirota, 1977)

A *probabilistic set* A is defined by a randomized membership function μ_A from $X \times \Omega$ to $[0, 1]$, where $\mu_A(x, \cdot)$ is measurable on the σ-algebra Ω.

The membership value $\mu_A(x, \omega)$ of x in A is a random variable built from the distribution p of ω, assumed independent of A. Fig. 5 depicts the "noised" fuzzy set A. p models the subjective imprecision of μ_A. Since p

does not depend on A, the set-theoretic operators $\cup$, $\cap$, $^-$ can be easily extended. Probabilistic sets in X form a pseudocomplemented distributive lattice. Probabilistic fuzzy sets, which are not L-fuzzy sets, are related to the result suggested by MacVicar-Whelan's experiment (Reference from IV.1, 1977), when asking several people to locate the boundary between membership and nonmembership, answers are randomly distributed in a given interval.

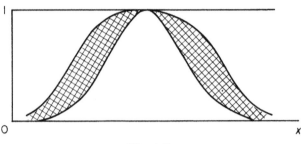

Figure 5

H.   MEASURES OF FUZZINESS

Various authors have proposed scalar indices to measure the degree of fuzziness of a fuzzy set. The degree of fuzziness is assumed to express on a global level the difficulty of deciding which elements belong and which do not belong to a given fuzzy set.

Mathematically, a measure of fuzziness is a mapping d from $\tilde{\mathcal{P}}$ (X) to $[0, +\infty)$ satisfying the conditions (De Luca and Termini, 1972b):

(1) $d(A) = 0$ iff A is an ordinary subset of X;
(2) $d(A)$ is maximum iff $\mu_A(x) = \frac{1}{2}$ $\forall x \in X$;
(3) $d(A^*) \leqslant d(A)$, where A^* is any sharpened version of A, that is, $\mu_{A^*}(x) \leqslant \mu_A(x)$ if $\mu_A(x) \leqslant \frac{1}{2}$ and $\mu_{A^*}(x) \geqslant \mu_A(x)$ if $\mu_A(x) \geqslant \frac{1}{2}$;
(4) $d(A) = d(\bar{A})$ ($\bar{A}$ is as fuzzy as A).

When X is finite, Loo (1977) has proposed a general mathematical form for d:

$$d(A) = F\left[\sum_{i=1}^{|X|} c_i f_i(\mu_A(x_i)) \right],$$

where $c_i \in \mathbb{R}^+$, $\forall i$; f_i is a real-valued function such that $f_i(0) = f_i(1) = 0$; $f_i(u) = f_i(1 - u)$ $\forall u \in [0, 1]$; and f_i is strictly increasing on $[0, \frac{1}{2}]$. F is a positive increasing function. d satisfies (1)–(4), but is not a priori the most general form. When F is linear, the following property holds:

$$d(A) + d(B) = d(A \cup B) + d(A \cap B).$$

Particular forms of d are:

Index of fuzziness (Kaufmann, 1975): F is the identity, $\forall i$, $c_i = 1$, $\forall i$, $f_i(u) = u$ when $u \in [0, \frac{1}{2}]$. $d(A)$ is the distance between A and the closest ordinary subset of X to A using a Hamming distance, i.e.,

$$d(A) = \sum_{i=1}^{|M|} |\mu_A(x_i) - \mu_{A_{1/2}}(x_i)|,$$

where $A_{1/2}$ is the $\frac{1}{2}$-cut of A.

Entropy (De Luca and Termini, 1972b): $F(u) = ku$, $k > 0$; $\forall i$, $c_i = 1$;

$$\forall i, \quad f_i(u) = -u\ln(u) - (1 - u)\ln(1 - u)$$

(Shannon function).

Note that measures of fuzziness evaluate A and $\overline{A}$ at the same time. They can be extended to evaluate a whole fuzzy partition in order to give a rating of the total amount of ambiguity that arises when deciding to which of $A_1, \ldots, A_m$ an element x belongs. We have $\sum_{i=1}^{m} \mu_{A_i}(x) = 1$. For such a fuzzy partition, the measure of fuzziness is (Capocelli and De Luca, 1973)

$$d(A_1, \ldots, A_m) = \sum_{i=1}^{|X|} \sum_{j=1}^{m} v(\mu_{A_j}(x_i)),$$

where v is any continuous and strictly concave function in $[0, 1]$. When $(A_1, \ldots, A_m)$ is an ordinary partition of X, $d(A_1, \ldots, A_m) = 0$. $d(A_1, \ldots, A_m)$ is maximum iff $\forall i$, $\forall j$, $\mu_{A_j}(x_i) = 1/m$ (maximum ambiguity).

Capocelli and De Luca (1973) have constructed a thermodynamics of fuzzy sets, introducing such concepts as absolute temperature, energy, . . . , even recovering Bose–Einstein and Fermi–Dirac distributions.

Entropy measures of a fuzzy set defined on a denumerable support are studied by De Luca and Termini (1977). The same authors extended this notion to L-fuzzy sets in a finite universe (De Luca and Termini, 1974).

Lastly, Knopfmacher (1975) gave a formulation of a measure of fuzziness, for fuzzy sets in a measurable universe, that satisfies (1)–(4):

$$d(A) = \frac{1}{P(X)} \int_X F(\mu_A(x)) \, dP(x),$$

where $F(u) = F(1 - u)$, $u \in [0, 1]$; $F(0) = F(1) = 0$; F is strictly increasing in $[0, \frac{1}{2}]$.

Remark   Instead of using a quantitative measure of fuzziness, we may simply employ a qualitative typology, as suggested by Kaufmann (1975, Vol. 3, p. 287 et seq.), in order to classify fuzzy sets in rough categories such as "slightly fuzzy," "almost precise," "very fuzzy."

REFERENCES

Backer, E. (1975). A non-statistical type of uncertainty in fuzzy events. (2nd Colloq. Keszthely, 1975) pp. 53–73. Colloq. Math. Soc. János Bolyai **16**, North-Holland, Amsterdam, 1977.

Bellman, R. E., and Giertz, M. (1973). On the analytic formalism of the theory of fuzzy sets. *Inf. Sci.* **5**, 149–157.

Brown, J. G. (1971). A note on fuzzy sets. *Inf. Control* **18**, 32–39.

Capocelli, R. M., and De Luca, A. (1973). Fuzzy sets and decision theory. *Inf. Control* **23**, 446–473.

Chapin, E. W. (1974). Set-valued set-theory. Part I. *Notre Dame J. Formal Logic* **4**, 619–634.

Chapin, E. W. (1975). Set-valued set-theory. Part II. *Notre Dame J. Formal Logic* **5**, 255–267.

De Luca, A., and Termini, S. (1972a). Algebraic properties of fuzzy sets. *J. Math. Anal. Appl.* **40**, 373–386.

De Luca, A., and Termini, S. (1972b). A definition of a non-probabilistic entropy in the setting of fuzzy sets theory. *Inf. Control* **20**, 301–312.

De Luca, A., and Termini, S. (1974). Entropy of *L*-fuzzy sets. *Inf. Control* **24**, 55–73.

De Luca, A., and Termini, S. (1977). On the convergence of entropy measures of a fuzzy set. *Kybernetes* **6**, 219–227.

Féron, R. (1976). Ensembles aléatoires flous. *C. R. Acad. Sci.* Paris, Sér. A, **282**, 903–906.

Fortet, R. and Kambouzia, M. (1976). Ensembles aléatoires et ensembles flous. *Publ. Econometr.* **9** No. 1.

Fung, L. W., and Fu, K. S. (1975). An axiomatic approach to rational decision-making in a fuzzy environment. In "Fuzzy Sets and Their Application to Cognitive and Decision Processes" (L. A. Zadeh, K. S. Fu, K. Tanaka, and M. Shimura, eds.), pp. 227–256. Academic Press, New York. (Reference from IV.3.)

Gentilhomme, Y. (1968). Les ensembles flous en linguistique. *Cah. Linguist. Theor. Appl.* (*Bucarest*) **5**, 47–63.

Giles, R. (1976). Lukasiewicz logic and fuzzy theory. *Int. J. Man-Mach. Stud.* **8**, 313–327. (Reference from III.1, 1976a.)

Goguen, J. A. (1967). *L*-fuzzy sets. *J. Math. Anal. Appl.* **18**, 145–174.

Goguen, J. A. (1974). Concept representation in natural and artificial languages: Axioms, extensions and applications for fuzzy sets. *Int. J. Man-Mach. Stud.* **6**, 513–561. (Reference from II.4.)

Gottinger, H. W. (1973). Towards a fuzzy reasoning in the behavioral sciences. *Cybernetica* **16**, No. 2, 113–135.

Grattan-Guiness, I. (1975). Fuzzy membership mapped onto interval and many-valued quantities. *Z. Math. Logik Grundlag. Math.* **22**, 149–160.

Hirota, K. (1977). Concepts of probabilistic sets. *Proc. IEEE Conf. Decision Control, New Orleans* pp. 1361–1366.

Jahn, K. U. (1975). Intervall-wertige Mengen. *Math. Nachr.* **68**, 115–132.

Kaufmann, A. (1975). Introduction to the Theory of Fuzzy Subsets, Vol. 1. Academic Press, New York. (Reference from I, 1975c.)

Knopfmacher, J. (1975). On measures of fuzziness. *J. Math. Anal. Appl.* **49**, 529–534.

Lake, J. (1976). Sets, fuzzy sets, multisets and functions. *J. London Math. Soc.* **12**, 323–326.

Loo, S. G. (1977). Measures of fuzziness. *Cybernetica* **20**, No. 3, 201–210.

Lowen, R. (1978). On fuzzy complements. *Inf. Sci.* **14**, 107–113.

Mizumoto, M., and Tanaka, K. (1976). Some properties of fuzzy sets of type 2. *Inf. Control* **31**, 312–340. (Reference from II.2.)

Negoita, C. V., and Ralescu, D. A. (1975a). Representation theorems for fuzzy concepts. *Kybernetes* **4**, 169–174.

Negoita, C. V., and Ralescu, D. A. (1975b). "Application of Fuzzy Sets to Systems Analysis," Chaps. 1 and 2. Birkaeuser, Basel. (Reference from I.)

Netto, A. B. (1968). Fuzzy classes. *Not. Am. Math. Soc.* p. 945 (68T-H28).

Ponasse, D. (1978). Algèbres floues et algèbres de Lukasiewicz. *Rev. Roum. Math. Pures Appl.* **23**, No. 1, 103–113.

Ponsard, C. (1975). "Contribution à une théorie des espaces économiques imprécis. *Publ. Econometr.* **8**, No. 2.

Radecki, T. (1977). Level-fuzzy sets. *J. Cybern.* **7**, 189–198.

Ragade, R. K., and Gupta, M. M. (1977). Fuzzy set theory: Introduction. In "Fuzzy Automata and Decision Processes" (M. M. Gupta, G. N. Saridis, and B. R. Gaines, eds.), pp. 105–131. North-Holland Publ., Amsterdam.

Rosenfeld, A. (1971). Fuzzy groups. *J. Math. Anal. Appl.* **35**, 512–517.

Sambuc, R. (1975). "Fonctions Φ-floues. Application a l'Aide au Diagnostic en Pathologie Thyroidienne." Thèse Univ. de Marseille, Marseille.

Sanchez, E., and Sambuc, R. (1976). Relations floues. Fonctions Φ-floues. Application à l'aide au diagnostic en pathologie thyroidienne. *Med. Data Process. Symp., I.R.I.A.,* Toulouse, Fr.

Trillas, E., and Riera, T. (1978). Entropies of finite fuzzy sets. *Inf. Sci.* **15**, No. 2, 159–168.

Willaeys, D., and Malvache, N. (1976). Utilisation d'un référentiel de sous-ensembles flous. Application à un algorithme flou. *Int. Conf. Syst. Sci.*, Wroclaw, Poland.

Zadeh, L. A. (1965). Fuzzy sets. *Inf. Control* **8**, 338–353.

Zadeh, L. A. (1966). Shadows of fuzzy sets. *Probl. Peredachi Inf.* **2**, No. 1, 37–44. (In Russ.) [Engl. trans., *Probl. Inf. Transm., (USSR)* **2**, No. 1, 29–34 (1966).]

Zadeh, L. A. (1975). The concept of a linguistic variable and its application to approximate reasoning. Parts 1, 2, and 3. *Inf. Sci.* **8**, 199–249; **8**, 301–357; **9**, 43–80. (Reference from II.2.)

Zadeh, L. A. (1977). "Theory of Fuzzy Sets," Memo UCB/ERL M77/1. Univ. of California, Berkeley. [Also in "Encyclopedia of Computer Science and Technology" (J. Belzer, A. Holzman, A. Kent, eds.) Marcel Dekker, New York, 1977.]

Chapter 2

EXTENSION PRINCIPLE, EXTENDED OPERATIONS, AND EXTENDED FUZZY SETS

The extension principle introduced by Zadeh is one of the most basic ideas of fuzzy set theory. It provides a general method for extending nonfuzzy mathematical concepts in order to deal with fuzzy quantities. Some illustrations are given including the notion of fuzzy distance between fuzzy sets. The extension principle is then systematically applied to real algebra: operations on fuzzy numbers are extensively developed. These operations generalize interval analysis and are computationally attractive. Although the set of real fuzzy numbers equipped with an extended addition or multiplication is no longer a group, many structural properties are preserved. Lastly, the extension principle is shown to be very useful for defining set-theoretic operations for higher order fuzzy sets.

A. EXTENSION PRINCIPLE

a. Definition

Let X be a Cartesian product of universes, $X = X_1 \times \cdots \times X_r$, and $A_1, \ldots, A_r$ be r fuzzy sets in $X_1, \ldots, X_r$, respectively. The *Cartesian product* of $A_1, \ldots, A_r$ is defined as

$$A_1 \times \cdots \times A_r = \int_{X_1 \times \cdots \times X_r} \min(\mu_{A_1}(x_1), \ldots, \mu_{A_r}(x_r))/(x_1, \ldots, x_r).$$

Let f be a mapping from $X_1 \times \cdots \times X_r$ to a universe Y such that

$y = f(x_1, \ldots, x_r)$. The extension principle (Zadeh, 1975) allows us to induce from r fuzzy sets A_i a fuzzy set B on Y through f such that

$$\mu_B(y) = \sup_{\substack{x_1, \ldots, x_r \\ y = f(x_1, \ldots, x_r)}} \min(\mu_{A_1}(x_1), \ldots, \mu_{A_r}(x_r)). \tag{1}$$

$$\mu_B(y) = 0 \quad \text{if} \quad f^{-1}(y) = \varnothing,$$

where $f^{-1}(y)$ is the inverse image of y. $\mu_B(y)$ is the greatest among the membership values $\mu_{A_1 \times \cdots \times A_r}(x_1, \ldots, x_r)$ of the realizations of y using r-tuples $(x_1, \ldots, x_r)$.

The special case when $r = 1$ was already solved by Zadeh (Reference from II.1, 1965). When f is one to one, (1) becomes $\mu_B(y) = \mu_A(f^{-1}(y))$ when $f^{-1}(y) \neq \varnothing$.

Zadeh usually writes (1) as

$$B = f(A_1, \ldots, A_r) = \int_{X_1 \times \cdots \times X_r} \min(\mu_{A_1}(x_1), \ldots, \mu_{A_r}(x_r))/f(x_1, \ldots, x_r),$$

where the sup operation is implicit.

b. Compatibility of the Extension Principle with α-Cuts

Denoting the image of $A_1, \ldots, A_r$, by $B = f(A_1, \ldots, A_r)$ the following proposition holds (Nguyen, 1976):

$$[f(A_1, \ldots, A_r)]_\alpha = f(A_{1\alpha}, \ldots, A_{r\alpha})$$

iff $\forall y \in Y, \quad \exists x_1^*, \ldots, x_r^*, \quad \mu_B(y) = \mu_{A_1 \times \cdots \times A_r}(x_1^*, \ldots, x_r^*) \quad (2)$

(the upper bound in (1) is attained).

Remark While a discretization of the valuation set generally commutes with the extension of function f, this is not true for the discretization of the universe ($X_i = \mathbb{R}$) as will be seen later (see Section B).

c. Other Extension Principles

Other extension principles can be considered.

Jain (1976) proposed replacing sup in (1) by the probabilistic sum $\hat{+}$ ($u \hat{+} v = u + v - uv$). The rationale behind this operator is that the membership of y in $f(A_1, \ldots, A_r)$ should depend on the number of r-tuples $(x_1, \ldots, x_r)$ such that $y = f(x_1, \ldots, x_r)$. This extension principle sounds more probabilistic than fuzzy, particularly if we also replace min by product. It has been pointed out by Dubois and Prade (1978a) that, in general $f(A_1, \ldots, A_r)$ is a classical subset of y when $X = \mathbb{R}$ (with min or product) and continuous membership functions are considered. So the result depends only on the supports of the A_i, which invalidates this principle as one of fuzzy extension.

Another extension principle can be obtained by just replacing min by product in (1). This principle implicitly assumes some "interactivity" or possible "compensation" between the A_i. The problem of interactivity will be considered later (see chapter 3). It does not seem that this latter principle has the same drawbacks as does that of Jain.

Note that $f_2(A_1, \ldots, A_r) \subseteq f_1(A_r, \ldots, A_r)$ where f_1 is the sup–min extended f and f_2 is the sup–· extended f.

d. Generality of the Extension Principle

Given this principle, it is possible to fuzzify any domain of mathematical reasoning based on set theory. As in Gaines (Reference from III.1, 1976b), "the fundamental change is to replace the precise concept that a variable has a *value* with the fuzzy concept that a variable has a *degree of membership to each possible value.*"

However, using the extension principle is not the only way of fuzzifying mathematical structures. Another way is just to replace ordinary sets by fuzzy sets (or the family of their α-cuts) in the framework of a given theory. For instance, fuzzy groups were defined in the previous chapter; their setting did not require an extension principle: a fuzzy group is nothing but a subgroup without sharp boundary. The group operation is still performed on the elements of the universe. Using the extension principle, however, we can extend the group operation to have it acting on fuzzy sets of the universe. The extended operation is not necessarily a group operation. This latter way of "fuzzifying a group" is radically different from Rosenfeld's (Reference from II.1) and will be investigated in Section B of this chapter.

e. Three Examples of Application of the Extension Principle

α. *Fuzzy Distance between Fuzzy Sets*

Let X be a metric space equipped with the pseudometric d, i.e.,

(1) d is a mapping from X^2 to $\mathbb{R}^+$;
(2) $d(x, x) = 0 \ \forall x$;
(3) $d(x_1, x_2) = d(x_2, x_1) \ \forall x_1, \forall x_2$;
(4) $d(x_1, x_3) \leqslant d(x_1, x_2) + d(x_2, x_3) \forall x_1, \forall x_2, \forall x_3$.

A fuzzy distance $\tilde{d}$ between fuzzy sets A and B on X is defined using (1) as

$$\forall \delta \in \mathbb{R}^+, \quad \mu_{\tilde{d}(A,B)}(\delta) = \sup_{\delta = d(u,v)} \min(\mu_A(u), \mu_B(v)).$$

$\tilde{d}(A,B)$ models a distance between fuzzy "spots." When A and B are connected subsets of X, $\tilde{d}(A,B)$ is an ordinary interval whose extremities are respectively the shortest and greatest distance between a point of A and a point of B. $\tilde{d}$ is a mapping from $[\tilde{\mathcal{P}}(X)]^2$ to the set of fuzzy sets on $\mathbb{R}^+$ (i.e., positive real fuzzy sets). $\tilde{d}(A,A)$ can be interpreted as the fuzzy diameter of A and $\mu_{\tilde{d}(A,A)}(0) = \text{hgt}(A)$. It is clear that we have $\tilde{d}(A,B) = \tilde{d}(B,A)$.

The question of knowing whether some triangular inequality like (4) still holds for $\tilde{d}$ is less straightforward. Let $A_\alpha, B_\alpha, C_\alpha$ be the α-cuts of three fuzzy sets on X. Let us respectively denote by u, v, w any element of $A_\alpha, B_\alpha, C_\alpha$. The following inequalities hold:

$$\sup_{u,w} d(u,w) = d(u^*,w^*) \leq d(u^*,v) + d(v,w^*) \leq \sup_v (d(u^*,v) + d(v,w^*));$$

$$\sup_v (d(u^*,v) + d(v,w^*)) \leq \sup_{u,v,w} (d(u,v) + d(v,w));$$

$$\sup_{u,v,w} (d(u,v) + d(v,w)) \leq \sup_{u,v} d(u,v) + \sup_{v,w} d(v,w),$$

$$\inf_{u,v} d(u,v) + \inf_{v,w} d(v,w) \leq \inf_{u,v,w} (d(u,v) + d(v,w)).$$

The sides of the two last inequalities correspond to two different fuzzifications:

$$\mu_{\tilde{d}(A,B,C)}(\delta) = \sup_{\substack{u,v,w \\ \delta = d(u,v)+d(v,w)}} \min(\mu_A(u), \mu_B(v), \mu_C(w)),$$

$$\mu_{\tilde{\Delta}(A,B,C)}(\Delta) = \sup_{\substack{\delta,\delta' \\ \Delta = \delta+\delta'}} \min\left[\sup_{\substack{u,v \\ \delta = d(u,v)}} \min(\mu_A(u), \mu_B(v)), \right.$$
$$\left. \sup_{\substack{v,w \\ \delta' = d(v,w)}} \min(\mu_B(v), \mu_C(w)) \right].$$

In $\mu_{\tilde{d}(A,B,C)}$ we consider all the paths between A and C with a detour in B, while in $\mu_{\tilde{\Delta}(A,B,C)}$ the arrival point in B is no longer constrained to be the departure point. Hence, we have, if $A_\alpha, B_\alpha, C_\alpha$ are connected and without holes, $\forall \alpha$, $\tilde{d}(A,B,C) \subseteq \tilde{\Delta}(A,B,C)$.

Let us denote by $\tilde{d}(A,B,C)_\alpha$ and $\tilde{\Delta}(A,B,C)_\alpha$ the α-cuts of $\tilde{d}(A,B,C)$ and $\tilde{\Delta}(A,B,C)$, respectively. It is easy to show that $\forall \alpha \in]0,1]$,

$$\inf(\tilde{d}(A,C)_\alpha) \leq \inf(\tilde{d}(A,B,C)_\alpha),$$
$$\sup(\tilde{d}(A,C)_\alpha) \leq \sup(\tilde{d}(A,B,C)_\alpha),$$
$$\sup(\tilde{d}(A,C)_\alpha) \leq \sup(\tilde{\Delta}(A,B,C)_\alpha),$$

and provided that these bounds are attained for some elements of X. Nothing can be said when comparing $\inf(\tilde{d}(A,C)_\alpha)$ and $\inf(\tilde{\Delta}(A,B,C)_\alpha)$. The deep reason is that when A,B,C are ordinary sets, the triangular inequality does not hold for the minimal distances between the subsets. The two first inequalities can be interpreted as a triangular inequality for fuzzy distances (see also B.d.ε). $\tilde{d}$ may also be viewed as a fuzzy measure of dissimilarity between fuzzy sets.

β. Compatibility of Two Fuzzy Sets (Zadeh, Reference from III.1, 1977a)

Given a fuzzy set A on X, $\mu_A(x)$ is the grade of membership of x in A. We may also call it the degree of *compatibility* of the fuzzy value A with the nonfuzzy value x. The extension principle allows us to evaluate the compatibility of the fuzzy value A with another fuzzy value B, taken as a reference.

Let τ be this compatibility. τ is a fuzzy set on $[0,1]$ since it is $\mu_A(B)$. Using (1),

$$\mu_\tau(u) = \sup_{x\,:\,u=\mu_A(x)} \mu_B(x) \qquad \forall u \in [0,1], \tag{3}$$

or, using Zadeh's notation,

$$\tau = \mu_A(B) = \int_X \mu_B(x)/\mu_A(x).$$

An example of computation of $\mu_\tau(u)$ is pictured in Fig. 1. When μ_A is one to one, $\mu_\tau = \mu_B \circ \mu_A^{-1}$, where $\circ$ is the composition of functions. When $A = B$, μ_τ is the identity function, $\mu_\tau(u) = u$. Remember that the converse proposition does not hold: A and B can be very different while $\mu_\tau(u) = u$.

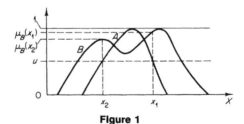

Figure 1

τ *is a normalized fuzzy set if B is.* To prove this, observe that if b is such that $\mu_B(b) = 1$, $\mu_\tau(\mu_A(b)) = 1$ also. The converse proposition is obvious provided that the sup is reached in (3). Q.E.D.

If μ_B has only one relative maximum b, $\mu_B(b) = \mathrm{hgt}(B)$, then μ_τ has only one relative maximum.

This is obvious from Zadeh's form of the extension principle.

From now on B is assumed to have only one global maximum b. $\mu_A(b)$ is the mean value of τ, i.e., the compatibility degree of A with respect to B is "approximately $\mu_A(b)$." $\mu_A(b)$ can be considered as a scalar inclusion index somewhat like consistency (cf. 1.E.c.β); instead of choosing $\mathrm{hgt}(A \cap B)$, we prefer here the membership value in A of the element that mostly belongs to B. Note that the mean value of τ is always less than $\mathrm{hgt}(A)$.

More generally, the compatibility of A with respect to B is a fuzzy inclusion index.

γ. *Fuzzy α-Cuts* (Zadeh, Reference from IV.6)

Let A be a fuzzy set on X and A_α its α-cut. A_α can be written $\mu_A^{-1}([\alpha, 1])$, i.e., the inverse image of the interval $[\alpha, 1]$. Let $\mu_{[\alpha, 1]}$ be the characteristic function of the interval $[\alpha, 1]$ in the universe $[0, 1]$. We get

$$\mu_{A_\alpha}(x) = \mu_{[\alpha, 1]}(\mu_A(x)) \qquad \forall x \in X. \tag{4}$$

A fuzzy α-cut can be understood as the set of elements whose membership values are greater than "approximately α," i.e., belong to a fuzzy interval $(\tilde{\alpha}, 1]$, where $\mu_{(\tilde{\alpha}, 1]}$ is a continuous nondecreasing function from $[0, 1]$ to $[0, 1]$ and $\mu_{(\tilde{\alpha}, 1]}(1) = 1$. Semantically, the fuzzy interval means something like "high." So it is natural to extend (4) into

$$\mu_{A_{\tilde{\alpha}}}(x) = \mu_{(\tilde{\alpha}, 1]}(\mu_A(x)) \qquad \forall x \in X \tag{5}$$

where $A_{\tilde{\alpha}}$ is the fuzzy α-cut of A.

Let us prove that (5) can be derived from the extension principle. Since $A_\alpha = \mu_A^{-1}([\alpha, 1])$, symbolically we also have $A_{\tilde{\alpha}} = \mu_A^{-1}((\tilde{\alpha}, 1])$; hence, we must extend μ_A^{-1} viewed as a multivalued function from $\mathcal{P}([0, 1])$ in X. Nevertheless, the extension principle can be generalized to deal with multivalued functions. In our example,

$$\mu_{A_{\tilde{\alpha}}}(x) = \sup_{\substack{\alpha \\ x \in \mu_A^{-1}([\alpha, 1])}} \mu_{(\tilde{\alpha}, 1]}^*([\alpha, 1]) \tag{6}$$

where $\mu_{(\tilde{\alpha}, 1]}^*([\alpha, 1]) = \mu_{(\tilde{\alpha}, 1]}(\alpha)$. Note that $\{\alpha, x \in \mu_A^{-1}([\alpha, 1])\} =]0, \mu_A(x)]$; and since $\mu_{(\tilde{\alpha}, 1]}^*$ is nondecreasing and continuous, $\mu_{A_{\tilde{\alpha}}}(x) = \mu_{(\tilde{\alpha}, 1]}^*([\mu_A(x), 1])$, which is the same as (5). In (6) $\mu_{A_{\tilde{\alpha}}}(x)$ is the greatest among the membership values of the sets $[\alpha, 1]$ whose images under μ_A^{-1} contain x; this contrasts with (1), where $=$ replaces $\in$.

B. EXTENDED REAL OPERATIONS

An important field of applications for the extension principle is given by algebraic operations such as addition and multiplication. More generally, given an *n*-ary composition law from X^n to X, it is possible to induce an

n-ary composition law in $\tilde{\mathscr{P}}(X)$. In this section we restrict ourselves to $X = \mathbb{R}$, the real line; so here we extend real algebra.

a. Operations on Fuzzy Numbers (Dubois and Prade, 1978c)

Some previous works related to operations on fuzzy numbers are those of Jain (1976), Nahmias (1978), Mizumoto and Tanaka (1976b, c), Baas and Kwakernaak (1977).

For simplicity, theorems and proofs will be stated for binary operations. However, they remain valid for n-ary operations (see Dubois and Prade, 1978c).

Definition A binary operation $*$ in $\mathbb{R}$ is said to be *increasing* iff:

$$\text{if} \quad x_1 > y_1 \quad \text{and} \; x_2 > y_2, \quad \text{then} \quad x_1 * x_2 > y_1 * y_2.$$

In the same way, $*$ is said to be *decreasing* iff $x_1 > y_1$ and $x_2 > y_2$ imply $x_1 * x_2 < y_1 * y_2$.

Using the extension principle, $*$ can be extended into $\circledast$ to combine two fuzzy numbers (i.e., convex and normalized fuzzy sets in $\mathbb{R}$) M and N. Moreover, μ_M and μ_N are assumed to be continuous functions on $\mathbb{R}$;

$$\mu_{M \circledast N}(z) = \sup_{z = x * y} \; \min(\mu_M(x), \mu_N(y)). \tag{7}$$

N.B.: Kaufmann (Reference from I, 1975c, pp. 290–295) considered a probabilistic method for extending addition to fuzzy numbers, by means of an ordinary convolution, $\mu_{M+N}(z) = \int_0^z \mu_M(x)\mu_N(z - x)\,dx$, for some particular kinds of μ_M and μ_N. See also Mareš (1977a, b).

From now on, $\mathfrak{N}(\mathbb{R})$ denotes the set of real fuzzy numbers.

Lemma 1 Let M and N be two continuous fuzzy numbers, and $*$ a continuous increasing binary operation. Let $[\lambda_M, \rho_M]$ and $[\lambda_N, \rho_N]$ be two intervals on nondecreasing parts of μ_M and μ_N, respectively (possibly $\lambda_M = \rho_M$ or $\lambda_N = \rho_N$) such that

$$\forall x \in [\lambda_M, \rho_M], \quad \forall y \in [\lambda_N, \rho_N], \quad \mu_M(x) = \mu_N(y) = \omega.$$

Then

$$\forall t \in [\lambda_M * \lambda_N, \rho_M * \rho_N], \mu_{M \circledast N}(t) = \omega.$$

Proof: Let x_M be an element of $[\lambda_M, \rho_M]$ and y_N an element of $[\lambda_N, \rho_N]$. We have $\min(\mu_M(x_M), \mu_N(y_N)) = \omega$.

Let $(x, y) \in \mathbb{R}^2$ be such that $x * y = x_M * y_N$. If $x \leqslant x_M$, then $\min(\mu_M(x),$ $\mu_N(y)) \leqslant \omega$ because μ_M is nondecreasing on $(-\infty, x_M]$, at least. If $x > x_M$, then $y \leqslant y_N$ because $*$ is increasing, and $\min(\mu_M(x), \mu_N(y)) \leqslant \omega$ since μ_N is nondecreasing on $(-\infty, y_N]$, at least. Hence $\mu_{M \odot N}(x_M * y_N) = \omega$. When x_M and y_N range over $[\lambda_M, \rho_M]$ and $[\lambda_N, \rho_N]$, respectively, $x_M * y_N$ ranges over $[\lambda_M * \lambda_N, \rho_M * \rho_N]$ since $*$ is increasing and continuous. Q.E.D.

A similar lemma holds when we consider the nonincreasing parts of μ_M and μ_N.

Lemma 2 Let M and N be two continuous fuzzy numbers such that μ_M is nondecreasing on $(-\infty, m]$ and nonincreasing on $[m, +\infty)$ and μ_N is nondecreasing on $(-\infty, n]$ and nonincreasing on $[n, +\infty)$. Let $*$ be a continuous increasing binary operation. Assume $\mu_M(\mathbb{R}') = \mu_N(\mathbb{R}') = [0, 1]$ where $\mathbb{R}' = \mathbb{R} \cup \{-\infty, +\infty\}$.

Then $\forall t \in]\inf_{x, y} x * y, \sup_{x, y} x * y[, \exists(x_M, y_N)$ such that:

Either $x_M \leqslant m$ and $y_N \leqslant n$, or $x_M \geqslant m$ and $y_M \geqslant n$;
$\mu_{M \odot N}(x_M * y_N) = \mu_M(x_M) = \mu_N(x_N) = \mu_{M \odot N}(t)$.

Proof: Note that since μ_M and μ_N are continuous and nondecreasing, they are locally either constant or strictly increasing on $(-\infty, m]$ and $(-\infty, n]$, respectively.

Let $\mu_{M_+}^{-1}$ be a function from $[0, 1]$ to the set of subintervals of $(-\infty, m]$, $\omega \mapsto [\lambda_M(\omega), \rho_M(\omega)] = \mu_{M_+}^{-1}(\omega)$, such that $x \in [\lambda_M(\omega), \rho_M(\omega)]$ iff $\mu_{M_+}(x) = \mu_M(x) = \omega$ (μ_{M_+} is the nondecreasing part of μ_M).
Similarly, $\mu_{N_+}^{-1}$: $\omega \mapsto [\lambda_N(\omega), \rho_N(\omega)] = \mu_{N_+}^{-1}(\omega)$.
Let $\bar{g}$ and $\underline{g}$ be functions from $[0, 1]$ to $[-\infty, m * n]$ defined as

$$\bar{g}(\omega) = \rho_M(\omega) * \rho_N(\omega), \qquad \underline{g}(\omega) = \lambda_M(\omega) * \lambda_N(\omega).$$

Since $*$ is increasing, $\bar{g}(\omega) \geqslant \underline{g}(\omega)$ and $\bar{g}, \underline{g}$ are nondecreasing. $\bar{g}(1) = m * n$, $\underline{g}(0) = \inf_{x, y}(x * y)$. Hence, $\forall t \in (\underline{g}(0), m * n], \exists \hat{\omega}$ such that $t \in [\underline{g}(\hat{\omega}), \bar{g}(\hat{\omega})]$. On the intervals $[\lambda_M(\hat{\omega}), \rho_M(\hat{\omega})], [\lambda_N(\hat{\omega}), \rho_N(\hat{\omega})], \mu_{M_+}$ and μ_{N_+} are constant, and their values are $\hat{\omega}$.
Hence, due to the continuity of $*$,

$$\exists(\hat{x}, \hat{y}) \in [\lambda_M(\hat{\omega}), \rho_M(\hat{\omega})] \times [\lambda_N(\hat{\omega}), \rho_N(\hat{\omega})]$$

such that $t = \hat{x} * \hat{y}$ and, by Lemma 1, $\mu_{M \odot N}(t) = \hat{\omega}$. When $t \geqslant m * n$, a similar proof holds. Q.E.D.

When μ_M and μ_N are strictly increasing respectively on $(-\infty, m]$ and on

$(-\infty, n]$, $\underline{g}(\omega) = \bar{g}(\omega) = g(\omega)$. g is bijective from $[0, 1]$ to $(g(0), m * n]$ and

$$
\mu_{M \circledast N} = g^{-1} = \begin{cases} \left(\mu_{M_+}^{-1} * \mu_{N_+}^{-1} \right)^{-1} & \text{on } \left[g(0), m * n \right], \\ \left(\mu_{M_-}^{-1} * \mu_{N_-}^{-1} \right)^{-1} & \text{on } \left[m * n, \sup_{x, y} x * y \right], \\ 0 & \text{otherwise.} \end{cases}
$$

(μ_{M_-} and μ_{N_-} denote the nonincreasing parts of μ_M and μ_N.)

We can now conclude:

Theorem 1 If M and N are continuous fuzzy numbers whose membership functions are onto and $*$ is a continuous increasing binary operation, then the extension $M \circledast N$ is a continuous fuzzy number whose membership function is onto. The effective construction of $M \circledast N$ can be performed separately on increasing and decreasing parts of μ_M and μ_N using the procedure given in Lemma 1.

When $*$ is a decreasing continuous binary operation, the same results hold; but we must use the decreasing parts of μ_M and μ_N to build the increasing part of $\mu_{M \circledast N}$ and vice versa.

Suppose the operation $*$ is for instance such that on $\mathbb{R}$

If $x_1 > y_1$ and $x_2 < y_2$, then $x_1 * x_2 > y_1 * y_2$.

It is easy to see that the operation $\perp$ defined by $x_1 \perp x_2 = x_1 * (-x_2)$ is increasing on $\mathbb{R}$. Theorem 1 applies to $\perp$ and hence to $*$. However, to perform $M \circledast N$ where M and N are fuzzy numbers, we must combine, by means of Lemma 1, the nondecreasing (resp. nonincreasing) parts of M with the nonincreasing (resp. nondecreasing) part of N. $*$ is in this case said to be "hybrid."

N.B.: Another approach to obtaining Theorem 1 is to use more explicitly α-cuts and their compatibility with the extension principle (2). In that framework some results may appear more intuitive.

Remark Baas and Kwakernaak (1977) have proved the following result: Let μ_i, $i = 1, n$ be n piecewise continuously differentiable membership functions with finite supports. Let f be a continuously differentiable mapping of $\mathbb{R}^n$ into $\mathbb{R}$. At points where the respective derivatives exist, we shall write

$$
\mu_i'(x_i) = d\mu_i(x_i)/dx_i, \qquad f_i(x) = \partial f(x_1, \ldots, x_n)/\partial x_i.
$$

Suppose that the point $\hat{x} = (\hat{x}_1, \hat{x}_2, \ldots, \hat{x}_n) \in \mathbb{R}^n$ satisfies the following conditions:

   (i) $\mu_i'(\hat{x}_i)$ and $f_i(\hat{x})$, $i = 1, n$, all exist and are all nonzero.
   (ii) $\mu_1(\hat{x}_1) = \mu_2(\hat{x}_2) = \cdots = \mu_n(\hat{x}_n)$.
  (iii) $\mu_i'(\hat{x}_i)/f_i(\hat{x})$ has the same sign for each $i \in \{1, 2, \ldots, n\}$.

Then $\hat{x}$ is a strict relative maximum point of the mathematical programming problem:

$$\text{Maximize} \quad \min_{i=1,n} \mu_i(x_i)$$
$$\text{subject to} \quad f(x_1, \ldots, x_n) = f(\hat{x}_1, \ldots, \hat{x}_n) = f(\hat{x}).$$

Note that this theorem gives only sufficient conditions for *relative* maximum points. Moreover, it is a local version of Lemma 1 with different hypotheses.

b. Properties of $\circledast$

   If $*$ is commutative, so is $\circledast$.
   If $*$ is associative, so is $\circledast$. (This is easy to check from the definition of $\circledast$.)

Distributivity of $\circledast$ *over* $\cup$,

$$\forall(M, N, P) \in \left[\tilde{\mathscr{P}}(\mathbb{R})\right]^3, \quad M \circledast (N \cup P) = (M \circledast N) \cup (M \circledast P)$$

(obvious). The result still holds for *n*-ary operations. On the contrary, $\cup$ is not distributive over $\circledast$ and $\circledast$ is not distributive over $\cap$.

Flattening effect   Let M, N be two fuzzy sets on an interval I of $\mathbb{R}$ such that $\mu_M(I) = [0, \omega_M]$ and $\mu_N(I) = [0, \omega_N]$; consider M' and N' such that $\mu_{M'}(x) = \min(\mu_M(x), \min(\omega_M, \omega_N))$ and $\mu_{N'}(y) = \min(\mu_N(y), \min(\omega_M, \omega_N))$; it is easy to see that $M' \circledast N' = M \circledast N$ for any operation $*$. Thus, M and N can be "flattened" into M' and N' that have the same height. When M and N are continuous convex fuzzy sets on I, Theorem 1 can be applied directly to M' and N' when $*$ is increasing or decreasing in I, replacing $[0, 1]$ by $[0, \min(\omega_M, \omega_N)]$.

   Moreover, if M and N are continuous fuzzy numbers on I such that $\mu_M(I) = [\epsilon_M, 1]$ and $\mu_N(I) = [\epsilon_N, 1]$, with $\epsilon_M \leqslant \epsilon_N$, denote $\{x \in I, \mu_M(x) \geqslant \epsilon_N\}$ by $[x_1, x_2]$. If $*$ is a continuous increasing operation such that

$$\forall t \in \,]\inf(I) * \inf(I), \sup(I) * \sup(I)[,$$

$$\exists y \in \,]\inf(I), \sup(I)[, \quad \exists x \in [x_1, x_2], \quad t = x * y$$

then, denoting by M'' the fuzzy number such that $\mu_{M''}(x) = \max(\mu_M(x), \epsilon_N)$, we have $M \circledast N = M'' \circledast N$, i.e., M has been flattened "from the bottom." This property can be applied to $(I, *) = (\mathbb{R}, +)$ or $(I, *) = (\mathbb{R}^+, \cdot)$, but not for instance to $(\mathbb{R}, \max)$ or $(\mathbb{R}, \min)$.

N.B.: If $\mu_M(I) = [0, \omega_N]$ and $\mu_N(I) = [\epsilon_N, 1]$ with $\epsilon_N > \omega_M$, then $\forall z$, $\mu_{M \odot N}(z) = \sup_{x \in Dz} \mu_M(x)$ with $Dz = \{x \in I, \exists y \in I, x * y = z\}$.

c. Outline of a General Algorithm for the Computation of Extended Operations (Dubois and Prade, 1978c)

We are now able to perform the exact computation of any extended continuous and increasing (or decreasing) n-ary operation between continuous fuzzy sets within the framework of the same algorithm. Any continuous fuzzy set can be decomposed into the union of convex, possibly nonnormalized, fuzzy sets whose membership functions are either strictly increasing or decreasing or constant in the only interval where they are not zero (see Fig. 2).

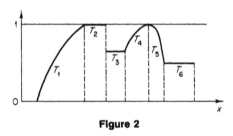

Figure 2

Owing to the distributivity of $\circledast$ over $\cup$, we can perform this operation on each part separately using the top-flattening effect and Theorem 1. The final result is the union of all the partial ones. Note that Theorem 1 can be extended to piecewise continuous convex fuzzy sets by considering any infinite-slope segment (discontinuity) as an increasing or a decreasing part of the fuzzy number, according to the context.

Description of the Algorithm   Each fuzzy set is assumed to be discretized into a finite number of membership levels ω_i, $i = 0, m$ ($\omega_0 = 0, \omega_m = 1$). To each level is assigned a set $P_i^k = \{p_{i1}^k, \ldots, p_{iJ}^k\}$ of real values such that $\mu_{M_k}(p_{ij}^k) = \omega_i$, $j = 1, J$, where M_k, $k = 1, n$, are the fuzzy sets considered and J is function of i and k. The P_i^k are assumed increasingly ordered.

Example   Two fuzzy numbers and a binary operation $*$, with $m = 3$:

$$M_1 = \omega_1/p_{11}^1 + \omega_2/p_{21}^1 + \omega_3/p_{31}^1 + \omega_2/p_{22}^1 + \omega_1/p_{12}^1,$$

$$M_2 = \omega_1/p_{11}^2 + \omega_2/p_{21}^2 + \omega_3/p_{31}^2 + \omega_2/p_{22}^2 + \omega_1/p_{12}^2,$$

then

$$M_1 \circledast M_2 = \omega_1/p_{11}^1 * p_{11}^2 + \omega_2/p_{21}^1 * p_{21}^2 + \omega_3/p_{31}^1 * p_{31}^2$$
$$+ \omega_2/p_{22}^1 * p_{22}^2 + \omega_1/p_{12}^1 * p_{12}^2.$$

The algorithm for an n-ary operation generally proceeds in four steps:

(1) Flattening: The n fuzzy sets are changed into fuzzy sets all having the same height.

(2) Decomposition of each fuzzy set as described above into two sets or pieces, the set of nondecreasing "parts" and the set of nonincreasing parts: The constant parts between two nondecreasing ones (resp. nonincreasing) belong to the nondecreasing (resp. nonincreasing) set. The constant parts, which are between parts of different kinds, belong to both. In Fig. 2, the nondecreasing set is $\{T_1, T_2, T_3, T_4\}$, and the nonincreasing set is $\{T_2, T_3, T_5, T_6\}$.

(3) Operation $*$: The operation $*$ is performed as in the above example for every n-tuple of parts (one part for each fuzzy set) all belonging to the same kind of sets (nondecreasing or nonincreasing). The flattening effect may be used.

(4) Union: For each n-tuple of parts a fuzzy set was built in step 3. The union of these fuzzy sets is the final result.

N.B.: The above algorithm can be easily adapted to deal with hybrid operations.

Example Consider the two fuzzy sets A and B pictured in Fig. 3. We want to calculate $C = A \oplus B$. $\oplus$ denotes the extended sum.

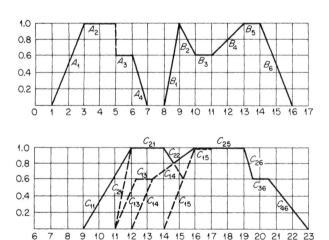

Figure 3

Step 1. A and B are normalized, no flattening is necessary.

Step 2.

A: nondecreasing set $\{A_1, A_2\}$,   nonincreasing set $\{A_2, A_3, A_4\}$;

B: nondecreasing set $\{B_1, B_3, B_4, B_5\}$,   nonincreasing set $\{B_2, B_3, B_5, B_6\}$.

Steps 3 and 4.   Let C_{ij} denote $A_i \oplus B_j$. The C_{ij}s will be calculated in lexicographic order:

$C_{11} = A_1 \oplus B_1$; $C \leftarrow C_{11}$.

C_{12} is not considered because A_1 and B_2 do not belong to the same kind of set.

$C_{13} = A_1 \oplus B_3$. Perform $C \leftarrow C \cup C_{13}$, the part of C_{13} between abscissas 11 and 12 is dropped.

$C_{14} = A_1 \oplus B_4$; $C \leftarrow C \cup C_{14}$.

$C_{15} = A_1 \oplus B_5$; $C \leftarrow C \cup C_{15}$. (The part of C_{15} between abscissas 14 and 16 is dropped.)

C_{16} is not considered.

$C_{21} = A_2 \oplus B_1$; $C \leftarrow C \cup C_{21}$. (The remaining part of C_{13} is dropped, and the part of C_{12} between abscissas 12 and 14 is dropped.)

And so on.

The final result is pictured in Fig. 3.

The above procedure is certainly not the most efficient one—a lot of redundancies remain that could be avoided through more careful analysis. We intend here only to indicate that the algebraic calculus on rather general fuzzy sets on $\mathbb{R}$ is practically possible.

Remark 1   It is clear that when discrete representations are used for continuous fuzzy sets, it is not suitable to perform a sup–min composition on the discrete data. The exact result is got by a direct performance of the operation $*$ using Theorem 1.

Example:

$$(0.5/4 + 1/5 + 0.5/6) \oplus (0.5/1 + 1/2 + 0.5/3)$$

$$= (0.5/5 + 1/7 + 0.5/9)$$

$$\neq (0.5/5 + 0.5/6 + 1/7 + 0.5/8 + 0.5/9),$$

where the latter was obtained by a direct application of sup–min composition. Here, $\oplus$ denotes the extended addition. ($+$ is, of course, an increasing operation.)

This is an illustration of the noncommutativity of support discretization and extended operations.

2 The decomposition of a convex fuzzy set into a union of convex, possibly nonnormalized fuzzy sets used in the above algorithm is very similar to the decomposition of a multimodal probability distribution considered as a mixture (linear convex combination) of unimodal ones.

3 The extension principle expresses generalized convolutions.

d. Usual Operations on Fuzzy Numbers (Dubois and Prade, 1978b, c)

α. Unary Operations

Let φ be a unary operation; the extension principle reduces to

$$\forall M \in \tilde{\mathcal{P}}(\mathbb{R}), \quad \mu_{\varphi(M)}(z) = \sup_{\substack{x \\ z = \varphi(x)}} \mu_M(x).$$

Opposite of a fuzzy number: $\varphi(x) = -x$. $\varphi(M)$ is denoted by $-M$ and is such that

$$\forall x \in \mathbb{R}, \quad \mu_{-M}(x) = \mu_M(-x).$$

M and $-M$ are symmetrical with respect to the axis $x = 0$.

Inverse of a fuzzy number: $\varphi(x) = 1/x$. $\varphi(M)$ is denoted by M^{-1} and is such that

$$\forall x \in \mathbb{R} - \{0\}, \quad \mu_{M^{-1}}(x) = \mu_M(1/x).$$

Let us call a fuzzy number M *positive* (resp. *negative*) if its membership function is such that $\mu_M(x) = 0$, $\forall x < 0$ (resp. $\forall x > 0$). This is denoted $M > 0$ (resp. $M < 0$).

If M is neither positive nor negative, M^{-1} is no longer convex, and generally does not vanish when $|x| \to \infty$ (see Fig. 4b). However, when M is positive or negative, M^{-1} is convex (Fig. 4a).

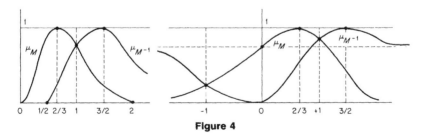

Figure 4

Scalar multiplication: $\mu_{\lambda \cdot M}(x) = \mu_M(x/\lambda)$, $\quad \forall \lambda \in \mathbb{R} - \{0\}$.

Exponential of a fuzzy number: $\varphi(x) = e^x$. $\varphi(M)$ is denoted e^M and is such that

$$\mu_{e^M}(x) = \begin{cases} \mu_M(\ln x), & x > 0, \\ 0, & \text{otherwise.} \end{cases}$$

e^M is a positive fuzzy number. Moreover, $e^{-M} = (e^M)^{-1}$.

Absolute value of a fuzzy number: The absolute value of M is denoted $\text{abs}(M)$;

$$\text{abs}(M) = \begin{cases} M \cup (-M) & \text{on} \quad \mathbb{R}^+, \\ 0 & \text{on} \quad \mathbb{R}^-. \end{cases}$$

$\text{abs}(M)$ is positive!

β. *Extended Addition and Multiplication*

Addition: Addition is an increasing operation. Hence, the extended addition ($\oplus$) of fuzzy numbers gives a fuzzy number. Note that $-(M \oplus N) = (-M) \oplus (-N)$. $\oplus$ is commutative and associative but has no group structure. The identity of $\oplus$ is the nonfuzzy number 0. But M has no symmetrical element in the sense of a group structure. In particular, $M \oplus (-M) \neq 0 \ \forall M \in \tilde{\mathscr{P}}(\mathbb{R}) - \mathbb{R}$.

Multiplication: Multiplication is an increasing operation on $\mathbb{R}^+$ and a decreasing operation on $\mathbb{R}^-$. Hence, the product of fuzzy numbers ($\odot$) that are all either positive or negative gives a positive fuzzy number. Note that $(-M) \odot N = -(M \odot N)$, so that the factors can have different signs. $\odot$ is commutative and associative. The set of positive fuzzy numbers is not a group for $\odot$: although $\forall M$, $M \odot 1 = M$, the product $M \odot M^{-1} \neq 1$ as soon as M is not a real number. M has no inverse in the sense of group structure.

The multiplication of ordinary fuzzy numbers can be performed by means of the general algorithm (see c) provided there is decomposition of each factor into a positive and a negative part. Note also that

$$\forall (M, N) \in [\tilde{\mathscr{P}}(\mathbb{R})]^2, \quad (M \odot N)^{-1} = (M^{-1}) \odot (N^{-1}).$$

Weak Distributivity of $\odot$ on $\oplus$

Theorem 2   Provided that M is either a positive or a negative fuzzy number and that N and P are *together* either positive or negative fuzzy numbers, then

$$M \odot (N \oplus P) = (M \odot N) \oplus (M \odot P) \tag{8}$$

Proof The membership functions of each side are, by definition and through an obvious reduction,

$$\mu_{M\odot(N\oplus P)}(x) = \sup_{z=x(y+t)} \min(\mu_M(x), \mu_N(y), \mu_P(t)), \tag{9}$$

$$\mu_{(M\odot N)\oplus(M\odot P)}(z) = \sup_{z=xy+tu} \min(\mu_M(x), \mu_N(y), \mu_M(u), \mu_P(t)). \tag{10}$$

Let $\varphi(x, y, t, u) = xy + tu$. When M, N, and P are positive, φ is increasing; and when they are all negative, decreasing. In both cases, using Theorem 1, the upper bound of the right-hand side of (10) is reached for $\mu_M(x) = \mu_N(y) = \mu_M(u) = \mu_P(t)$ either in the increasing or the decreasing parts of the membership functions. Hence, $x = u$ and the right-hand sides of (9) and (10) are equal. When M and the pair (N, P) have opposite signs, we apply the same method to

$$-[(-M)\odot(N \oplus P)] = M\odot(N \oplus P). \qquad \text{Q.E.D.}$$

N.B.:   1.   When N and P have opposite signs, (8) no longer holds. A counterexample will be provided later (see f). However, note that $M\odot(N \oplus P) \subseteq (M\odot N)\oplus(M\odot P)$ always holds, i.e. the right-hand side is fuzzier.

2.   Zadeh (1975, Part 1) gives a demonstration of the nondistributivity of $\odot$ on $\oplus$, in the general case, for discrete support fuzzy sets. (See also Mizumoto and Tanaka, 1976b.)

3.   Because of the nondistributivity of $\odot$ on $\oplus$, some nonincreasing operations involving sum and product cannot be extended using $\odot$ and $\oplus$. For instance, consider $\varphi(x, y, z, t) = xy + ty + xz$; $\varphi(M, N, P, Q)$ is neither $(M\odot N)\oplus(Q\odot N)\oplus(M\odot P)$ nor $[(M \oplus Q)\odot N]\oplus(M\odot P)$ nor $[M\odot(N \oplus P)]\oplus(Q\odot N)$.

A property of the fuzzy exponential:

$$e^M\odot e^N = e^{M\oplus N} \qquad \forall(M, N) \in [\tilde{\mathscr{P}}(\mathbb{R})]^2. \tag{11}$$

This is obvious since e^{x+y} is an increasing binary operation.

γ. *Extended Subtraction* ($\ominus$)

Subtraction is neither increasing nor decreasing. However, it is easy to check that $M \ominus N = M \oplus(-N) \; \forall(M, N) \in [\tilde{\mathscr{P}}(\mathbb{R})]^2$, so that $M \ominus N$ is a fuzzy number whenever M and N are.

δ. *Extended Division* ($\oslash$)

Division is neither increasing nor decreasing. But, since $M\oslash N = M\odot(N^{-1}) \; \forall(M, N) \in [\tilde{\mathscr{P}}(\mathbb{R}^+) \cup \tilde{\mathscr{P}}(\mathbb{R}^-)]^2$, $M\oslash N$ is a fuzzy number when M and N are positive or negative fuzzy numbers. The division of ordinary fuzzy numbers can be performed similarly to multiplication, by decomposition.

ε. *Extended max and min*

Max and min are increasing operations in $\mathbb{R}$. The maximum (resp. minimum) of n fuzzy numbers $M_1, \ldots, M_n$, denoted $\widetilde{\max}(M_1, \ldots, M_n)$ (resp. $\widetilde{\min}(M_1, \ldots, M_n)$), is a fuzzy number. A direct application of theorem 1 gives a practical rule for construction of $\widetilde{\max}(M_1, \ldots, M_n)$ and $\widetilde{\min}(M_1, \ldots, M_n)$, already stated in Dubois and Prade (1978b): the maximum (resp. minimum), $\widetilde{\max}$ (resp. $\widetilde{\min}$) is the dual operation with respect to union (resp. intersection) because $M_1 \cup M_2 \cup \cdots \cup M_n$ (resp. $M_1 \cap M_2 \cap \cdots \cap M_n$) is obtained by considering the nonfuzzy maximum (resp. minimum) of the n membership functions. And $\widetilde{\max}(M_1, \ldots, M_n)$ (resp. $\widetilde{\min}(M_1, \ldots, M_n)$) is similarly obtained provided that we exchange the coordinate axes $0x$ and $0y$ and that we consider increasing and decreasing parts separately (see Fig. 5).

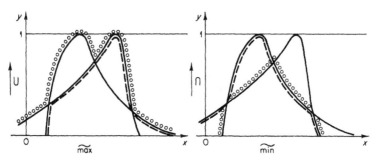

Figure 5 $\widetilde{\max}$(---)$\widetilde{\min}$(---).

Let M, N, P be three fuzzy numbers (i.e., convex normalized fuzzy sets of $\mathbb{R}$). The following properties hold:

$\widetilde{\max}$ and $\widetilde{\min}$ are commutative and associative operations; they are mutually distributive,

$$\widetilde{\min}(M, \widetilde{\max}(N, P)) = \widetilde{\max}[\widetilde{\min}(M, N), \widetilde{\min}(M, P)],$$

$$\widetilde{\max}(M, \widetilde{\min}(N, P)) = \widetilde{\min}[\widetilde{\max}(M, N), \widetilde{\max}(M, P)];$$

absorption laws,

$$\widetilde{\max}(M, \widetilde{\min}(M, N)) = M, \qquad \widetilde{\min}(M, \widetilde{\max}(M, N)) = M;$$

De Morgan laws,

$$1 \ominus \widetilde{\min}(M, N) = \widetilde{\max}(1 \ominus M, 1 \ominus N),$$

$$1 \ominus \widetilde{\max}(M, N) = \widetilde{\min}(1 \ominus M, 1 \ominus N);$$

note that $1 \ominus M$ is the "dual" of $\overline{M}$: indeed, $1 \ominus (1 \ominus M) = M$: idempotence, $\widetilde{\max}(M, M) = M = \widetilde{\min}(M, M)$;
$M \oplus \widetilde{\max}(N, P) = \widetilde{\max}(M \oplus N, M \oplus P)$; the same with $\widetilde{\min}$;
$\widetilde{\max}(M, N) \oplus \widetilde{\min}(M, N) = M \oplus N$.
$(\widetilde{\mathfrak{N}}(\mathbb{R}), \widetilde{\max}, \widetilde{\min})$ is thus a noncomplemented distributive lattice.

Lastly, from the results of A.e.α, we infer the following equality

$$\widetilde{\max}(\tilde{d}(A, C), \tilde{d}(A, B, C)) = \tilde{d}(A, B, C)$$

where $\tilde{d}$ is the fuzzy distance introduced in A.e.α. This is a compact presentation of a triangular inequality for fuzzy distances.

ζ. *Extended Power Function*

x^y is defined when $x > 0$. We consider only this case. x^y is increasing when $x \in [1, +\infty)$ and $y \in [0, +\infty)$ and decreasing when $x \in]0, 1]$ and $y \in (-\infty, 0]$. So it is possible to show that

$$\forall M \in \tilde{\mathcal{P}}([1, +\infty)), \quad \forall \Lambda > 0, \quad \forall P > 0, \quad M^{\Lambda} \odot M^{P} = M^{\Lambda \oplus P}$$

and

$$\forall M \in \tilde{\mathcal{P}}(]0, 1]), \quad \forall \Lambda < 0, \quad \forall P < 0, \quad M^{\Lambda} \odot M^{P} = M^{\Lambda \oplus P}$$

(because $x^y t^u$ and x^{y+u} are increasing operations for $x \geqslant 1$, $t \geqslant 1$, $y \geqslant 0$, and $u \geqslant 0$ and decreasing for $0 < x \leqslant 1$, $0 \leqslant t < 1$, $y \leqslant 0$, and $u \leqslant 0$; hence, Theorem 1 can be applied). Owing to $(M \odot N)^{-1} = (M^{-1}) \odot (N^{-1})$ and $\forall M > 0$, $(M^{\Lambda})^{-1} = M^{(-\Lambda)}$, the formula $M^{\Lambda} \odot M^{P} = M^{\Lambda \oplus P}$ holds as soon as $M \in \tilde{\mathcal{P}}(]0, 1]) \cup \tilde{\mathcal{P}}([1, +\infty))$ and both Λ and P are positive or negative. When m, λ, ρ are just ordinary real numbers, we have

$$\forall(\Lambda, P) \in [\tilde{\mathcal{P}}(\mathbb{R})]^2, \quad m^{\Lambda} \odot m^{P} = m^{\Lambda \oplus P},$$

$$\forall M < 0 \text{ or } \forall M > 0, \quad M^{\lambda} \odot M^{\rho} = M^{\lambda + \rho}.$$

N.B.: Here, M^{λ} does not denote the λth power of the fuzzy set M in the sense of II.1.B.f.

e. **Fast Computation Formulas**

α. *L-R Representation of Fuzzy Numbers* (Dubois and Prade, 1978b, c)

A function, usually denoted L or R, is a reference function of fuzzy numbers iff (1) $L(x) = L(-x)$; (2) $L(0) = 1$; (3) L is nonincreasing on

$[0, +\infty)$. For instance, $L(x) = 1$ for $x \in [-1, +1]$ and 0 outside; $L(x) = \max(0, 1 - |x|^p), p \geqslant 0$; $L(x) = e^{-|x|^p}, p \geqslant 0$; $L(x) = 1/(1 + |x|^p), p \geqslant 0$.

A fuzzy number M is said to be an L-R type fuzzy number iff

$$\mu_M(x) = \begin{cases} L((m - x)/\alpha) & \text{for } x \leqslant m, \alpha > 0, \\ R((x - m)/\beta) & \text{for } x \geqslant m, \beta > 0. \end{cases}$$

L is for left and R for right reference. m is the *mean value* of M. α and β are called *left* and *right spreads*, respectively. When the spreads are zero, M is a nonfuzzy number by convention. As the spreads increase, M becomes fuzzier and fuzzier. Symbolically, we write

$$M = (m, \alpha, \beta)_{LR}.$$

β. Addition

Let us consider the increasing parts of two fuzzy numbers $M = (m, \alpha, \beta)_{LR}$ and $N = (n, \gamma, \delta)_{LR}$. Let x and y be the unique real numbers such that

$$L((m - x)/\alpha) = \omega = L((n - y)/\gamma),$$

where ω is a fixed value in $[0, 1]$. This is equivalent to

$$x = m - \alpha L^{-1}(\omega), \qquad y = n - \gamma L^{-1}(\omega),$$

which implies

$$z = x + y = m + n - (\alpha + \gamma)L^{-1}(\omega) \qquad \text{and}$$

$$L\left(\frac{m + n - z}{\alpha + \gamma}\right) = \omega.$$

The same reasoning holds on decreasing parts of M and N and

$$R\left(\frac{z - (m + n)}{\beta + \delta}\right) = \omega.$$

Using Theorem 1, we prove

$$(m, \alpha, \beta)_{LR} \oplus (n, \gamma, \delta)_{LR} = (m + n, \alpha + \gamma, \beta + \delta)_{LR}. \tag{12}$$

More generally,

$$(m, \alpha, \beta)_{LR} \oplus (n, \gamma, \delta)_{L'R'} = (m + n, 1, 1)_{L''R''}$$

with

$$L'' = (\alpha L^{-1} + \gamma L'^{-1})^{-1}, \qquad R'' = (\beta L^{-1} + \delta L'^{-1})^{-1}.$$

The formula for the opposite of a fuzzy number is

$$-(m, \alpha, \beta)_{LR} = (-m, \beta, \alpha)_{RL}. \tag{13}$$

Note that the references are exchanged.

From (12) and (13) we deduce the formula for subtraction

$$(m, \alpha, \beta)_{LR} \ominus (n, \gamma, \delta)_{RL} = (m - n, \alpha + \delta, \beta + \gamma)_{LR}. \qquad (14)$$

γ. Multiplication

Using the same reasoning as above, for positive fuzzy numbers, we get

$$z = x \cdot y = m \cdot n - (m\gamma + n\alpha)L^{-1}(\omega) + \alpha\gamma\big(L^{-1}(\omega)\big)^2.$$

Without any approximation, this second-order equation in $L^{-1}(\omega)$, whose discriminant is $(m\gamma - n\alpha)^2 + \alpha\gamma z \geq 0$, always has one positive root when $z \leqslant mn$. Using Theorem 1, we could deduce explicitly $\mu_{M \odot N}$. Usually, $M \odot N$ will not be an L-R type fuzzy number.

Nevertheless, if we neglect the term $\alpha\gamma(L^{-1}(\omega))^2$, provided that α and γ are small compared with m and n, and/or ω is in the neighborhood of 1, the above equation becomes simpler, and we infer the approximation formula $(M > 0, N > 0)$

$$(m, \alpha, \beta)_{LR} \odot (n, \gamma, \delta)_{LR} \simeq (mn, m\gamma + n\alpha, m\delta + n\beta)_{LR}. \qquad (15)$$

When $M < 0$ and $N > 0$, (15) becomes

$$(m, \alpha, \beta)_{RL} \odot (n, \gamma, \delta)_{LR} \simeq (mn, n\alpha - m\delta, n\beta - m\gamma)_{RL}. \qquad (16)$$

When $M < 0$ and $N < 0$, (15) becomes

$$(m, \alpha, \beta)_{LR} \odot (n, \gamma, \delta)_{LR} \simeq (mn, -n\beta, -m\delta, -n\alpha - m\gamma)_{RL}. \qquad (17)$$

When spreads are not small compared with mean values, other approximation formulas can be used to give the rough shape of $\mu_{M \odot N}$; for instance, when $M > 0$ and $N > 0$,

$$(m, \alpha, \beta)_{LR} \odot (n, \gamma, \delta)_{LR} \simeq (mn, m\gamma + n\alpha - \alpha\gamma, m\delta + n\beta + \beta\delta)_{LR}. \qquad (18)$$

The membership function defined on the right-hand side of (18) coincides with $\mu_{M \odot N}$ at at least three points: $(mn, 1)$, $[(m - \alpha)(n - \gamma), L(1)]$, $[(m + \beta)(n + \delta), R(1)]$. When more precision is required, it is always possible to get more points of $\mu_{M \odot N}$, such as $(x_L \cdot y_L, \omega)(x_R \cdot y_R, \omega)$, where $\mu_M(x_L) = \mu_N(y_L) = \mu_M(x_R) = \mu_N(y_R) = \omega$, and x_L, y_L (resp. x_R, y_R) are on the left (resp. right) parts of μ_M and μ_N.

N.B.: *Scalar multiplication.* Obviously, from d.α,

$$\forall \lambda > 0, \quad \lambda \in \mathbb{R}, \quad \lambda \odot (m, \alpha, \beta)_{LR} = (\lambda m, \lambda\alpha, \lambda\beta)_{LR},$$

$$\forall \lambda < 0, \quad \lambda \in \mathbb{R}, \quad \lambda \odot (m, \alpha, \beta)_{LR} = (\lambda m, -\lambda\beta, -\lambda\alpha)_{RL}.$$

δ. Inverse of a Fuzzy Number

We know that $\mu_{M^{-1}}(x) = \mu_M(1/x) \; \forall x \neq 0, \; \forall M \in \tilde{\mathscr{P}}(\mathbb{R} - \{0\})$. Let M be a positive L-R type fuzzy number. The equation of the right part of M^{-1}

is

$$\mu_{M^{-1}}(x) = L\left(\frac{1 - mx}{\alpha x}\right), \qquad x \geq 1/m,$$

when $M = (m, \alpha, \beta)_{LR}$. Note that the right part of M^{-1} is built with the left part of M. Moreover M^{-1} is neither a L-R type fuzzy number, nor an R-L type. But if we consider only a neighborhood of $1/m$, $(1 - mx)/\alpha x$ $\simeq ((1/m) - x)/(\alpha/m^2)$ and M^{-1} is approximately of R-L type:

$$(m, \alpha, \beta)_{LR}^{-1} \simeq (m^{-1}, \beta m^{-2}, \alpha m^{-2})_{RL}. \tag{19}$$

A similar formula holds when $M < 0$ since $-(M^{-1}) = (-M)^{-1}$.

ϵ. Division of Fuzzy Numbers

Using the identity $M \oslash N = M \odot N^{-1}$, and (15), (19) for positive L-R and R-L type fuzzy numbers, the following approximate result can be found:

$$(m, \alpha, \beta)_{LR} \oslash (n, \gamma, \delta)_{RL} \simeq \left(m/n, \frac{\delta m + \alpha n}{n^2}, \frac{\gamma m + \beta n}{n^2}\right)_{LR}. \tag{20}$$

Similar formulas could be given when M and/or N are negative.

ζ. Maximum and Minimum of Fuzzy Numbers

Figure 5 shows that when M and N are L-R type fuzzy numbers, $\widetilde{\max}(M, N)$ and $\widetilde{\min}(M, N)$ are not always such since they may be built with parts of both M and N. This happens when μ_M and μ_N have more than one intersection point. More precisely, if M and N have at most one intersection point,

$$\widetilde{\min}(M, N) = M, \quad \widetilde{\max}(M, N) = N \qquad \text{iff} \quad m < n. \tag{21}$$

If M and N have two or three intersection points, x_i, $i = 1, 3$, they are always such that $x_1 < m \leq x_2 \leq n < x_3$ when $m \leq n$ and left (resp. right) parts of M and N are strictly increasing (resp. decreasing). Moreover, $\mu_M(x_2) \geq \max(\mu_M(x_1), \mu_M(x_3))$.

When $n \gg m$ and $\max(\mu_M(x_1), \mu_M(x_3))$ is low, (21) still holds approximately. When $|x_1 - x_3|$ is small with respect to m and n, we can use the formulas

$$\widetilde{\max}((m, \alpha, \beta)_{LR}, (n, \gamma, \delta)_{LR}) \simeq (\max(m, n), \min(\alpha, \gamma), \max(\beta, \delta))_{LR}, \tag{22}$$

$$\widetilde{\min}((m, \alpha, \beta)_{LR}, (n, \gamma, \delta)_{LR}) \simeq (\min(m, n), \max(\alpha, \gamma), \min(\beta, \delta))_{LR}. \tag{23}$$

When $m = n$, (22) and (23) exactly hold.

When more than three intersection points exist, no approximation formula seems available. When M and N have intervals of constant membership, the above reasoning holds replacing "intersection points" by "intersection intervals"; formulas will be given below.

η. Flat Fuzzy Numbers

A *flat fuzzy number* (see Fig. 6) is a fuzzy number M such that $\exists (m_1, m_2) \in \mathbb{R}$, $m_1 < m_2$, and $\mu_M(x) = 1 \; \forall x \in [m_1, m_2]$. A flat fuzzy number can model a fuzzy interval. An L-R type flat fuzzy number M is defined as

$$\mu_M(x) = L((m_1 - x)/\alpha) \qquad x \leqslant m_1, \quad \alpha > 0,$$
$$= R((x - m_2)/\beta) \qquad x \geqslant m_2, \quad \beta > 0,$$
$$= 1 \qquad \text{otherwise.}$$

More briefly, we denote $(m_1, m_2, \alpha, \beta)_{LR}$ by M where L and R are reference functions.

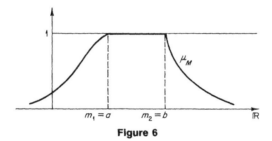

Figure 6

N.B.: A flat fuzzy number could be represented with only three parameters, with flat references, but it would be less convenient than four because the size of the flat part would depend on the values of the spreads. Hence, the four-parameter representation is more general since it avoids this dependency. Formulas for L-R type fuzzy numbers are easily constructed; for instance,

$$(m_1, m_2, \alpha, \beta)_{LR} \oplus (n_1, n_2, \gamma, \delta)_{LR} = (m_1 + n_1, m_2 + n_2, \alpha + \gamma, \beta + \delta)_{LR},$$
$$(24)$$

$$(m_1, m_2, \alpha, \beta)_{LR} \odot (n_1, n_2, \gamma, \delta)_{LR} \simeq (m_1 n_1, m_2 n_2, m_1 \gamma + n_1 \alpha, m_2 \delta + n_2 \beta)_{LR},$$
$$(25)$$

(25) holds for M and $N > 0$.

f. Interpretation and Comments

We already hinted that a fuzzy number M can model an ill-known quantity whose value is "approximately m" and that a flat fuzzy number can be an interval whose boundaries are not sharp, for instance, a fuzzy tolerance interval. A fuzzy set of $\mathbb{R}$ having distinct maxima whose membership values are 1 can model a set of imprecise measures of a given phenomenon. When the maxima have different membership values, they may express the degree of quality of the information inherent in these maxima.

Hence, the maximum membership value of a fuzzy number is interpreted as a grade of reliability, and its spreads model the imprecision of a measurement. The flattening effect supports this interpretation: the reliability of $M \circledast N$ is the least of the reliabilities of M and N.

The distributivity of any extended operation on the union of fuzzy sets is easily interpreted: for instance, it seems quite natural that ("approximately 2" or "approximately 3") + "approximately 1" gives "approximately 3" or "approximately 4." On the contrary, a number whose value is "approximately 1" and "approximately 3" has less meaning: this is consistent with the nondistributivity of any extended operation on the intersection of fuzzy sets; such a number results from conflicting sources of information.

The problem of identification of a membership function is considered at the beginning of Part IV.

Note that our interpretation of fuzzy numbers in the framework of tolerance analysis is supported by the fact that formulas (12) et seq. generalize those of nonfuzzy tolerance analysis. The algebra of real intervals as developed by Moore (NF 1966) is entirely consistent with our results. In particular, Moore points out the nondistributivity of the product of intervals over subtraction of intervals of $\mathbb{R}^+$, i.e., $\Delta(a(b - c)) \neq \Delta(ab - ac)$ where Δx denotes the absolute error in x, and a, b, and c are positive.

The main appeal of formulas (12) et seq. is to extend tolerance analysis to fuzzy intervals, without increasing the amount of computation too much, which makes it possible on a practical level: to be represented a nonfuzzy interval needs two parameters, a fuzzy number requires three, a fuzzy interval four. A manipulation of these parameters is enough to obtain the final membership function. Lastly, the L-R representation is general enough to encompass many shapes of membership functions.

NB.:   Formula (12), which is for addition, allows an empirical comparison between the sup–min and sup–product extension principles: Let M and N be two L-L fuzzy numbers, with $L(x) = e^{-x^2}$, $M = (m, \alpha, \beta)_{LL}$, $N = (n, \gamma, \delta)_{LL}$. Using the max–product extension principle, it is very easy to show that, using this particular reference,

$$M \boxplus N = \left(m + n, \sqrt{(\alpha^2 + \gamma^2)} , \sqrt{(\beta^2 + \delta^2)} \right)_{LL} \qquad (26)$$

where $\boxplus$ denotes the sup–product extended addition.[†] Since $M \oplus N$ $= (m + n, \alpha + \gamma, \beta + \delta)_{LL}$, $M \boxplus N \subseteq M \oplus N$; but this result holds for any extended operation as well, and any real fuzzy sets.

g. Comparison of Fuzzy Numbers

When comparing fuzzy numbers, two kinds of questions may arise:

(1) What is the fuzzy value of the least or the greatest number from a family of fuzzy numbers?

(2) Which is the greatest or the least among several fuzzy numbers?

The answer to the first question is given by the use of the operations $\widetilde{\max}$ and $\widetilde{\min}$. The above two questions are not simultaneously answered because, given a family $M_1, \ldots, M_n$ of fuzzy numbers, $\widetilde{\max}(M_1, \ldots, M_n)$ (resp. $\widetilde{\min}(M_1, \ldots, M_n)$) is not necessarily one of the M_i.

Hence, another method is required to answer question 2. We must evaluate the degree of possibility for $x \in \mathbb{R}$, fuzzily restricted to belong to $M \in \widetilde{\mathcal{P}}(\mathbb{R})$, to be greater than $y \in \mathbb{R}$ fuzzily restricted to belong to $N \in \widetilde{\mathcal{P}}(\mathbb{R})$.

The degree of possibility of $M \geqslant N$ is defined as

$$v(M \geqslant N) = \sup_{x, y : x \geqslant y} \min(\mu_M(x), \mu_N(y)). \tag{27}$$

This formula is an extension of the inequality $x \geqslant y$ according to the extension principle. It is a degree of possibility in the sense that when a pair (x, y) exists such that $x \geqslant y$ and $\mu_M(x) = \mu_N(y) = 1$, then $v(M \geqslant N) = 1$.

Since M and N are convex fuzzy numbers, it can be seen on Fig. 7 that

$$v(M \geqslant N) = 1 \quad \text{iff} \quad m \geqslant n,$$

$$v(N \geqslant M) = \text{hgt}(M \cap N) = \mu_M(d)$$

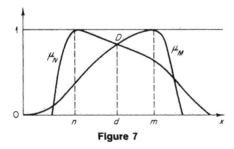

Figure 7

[†] $M \boxplus N$ has the same reference function L as M or N only because we use an exponential function; more generally, this is not true.

where d is the ordinate of the highest intersection point D between μ_M and μ_N. Note that $\text{hgt}(M \cap N)$ is a good separation index for two fuzzy numbers—the closer to 1 is $\text{hgt}(M \cap N)$, the harder it is to know whether M is either greater or less than N. ($\geqslant$ is actually a fuzzy relation (see Chapter 3) between fuzzy numbers.)

When $M = (m, \alpha, \beta)_{LR}$, $N = (n, \gamma, \delta)_{RL}$, the ordinate of D is given by the equation $L((n - d)/\delta) = L((m - d)/\alpha) = \mu_M(d)$, i.e., $\mu_M(d) = L((m - n)/(\alpha + \delta))$ if $m \geqslant n$. Note that the type of M is LR when that of N is RL.

To compare M and N, we need both $v(M \geqslant N)$ and $v(N \geqslant M)$. If, for instance, $v(M \geqslant N) = 1$, we know that either $M \geqslant N$, or M and N are too close to be separated. We may then choose a threshold θ and admit that $M \geqslant_\theta N$ as soon as $v(N \geqslant M) < \theta$.[†] For L-R and R-L type fuzzy numbers, this latter rule reduces to

$$\begin{cases} N \geqslant_\theta M & \text{iff} \quad n - m > \beta + \gamma \quad (\theta = R(1)), \\ M \geqslant_\theta N & \text{iff} \quad m - n > \alpha + \delta \quad (\theta = L(1)). \end{cases}$$

When $\min(v(M \geqslant N), v(N \geqslant M)) \geqslant \theta$, we shall say that M and N are approximately equal, in the sense that they may be very close after a learning process. This is a very weak equality between fuzzy numbers. Stronger equalities could be defined using similarity indices defined in 1.E.c.β. The consistency of fuzzy numbers works much better as a separation index than as a similarity index.

N.B.: 1. Comparison of M and N is equivalent to that of $M \ominus N$ and 0 because $\mu_{M \ominus N}(0) = \text{hgt}(M \cap N) \; \forall M, N \in \tilde{\mathscr{P}}(\mathbb{R})$. Hence if M and N are convex and $v(M \geqslant N) = 1$, $v(N \geqslant M) = \mu_{M \ominus N}(0)$.

2. All results hold for flat fuzzy numbers.

3. We could define $M \succ N$ by $\widetilde{\max}(M, N) = M$ and/or $\widetilde{\min}(M, N) = N$. Such a definition is not very good because M can be very close to N, and still $M \succ N$ can be true while neither $M \geqslant_\theta N$ nor $N \geqslant_\theta M$ holds.

h. Fuzzy Equations

A *fuzzy equation* is an equation whose coefficients and/or variables are fuzzy sets of $\mathbb{R}$. The concept of equation can be extended to deal with fuzzy quantities in several ways. Consider the very simple equation $a * x = b$ where $(a, b) \in \mathbb{R}^2$, x is a real variable, and $*$ is a group operation on $\mathbb{R}$, so that the unique solution is $x = a^{-1} * b$ where a^{-1} is the inverse of a.

α. $A \circledast x = B, \; A, B \in \tilde{\mathscr{P}}(\mathbb{R}), \; x \in \mathbb{R}$

The above equation means that the fuzzy set $A \circledast x$ is the same as B. Note that it is forbidden to shift terms from one side to another. For

---

[†] $\geqslant_\theta$ is a crisp relation and $M \geqslant_\theta N$ means "M is greater than N at level θ."

instance, the equation $A \circledast x = B$ is not equivalent to $(A \circledast x) \ominus B = 0$: the first may have solutions, while the second surely does not, since $(A \circledast x) \ominus B$ is fuzzy and 0 is a scalar.

β. $A \circledast x \subseteq B$

The above equation is a relaxed version of α. The fuzzy set $A \circledast x$ must be contained in B. B could be for example a prescribed tolerance constraint on the output of a given device, A a known error rate of its input, and x a characteristic of the device. The solution of the equation is a tolerance interval for x.

γ. $A \circledast x \simeq B$

The above equation is another relaxed version of α. The fuzzy set $A \circledast x$ must be approximately equal to B. $\simeq$ is defined as a weak equality in the sense of 1.E.c.α or an ϵ-equality in the sense of 1.E.c.β. Besides, we can interpret $A \circledast x \simeq B$ as "neither $A \circledast x \geqslant_\theta B$ nor $B \geqslant_\theta A \circledast x$ hold." Once more the range of x is generally an interval of $\mathbb{R}$.

δ. $A \circledast X \equiv b$, $X \in \mathscr{P}(\mathbb{R})$, $b \in \mathbb{R}$

The above equation is not related to the preceding ones. It means, is there a real fuzzy set X such that $\forall \omega \in [0, 1]$, $\forall a \in \mathbb{R}$ satisfying $\mu_A(a) = \omega$, $\exists x \in \mathbb{R}$, $\mu_X(x) = \omega$, and $a * x = b$? Here, since $*$ is a group operation, it is easy to check that $X = A^{-1} \circledast b$. This type of fuzzy equation could be generalized with a fuzzy right-hand side B. It is consistent with the extension principle.

Equations of type δ may be interpreted in the following way: knowing the fuzzy tolerance interval A of a quantity a, what is the fuzzy tolerance interval X of the quantity x that must satisfy the requirement $a * x = b$?

N.B.: α, β, γ can be generalized to fuzzy variables.

When coefficients are L-R type fuzzy numbers, the actual solution of all fuzzy equations is made much easier. Usually fuzzy equations will be equivalent to a system of nonfuzzy equations (see III.4).

i. Fuzzy Matrices

A fuzzy matrix is a rectangular array of fuzzy numbers. Obviously, there is no difficulty in performing additions on fuzzy matrices. But the product of fuzzy matrices is no longer associative because of the lack of complete distributivity of $\odot$ over $\oplus$. A sufficient condition to preserve associativity is to work only with positive fuzzy matrices (i.e., matrices all of whose elements are positive fuzzy numbers), only with negative fuzzy matrices, or only with diagonal fuzzy matrices.

The problem of the inversion of a fuzzy square matrix M in the sense of h.δ (find M' such that $M \odot M' \equiv I$ where I is the *ordinary* identity matrix) was approximately solved by Moore (NF 1966) when the fuzzy numbers are just ordinary intervals. He used an algorithm based on Hansen's method (Hansen, NF 1965; Hansen and Smith, NF 1967).

The problem of finding the fuzzy eigenvalues of a fuzzy matrix can be solved in the framework of systems of linear fuzzy equations (III.4.A.b).

j. Entropy of a Fuzzy Number

Let M be an L-R fuzzy number and s an entropy function that satisfies the requirements of (1.H). $M = (m, \alpha, \beta)_{LR}$ where L, R are such that

$$\int_{-\infty}^{0} s(L(x))\,dx = k_L < +\infty \qquad \text{and} \qquad \int_{0}^{+\infty} s(R(x))\,dx = k_R < +\infty.$$

The entropy $d(M)$ of M is

$$d(M) = \int_{-\infty}^{m} s\left(L\left(\frac{m-x}{\alpha}\right)\right) dx + \int_{m}^{+\infty} s\left(R\left(\frac{x-m}{\beta}\right)\right) dx$$

$$= \alpha \int_{-\infty}^{0} s(L(x))\,dx + \beta \int_{0}^{+\infty} s(R(x))\,dx,$$

or

$$d(M) = k_L \alpha + k_N \beta. \tag{28}$$

Thus, for a given L-R type, the entropy of a fuzzy number is a linear function of its spreads. It is easy to check that

$$d(M \oplus N) = d(M) + d(N), \qquad d(M \odot N) \simeq m\,d(N) + n\,d(M)$$

(n denotes the mean value of N).

C. EXTENDED FUZZY SETS

a. Type m or Level p Fuzzy Sets

Let A be an ordinary fuzzy set on a universe X. In Zadeh's notation, $A = \int_X \mu_A(x)/x$.

Zadeh (1972) called "fuzzification" of A the change of x or $\mu_A(x)$ into a fuzzy set on X or $[0, 1]$, respectively, for every $x \in X$. When $\mu_A(x)$ becomes fuzzy, A becomes a type 2 fuzzy set (1.G.d). This transformation of an ordinary fuzzy set into a type 2 fuzzy set by blurring the grades of membership is called g-fuzzification (Zadeh, 1972). When x is blurred into a fuzzy set $\tilde{x}$ on X, A is a fuzzy set on $\tilde{\mathcal{P}}(X)$, and is said to be a level 2 fuzzy set (Zadeh, Reference from IV.2, 1971).

More generally, A is said to be a *level p fuzzy set* iff it is a fuzzy set on $\tilde{\mathcal{P}}^{p-1}(X)$ recursively defined as

$$\tilde{\mathcal{P}}^1(X) = \tilde{\mathcal{P}}(X), \qquad \tilde{\mathcal{P}}^p(X) = \tilde{\mathcal{P}}(\tilde{\mathcal{P}}^{p-1}(X)), \quad p > 1.$$

A level p fuzzy set can be viewed as a hierarchy of fuzzy sets.

If X is finite ($|X| = n$) and $[0, 1]$ discretized into k grades of membership, the number of level 2 fuzzy sets of X is $k^{[k^n]}$, that of type 2 fuzzy sets of X is $[k^k]^n$. Since $k^{[k^n]} \geqslant [k^k]^n$ as soon as $k \geqslant 2$ (strict inequality when $k > 2, n > 2$), there are always more level 2 fuzzy sets on X than those of type 2. More precisely, there is no bijection between $\tilde{\mathcal{P}}_2(X)$ and $\tilde{\mathcal{P}}^2(X)$. The two notions are not equivalent: a type 2 fuzzy set is a fuzzy-valued fuzzy set, a level 2 fuzzy set is a fuzzy set of fuzzy sets.

Zadeh (1972) introduced the notion of s-fuzzification (s for support). In this fuzzification each singleton of X, denoted by $1/x$, is changed into a fuzzy set $K(x)$, clustered around x; the mapping K from X to $\tilde{\mathcal{P}}(X)$ is called the *kernel*. The result of an s-fuzzification is a fuzzy set on X:

$$F(A, K) = \bigcup_{x \in X} \mu_A(x) K(x) \tag{29}$$

where $\mu_A(x) K(x) = \int_X \mu_A(x) \mu_{K(x)}(x')/x' \; \forall x \in X$. Note that a level 2 fuzzy set can always be reduced to an ordinary one in a similar way:

$$A = \sum_{i=1}^{k} \mu_i / A_i, \qquad A_i \in \tilde{\mathcal{P}}(X)$$

is changed into $\bigcup_{i=1}^{k} \mu_i A_i$.

The effect of an s-fuzzification is to make a fuzzy set more fuzzy.

b. Extended Set-Theoretic Operations for Type 2 Fuzzy Sets

In 1.G.d we defined set-theoretic operations on fuzzy sets of type 2, by induction from the lattice structure of $(\tilde{\mathcal{P}}([0, 1]), \cup, \cap, ^-)$. This definition proved to be semantically very poor.

Now, since a fuzzy set of type 2 is obtained by assigning fuzzy membership values to elements of X, we can, following the same idea, extend the set-theoretic operations of ordinary fuzzy set theory to allow them to deal with fuzzy grades of membership; this is done using the extension principle. Let A and B be fuzzy sets of type 2 of X. $\mu_A(x)$ and $\mu_B(x)$ belong to $\tilde{\mathcal{P}}([0, 1])$. We write

$$\mu_{A \sqcup B}(x) = \widetilde{\max}(\mu_A(x), \mu_B(x)) \qquad \forall x \in x, \tag{30}$$

$$\mu_{A \sqcap B}(x) = \widetilde{\min}(\mu_A(x), \mu_B(x)) \qquad \forall x \in X, \tag{31}$$

$$\mu_{\bar{A}}(x) = 1 \ominus \mu_A(x) \qquad \forall x \in X. \tag{32}$$

$\sqcap$, $\sqcup$, $^\dashv$ were first proposed by Mizumoto and Tanaka (1976a). These operators allow overcoming the paradox quoted in 1.G.d: when $\mu_A(x)$ is approximately $l_A(x)$, $\mu_B(x)$ approximately $l_B(x)$, and $l_A(x) > l_B(x)$, we know from B.d that $\mu_{A \sqcup B}(x)$ will be "approximately $l_A(x)$."

On the contrary, the structure of $\tilde{\mathcal{P}}([0,1])$ is poorer: $(\tilde{\mathcal{P}}((0,1)), \sqcup, \sqcap, ^\dashv)$ is only a pseudocomplemented structure. The structure is pseudocomplemented because $M \in \tilde{\mathcal{P}}([0,1])$ iff $1 \ominus M \in \tilde{\mathcal{P}}([0,1])$. However, among the properties listed in B.d, distributivity does not hold for fuzzy sets of $[0,1]$, as proved by Mizumoto and Tanaka (1976a), neither does absorption hold, existence of a 0 and a 1, identity, and excluded middle laws.

If we restrict ourselves to the set $\mathcal{N}([0,1])$ of fuzzy numbers of $[0,1]$, the structure is richer: $(\mathcal{N}([0,1]), \sqcup, \sqcap, ^\dashv)$ is a pseudocomplemented complete distributive lattice, and all the properties of ordinary set-theoretic operators on fuzzy sets are satisfied, i.e., commutativity, associativity, idempotency, distributivity, identity, $A \sqcap \emptyset = \emptyset$, $A \sqcup X = X$, absorption, De Morgan's laws, involution (see 1.B.d). Only the equivalence and symmetrical difference formulas fail to hold any longer. This result stems from (see Dubois and Prade, Reference from III.1, 1978b):

$$\widetilde{\max}[\widetilde{\min}(M,N), \min(1 \ominus M, 1 \ominus N)] \neq \widetilde{\min}[\widetilde{\max}(M, 1 \ominus N),$$
$$\widetilde{\max}(1 \ominus M, N)] \text{ for some } M, N \in \mathcal{N}([0,1]).$$
$$\widetilde{\min}[\widetilde{\max}(M,N), \widetilde{\max}(1 \ominus M, 1 \ominus N)] \neq \widetilde{\max}[\widetilde{\min}(M, 1 \ominus N),$$
$$\min(1 \ominus M, N)] \text{ for some } M, N \in \mathcal{N}([0,1]).$$

Note that fuzzy-number-valued grades of membership are intuitively appealing since they may model our imprecise knowledge of these grades. Fuzzy numbers of $[0,1]$ are also easily combined thanks to the algebraic formulas provided in section B. Thus, intuitive meaning and practical reasons induce us to adopt $\mathcal{N}([0,1])$ as the best valuation set for type 2 fuzzy sets.

To define inclusion of fuzzy-number-valued fuzzy sets, we must compare fuzzy grades of membership, in order to be consistent with the extension principle. For instance, we may write

$$A \subseteq B \quad \text{iff} \quad \forall x \in X, \quad \widetilde{\min}(\mu_A(x), \mu_B(x)) = \mu_A(x)$$
$$\text{and} \quad \widetilde{\max}(\mu_A(x), \mu_B(x)) = \mu_B(x).$$

This definition is somewhat rigid (see B.g, N.B.3). We may choose

$$A \subseteq_\theta B \quad \text{iff} \quad \forall x \in X, \quad v(\mu_B(x) \geqslant \mu_A(x)) = 1,$$
$$v(\mu_A(x) \geqslant \mu_B(x)) < \theta$$

where θ is a threshold (see B.g). Set equality may be also very strict: $(A = B \Leftrightarrow \mu_A(x) = \mu_B(x) \; \forall x \in X)$ can be relaxed using the similarity indices given in 1.E.

Remarks 1 (30), (31), (32) restricted to act on interval-valued membership grades give operators for Φ-fuzzy sets (Sambuc, Reference from II.1) and multivalued quantities (Grattan-Guiness, Reference from II.1): Example:

$$\widetilde{\max}([a,b],[c,d]) = [\max(a,c),\max(b,d)],$$
$$\widetilde{\min}([a,b],[c,d]) = [\min(a,c),\min(b,d)],$$
$$1 \ominus [a,b] = [1-b, 1-a]; \quad \forall [a,b],[c,d] \subseteq [0,1].$$

2 The set of intervals of $[0,1]$ is only partially ordered under $\widetilde{\max}$ and $\widetilde{\min}$. In the context of an application, Ponsard (Reference from V, 1977a) introduced an inclusion of fuzzy sets so as to recover a linear ordering. The corresponding inequality of membership values is $[a,b] \leqslant [c,d]$ iff either $b < d$ or ($b = d$ and $a \leqslant c$). An alternative definition is $[a,b] \leqslant [c,d]$ iff either $a < c$ or ($a = c$ and $b \leqslant d$). Using one of these inclusions, the set of intervals of $[0,1]$ is linearly ordered. The union of Φ-fuzzy sets is now defined by means of the operator W on the interval-valued membership grades:

$$[a,b]W[c,d] = \begin{cases} [a,b] & \text{iff} & [c,d] \leqslant [a,b] \\ [c,d] & \text{iff} & [a,b] \leqslant [c,d]. \end{cases}$$

The intersection of Φ-fuzzy sets using the operator M is

$$[a,b]M[c,d] = \begin{cases} [a,b] & \text{iff} & [a,b] \leqslant [c,d], \\ [c,d] & \text{iff} & [c,d] \leqslant [a,b]. \end{cases}$$

Under W and M the set of intervals of $[0,1]$ is a distributive linear ordered lattice. W and M are associative, commutative, idempotent, and satisfy the law of absorption. Lastly,

$$[a,b]M[c,d] = [a,b]W[c,d] \quad \text{iff} \quad [a,b] = [c,d].$$
$$[a,b]M[c,d] = [c,d] \quad \text{iff} \quad [a,b]W[c,d] = [a,b].$$

c. Some Further Operations on Fuzzy Sets of Type 2

mth power of a fuzzy set of type 2 (Zadeh, 1975): Let A be a fuzzy set of type 2 on X. $\mu_A(x)$ is a fuzzy set on $[0,1]$:

$$\mu_A(x) = \int_{[0,1]} \lambda(t)/t.$$

Just as for the definition of union, intersection, or complementation of type 2 fuzzy sets, the extension principle provides us a way to define the

mth power $A^{\boxed{m}}$ of A as

$$\mu_{A^{\boxed{m}}}(x) = \int_{[0,\,1]} \lambda(t)/t^m = (\mu_A(x))^m \text{(see II.2.B.d.}\zeta).$$

For instance, for $m = 2$, if $\mu_A(x)$ is a fuzzy number, $\mu_{A^{\boxed{2}}}(x)$ is also a fuzzy number that is less than $\mu_A(x)$ in the sense of min. It must be clear that $\mu_{A^{\boxed{2}}}(x)$ is completely different from $\int_{[0,\,1]} (\lambda(t))^2/t = \mu_{A'}(x)$ (see Fig. 8), the second power of $\mu_A(x)$ in the sense of II.1.B.f.

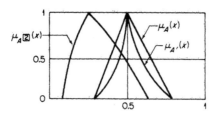

Figure 8

Addition: Let us consider the following level 2 fuzzy set $\mathfrak{M}$ on $\mathbb{R}$: $\mathfrak{M}$ is a fuzzy set of L-R fuzzy numbers M_λ. $\mathfrak{M} = \int \lambda/M_\lambda$ where $M_\lambda = (m, \alpha_\lambda, \beta_\lambda)_{LR}$. Symbolically, we write $\mathfrak{M} = (m, \tilde{\alpha}, \tilde{\beta})_{LR}$ where $\mu_{\tilde{\alpha}}(\alpha_\lambda) = \lambda = \mu_{\tilde{\beta}}(\beta_\lambda)$. $\mathfrak{M}$ has an ordinary mean value but fuzzy spreads. $\mathfrak{M}$ is represented in Fig. 9. $\mathfrak{M}$ may be also viewed as a type 2 fuzzy number: $\mu_{\mathfrak{M}}(m_0)$ is sketched in the right part of Fig. 9.

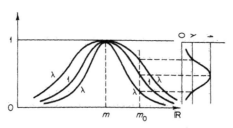

Figure 9

If we suppose that the spreads $\tilde{\alpha}$ and $\tilde{\beta}$ are l-r type fuzzy numbers, addition can easily be extended to level 2 fuzzy sets of $\mathbb{R}$, like $\mathfrak{M} = (m, \tilde{\alpha}_{lr}, \tilde{\beta}_{lr})_{LR}$ and $\mathfrak{N} = (n, \tilde{\gamma}_{lr}, \tilde{\delta}_{lr})_{LR}$, through the formula

$$\mathfrak{M} \oplus \mathfrak{N} = (m + n, \tilde{\alpha}_{lr} \oplus \tilde{\gamma}_{lr}, \tilde{\beta}_{lr} \oplus \tilde{\delta}_{lr})_{LR}. \tag{33}$$

Such fuzzy–fuzzy numbers can model situations where only the rough shape of the characteristic functions of the fuzzy numbers is known.

REFERENCES

Baas, S. M., and Kwakernaak, H. (1977). Rating and ranking of multiple-aspect alternatives using fuzzy sets. *Automatica* **13**, 47–58. (Reference from IV.3.)

Borghi, O., Marchi, E., and Zo, F. (1976). A note on embedding of fuzzy sets in a normed space. *Rev. Union Mat. Argent. Assoc. Fis. Argent.* **28**, No. 1, 36–41.

Dubois, D., and Prade, H. (1978a). Comment on "Tolerance analysis using fuzzy sets" and "A procedure for multiple aspect decision making." *Int. J. Syst. Sci.* **9**, No. 3, 357–360.

Dubois, D., and Prade, H. (1978b). Operations on fuzzy numbers. *Int. J. Syst. Sci.* **9**, No. 6, 613–626.

Dubois, D., and Prade, H. (1978c). Fuzzy real algebra: Some results. In "Fuzzy Algebra, Analysis, Logics," Tech. Rep. TR-EE 78/13. Purdue Univ., Lafayette, Indiana. [*Int. J. Fuzzy Sets Syst.* **2**, 327–348 (1979).]

Dubois, D., and Prade, H. (1978d). Systems of linear fuzzy constraints. In "Fuzzy Algebra, Analysis, Logics," Tech. Rep. TR-EE 78/13. Purdue Univ., Lafayette, Indiana. [*Int. J. Fuzzy Sets Syst.* **3**, 37–48 (1980).] (Reference from III.4, 1978b.)

Jain, R. (1976). Tolerance analysis using fuzzy sets. *Int. J. Syst. Sci.* **7**, No. 12, 1393–1401.

Mareš, M. (1977a). How to handle fuzzy-quantities. *Kybernetika* (Prague) **13**, No. 1, 23–40.

Mareš, M. (1977b). On fuzzy-quantities with real and integer values. *Kybernetika* (Prague) **13**, No. 1, 41–56.

Mizumoto, M., and Tanaka, K. (1976a). Some properties of fuzzy sets of type 2. *Inf. Control* **31**, 312–340.

Mizumoto, M., and Tanaka, K. (1976b). The four operations of arithmetic on fuzzy numbers. *Syst.—Comput.—Controls* **7**, No. 5, 73–81.

Mizumoto, M., and Tanaka, K. (1976c). Algebraic properties of fuzzy numbers. *Int. Conf. Cybern. Soc., Washington, D.C.*

Nahmias, S. (1978). Fuzzy variables. *Int. J. Fuzzy Sets Syst.* **1**, No. 2, 97–111.

Nguyen, H. T. (1976). "A Note on the Extension Principle for Fuzzy Sets," UCB/ERL Memo M-611. Univ. of California, Berkeley. [Also in *J. Math. Anal. Appl.* **64**, No. 2, 369–380 (1978).]

Nieminen, J. (1977). On the algebraic structure of fuzzy sets of type 2. *Kybernetika* (Prague) **13**, No. 4, 261–273.

Zadeh, L. A. (1972). A fuzzy set theoretic interpretation of linguistic hedges. *J. Cybern.* **2**, No. 3, 4–34.

Zadeh, L. A. (1975). The concept of a linguistic variable and its application to approximate reasoning. Parts 1, 2, and 3. *Inf. Sci.* **8**, 199–249; **8**, 301–357; **9**, 43–80.

FUZZY RELATIONS

The concept of a fuzzy relation is introduced naturally, as a generalization of crisp relations, in fuzzy set theory. It can model situations where interactions between elements are more or less strong. Fuzzy relations can be composed, and this composition is closely related to the extension principle.

A great deal of work has been done in the field of binary fuzzy relations. Notions such as equivalence and ordering have been generalized to fuzzy similarity and fuzzy ordering. However, it has been made clear that most of the mathematical tools that have been developed concerning this topic are not new. Similarities are very connected to distances. Fuzzy preorderings still contain undominated and undominating elements.

More original are the equations of fuzzy relations. Moreover, their resolution may prove to be useful in the framework of computerized diagnosis.

More sophisticated fuzzy relations are briefly outlined at the end of this chapter.

A. *n*-ARY FUZZY RELATIONS

a. Fuzzy Relations and Fuzzy Restrictions

Let $X_1, \ldots, X_n$ be n universes. An n-ary fuzzy relation R in $X_1 \times \cdots \times X_n$ is a fuzzy set on $X_1 \times \cdots \times X_n$ (Zadeh, Reference from II.1, 1965). An ordinary relation is a particular case of fuzzy relations.

Example $n = 2$. $X_1 = X_2 = \mathbb{R}^+ - \{0\}$. $R =$ "much greater than" may be defined by $\mu_R(x, y) = 0$ iff $x \leqslant y$ and $\mu_R(x, y) = \min(1, (x - y)/9y)$ iff $x \geqslant y$; $\mu_R(x, y) = 1$ as soon as $x \geqslant 10y$.

Let $v = (v_1, \ldots, v_n)$ be a variable on $X = X_1 \times \cdots \times X_n$. A *fuzzy restriction*, denoted $R(v)$, is a fuzzy relation R that acts as an elastic constraint on the values, elements of X, that may be assigned to a variable v (Zadeh, 1975a). In this context a *variable* is viewed as a 3-tuple $(v, X, R(v))$; v is the name of the variable.

A fuzzy relation R is normalized iff the fuzzy set R is normalized.

The *projection* of a fuzzy relation R on $X_{i_1} \times \cdots \times X_{i_k}$, where $(i_1, \ldots, i_k)$ is a subsequence of $(1, 2, \ldots, n)$, is a relation on $X_{i_1} \times \cdots \times X_{i_k}$ defined by (Zadeh, 1975a):

$$\text{proj}\big[R; X_{i_1}, \ldots, X_{i_k} \big]$$
$$= \int_{X_{i_1} \times \cdots \times X_{i_k}} \sup_{x_{j_1}, \ldots, x_{j_l}} \mu_R(x_1, \ldots, x_n)/(x_{i_1}, \ldots, x_{i_k}) \qquad (1)$$

where $(j_1, \ldots, j_l)$ is the subsequence complementary to $(i_1, \ldots, i_k)$ in $(1, \ldots, n)$.

N.B.:   Projections are also called *marginal fuzzy restrictions*.

Conversely, if R is a fuzzy set in $X_{i_1} \times \cdots \times X_{i_k}$, then its *cylindrical extension* in $X_1 \times \cdots \times X_n$ is a fuzzy set $c(R)$ on $X_1 \times \cdots \times X_n$ defined by (Zadeh, 1975a)

$$c(R) = \int_{X_1 \times \cdots \times X_n} \mu_R(x_{i_1}, \ldots, x_{i_k})/(x_1, \ldots, x_n). \qquad (2)$$

Let R and S be two fuzzy relations on $X_1 \times \cdots \times X_r$ and $X_s \times \cdots \times X_n$, respectively, with $s \leqslant r + 1$: the *join* of R and S is $c(R) \cap c(S)$, where $c(R)$ and $c(S)$ are cylindrical extensions on $X_1 \times \cdots \times X_n$.

Example $n = 3$. $X_i = \mathbb{R}$, $i = 1, 3$. (x, y, z) is fuzzily restricted to be on a sphere ($\mu_R(x, y, z) = e^{-k^2|x^2 + y^2 + z^2 - R^2|}$). The projection of R on the (x, y) plane has membership function

$$\mu_R(x, y) = e^{-k^2|x^2 + y^2 - R^2|} \qquad \text{iff} \quad x^2 + y^2 \geqslant R^2,$$

$\mu_R(x, y) = 1$ otherwise. We obtain a fuzzy disk. The cylindrical extension of this fuzzy disk is the fuzzy cylindrical volume whose base is the fuzzy disk and which contains the fuzzy sphere.

N.B.:   A *section* of a fuzzy relation R is obtained by assigning constant values to some of the variables fuzzily restricted by R. In the above example a section of the fuzzy sphere is a fuzzy circle.

In terms of their cylindrical extensions the *composition* of two fuzzy relations R and S respectively on $X_1 \times \cdots \times X_r$ and on $X_s \times \cdots \times X_n$ with $s \leqslant r$ is expressed by (Zadeh, 1975a)

$$R \circ S = \operatorname{proj}\left[c(R) \cap c(S); X_1 \times \cdots \times X_{s-1} \times X_{r+1} \times \cdots \times X_n\right]. \quad (3)$$

$R \circ S$ is a fuzzy relation in the symmetrical difference of the universes of R and S.

b. Interactivity (Zadeh, 1975b)

An n-ary fuzzy restriction $R(v_1, \ldots, v_n)$ is said to be *separable* iff $R(v_1, \ldots, v_n) = R(v_1) \times \cdots \times R(v_n)$ where $\times$ denotes the cartesian product (2.A.a) and $R(v_i)$ is the projection of R on X_i, i.e.,

$$\mu_R(x_1, \ldots, x_n) = \min_{i=1, n} \mu_{\operatorname{proj}[R; X_i]}(x_i).$$

Note that in terms of cylindrical extension, the above formula can be written

$$R = \bigcap_{i=1, n} c(\operatorname{proj}[R; X_i]). \quad (4)$$

R is separable iff it is the join of its projections. If R is separable, so are all its marginal fuzzy restrictions. The variables $v_1, \ldots, v_n$ are said to be *noninteractive* iff their restriction $R(v_1, \ldots, v_n)$ is a separable fuzzy restriction. It is easy to check that

$$R(v_1, \ldots, v_n) \subseteq R(v_1) \times \cdots \times R(v_n) = \bigcap_{i=1, n} c(\operatorname{proj}[R(v_1, \ldots, v_n); X_i]).$$

$$(5)$$

Figure 1 sketches two binary nonfuzzy relations. On the left-hand side the choice of a given value in $R(v_1)$ for v_1 does not at all restrict the choice of a value in $R(v_2)$ for v_2. This pair of values will always satisfy the relation R. On the contrary, the choice of a value for v_2 depends upon the value of v_1 and conversely, in order to satisfy the relation of the right-hand part of Fig. 1. It is an example of noninteractivity and interactivity, respectively, for nonfuzzy relations.

Note that the ordinary product of projections of a given relation R, $\operatorname{proj}[R; X_1] \cdot \operatorname{proj}[R; X_2] \cdot \ldots \cdot \operatorname{proj}[R; X_n]$ (see 1.B.e) is an interactive relation contained in $\bigcap_{i=1, n} c(\operatorname{proj}[R; X_i])$. As a matter of fact, the separable restriction $R(v_1) \times \cdots \times R(v_n)$ is associated with the greatest (in the sense of Zadeh's inclusion of fuzzy sets) of the relations whose projections are $\operatorname{proj}[R(v_1, \ldots, v_n); X_i] = R(v_i)$.

Interactivity must be considered when extending a given function, in the sense of 2.A. For instance, the nondistributivity of $\odot$ over $\oplus$ (2.B.d.β) can

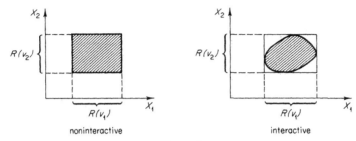

noninteractive   interactive

Figure 1

be explained in terms of interactivity: while there is no problem in the extension of the function $z(x + y)$, the situation is different for the extension of $zx + ty$ forgetting the constraint of interactivity $z = t$.

More generally, given a function from $X_1 \times \cdots \times X_n$ to Y and a fuzzy restriction R on the arguments of f, the extension principle becomes

$$f(A_1, \ldots, A_n)$$

$$= \int_{X_1 \times \cdots \times X_n} \min(\mu_{A_i}(x_1), \ldots, \mu_{A_n}(x_n), \mu_R(x_1, \ldots, x_n))/f(x_i, \ldots, x_n).$$

$$(6)$$

where A_i is a fuzzy set on X_i.

When R is an ordinary separable relation on $\mathbb{R}^n$, the associated restriction means $\forall i = 1, n$, $\exists I_i$, $x_i \in I_i$ where I_i is a union of disjoint intervals. The constraint is implicitly satisfied as soon as $\mu_{A_i}(x_i) = 0$, $\forall x_i \notin I_i$. An example of interactivity where R is an ordinary nonseparable relation on $\mathbb{R}^2$ is given in the following paragraph.

Calculate the fuzzy restriction H of $ax + by$ (a fuzzy set is a unary relation) when x and y are restricted by fuzzy sets M and N, respectively, and by the constraint $x + y = 1$, $(a, b) \in \mathbb{R}^2$:

$$\mu_H(z) = \sup_{\substack{z = ax + by \\ x + y = 1}} \min(\mu_M(x), \mu_N(y));$$

hence,

$$\mu_H(z) = \min\left(\mu_M\left(\frac{z - b}{a - b} \right), \mu_N\left(\frac{a - z}{a - b} \right) \right) \qquad \text{if} \quad a \neq b;$$

and if $a = b$,

$$\mu_H(z) = \begin{cases} \mu_{M \oplus N}(1) & \text{if} \quad z = a, \\ 0 & \text{otherwise.} \end{cases}$$

However, the existence of a nonseparable restriction R does not always simplify the computation of $f(A_1, \ldots, A_n)$ as above, but can make it totally unwieldy.

N.B.: Interactivity in the sense of this section was called β-interactivity by Zadeh (1975b). Another kind of interactivity will be introduced in IV.2.

c. Extension Principle and Composition of n-ary Fuzzy Relations

The extension principle can be written (see 2.A.a)

$$\mu_B(y) = \sup_{\substack{x_1, \ldots, x_n \\ y = f(x_1, \ldots, x_n)}} \min(\mu_{A_1}(x_1), \ldots, \mu_{A_n}(x_n))$$

where $B = f(A_1, \ldots, A_n)$. By denoting $R = c(A_1) \cap \cdots \cap c(A_n) = A_1 \times \cdots \times A_n$ and letting S be the ordinary relation defined by $\mu_S(x_1, \ldots, x_n, y) = 1$ iff $y = f(x_1, \ldots, x_n)$, we have $B = R \circ S$, and the extension principle appears as a particular case of composition of fuzzy relations. When a restriction T on $(x_1, \ldots, x_n)$ is added, B becomes $B = (R \cap T) \circ S$.

Remark From a computational point of view it may be interesting to solve the equation $y = f(x_1, \ldots, x_n)$ (or the corresponding system if nonfuzzy restrictions on $(x_1, \ldots, x_n)$ exist) and to introduce the calculated x_i in μ_{A_i}. Once more the formula becomes a composition of fuzzy relations.

B. BINARY FUZZY RELATIONS

Binary relations have received much attention in the literature because the notion of a link between two elements belonging to the same universe or two different universes is fundamental in systems theory. Some classical definitions follow.

a. Definitions

Let R be a fuzzy relation on $X \times Y$. The *domain* of R, denoted dom(R), and the *range* of R, denoted ran(R), are respectively defined by

$$\mu_{\text{dom}(R)}(x) = \sup_y \mu_R(x, y) \quad \forall x \in X$$

and

$$\mu_{\text{ran}(R)}(y) = \sup_x \mu_R(x, y) \quad \forall y \in Y.$$

The *inverse* of R, denoted R^{-1}, is the fuzzy relation on $Y \times X$ defined by $\mu_{R^{-1}}(y,x) = \mu_R(x,y)$ (Yeh, 1973).

Yeh (1973) has extended to fuzzy relations definitions which are rather specific of functions. R is:

ϵ-*determinate* iff $\forall x \in X$, $\exists$ at most one $y \in Y$, such that $\mu_R(x,y) \geqslant \epsilon$;
ϵ-*productive* iff $\forall x \in X$, $\exists y$, $\mu_R(x,y) \geqslant \epsilon$;
An ϵ-*function* iff R is both ϵ-determinate and ϵ-productive; a 1-function is an ordinary function when restricted to its 1-cut;
ϵ-*onto* iff $\forall y \in Y$, $\exists x \in X$, $\mu_R(x,y) \geqslant \epsilon$;
ϵ-*injective* iff R is an ϵ-function and R^{-1} is ϵ-determinate.
ϵ-*bijective* iff R and R^{-1} are both ϵ-functions.

In the following definitions $X = Y$. Now we give the fuzzy version of well-known possible properties of relations in a universe X.

Three extensions of reflexivity have been proposed. R is:

reflexive iff $\forall x \in X$, $\mu_R(x,x) = 1$ (Zadeh, 1971);
ϵ-*reflexive* iff $\forall x \in X$, $\mu_R(x,x) \geqslant \epsilon$ (Yeh, 1973);
weakly reflexive iff $\forall x \in X$, $\forall y \in X$, $\mu_R(x,x) \geqslant \mu_R(x,y)$ (Yeh, 1973).

Symmetry is defined by: R is *symmetric* iff $\forall x \in X$, $\forall y \in X$, $\mu_R(x,y) = \mu_R(y,x)$.

b. Composition of Binary Fuzzy Relations

α. Properties

The composition of fuzzy relations has already been introduced in A.a. In the particular case of binary relations the composition of R and S on $X \times Y$ and $Y \times Z$ respectively can be written

$$\mu_{R \circ S}(x,y) = \sup_{y \in Y} \min(\mu_R(x,y), \mu(y,z)) \qquad \forall x \in X, \quad \forall z \in Z. \quad (7)$$

There are some properties that are common to binary relations. They can be proven without difficulty. Let U be an extra relation on $Z \times W$ and T on $Y \times Z$. Then:

associativity: $R \circ (S \circ U) = (R \circ S) \circ U$;
distributivity over union: $R \circ (S \cup T) = (R \circ S) \cup (R \circ T)$;
weak distributivity over intersection: $R \circ (S \cap T) \subseteq (R \circ S) \cap (R \circ T)$;
monotonicity $S \subseteq T$ implies $R \circ S \subseteq R \circ T$;
symmetrization: $R \circ R^{-1}$ is a weakly reflexive and symmetric relation on $X \times X$.

A nonzero fuzzy relation Q on X is weakly reflexive and symmetric iff there is a universe Y and a fuzzy relation R on $X \times Y$ such that $Q = R \circ R^{-1}$ (Yeh, 1973).

β. Interpretations

(7) can be interpreted in the following way: $\mu_{R \circ S}(x, z)$ is the strength of a set of chains linking x to z. Each chain has the form x-y-z. The strength of such a chain is that of the weakest link. The strength of the relation between x and z is that of the strongest chain between x and z.

Let A be a fuzzy set in X: (7) can be rewritten

$$\mu_{A \circ R}(y) = \sup_{x} \min(\mu_A(x), \mu_R(x, y)).$$

We say that $B = A \circ R$ is a fuzzy set *induced* from A through R. This induction generalizes a well-known nonfuzzy rule: if $x = a$ and $y = f(x)$, then $y = f(a)$—as shown in Fig. 2 (Zadeh, 1975b): We have $B = \text{proj}[c(A) \cap R; Y]$.

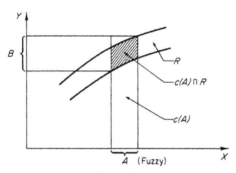

Figure 2

γ. Representation of a Fuzzy Relation on Finite Universes

When the related universes X and Y are finite, a fuzzy relation R on $X \times Y$ can be represented as a matrix $[R]$ whose generic term $[R]_{ij}$ is $\mu_R(x_i, y_j) = r_{ij}$, $i = 1, n, j = 1, m$, where $|X| = n$ and $|Y| = m$.

The composition of finite fuzzy relations can thus be viewed as a matrix product. With $[S]_{jk} = s_{jk}$, $k = 1, p$, $p = |Z|$,

$$[R \circ S]_{ik} = \sum_{j} r_{ij} s_{jk}$$

where $\sum$ is in fact the operation max and product the operation min.

δ. *Convergence of Powers of Fuzzy Relations on a Finite Universe* (Thomason, 1977)

Let R be a fuzzy relation on $X \times X$ where $|X| = n$. The mth power of a fuzzy relation is defined as $R^m = R \circ R^{m-1}$, $m > 1$, and $R^1 = R$. The following propositions hold:

the power of R either converges to idempotent R^c for a finite c or oscillates with finite period (if R^m does not converge, then it must oscillate with a finite period since $|X|$ is finite and the composition is deterministic and cannot introduce numbers not in R originally);

if $\forall i, j$, $\exists k$ such that $r_{ij} \leqslant \min(r_{ik}, r_{kj})$, then R converges to R^c where $c \leqslant n - 1$. (See Thomason, 1977, for a proof.)

Other results in more particular cases can be found in Thomason (1977).

ε. *Other Compositions*

Since $R \circ S$ can be written $\text{proj}[c(R) \cap c(S); X \times Z]$ where R and S are respectively on $X \times Y$ and $Y \times Z$, other compositions may be introduced by modifying the operator used for the intersection.

Changing min to $*$, we define $R \boxdot S$ through

$$\mu_{R \boxdot S}(x, z) = \sup_{y} (\mu_R(x, y) * \mu_S(y, z)).$$

Zadeh (1971) proved that when $*$ is associative, and nondecreasing with respect to each of its arguments, the sup-$*$ composition satisfies associativity, distributivity over union, and monotonicity.

Examples of such operations are product and bold intersection (1.B.e, formula (12)).

We may encounter another kind of alternative compositions, inf–max composition. The following property holds: $\overline{R \circ S} = \overline{R} \,\overline{\circ}\, \overline{S}$ where $\overline{\circ}$ denotes inf–max composition.

c Transitivities

α. *Max–Min Transitivity*

The idea behind transitivity is that the shorter the chain, the stronger the relation. In particular, the strength of the link between two elements must be greater than or equal to the strength of any indirect chain (i.e., involving other elements).

Let R be a fuzzy relation on $X \times X$, R is max–min transitive iff $R \circ R \subseteq R$, or more explicitly (Zadeh, 1971)

$$\forall(x, y, z) \in X^3, \quad \mu_R(x, z) \geqslant \min(\mu_R(x, y), \mu_R(y, z)).$$

Write $R^m = R \circ R^{m-1}$ with $m > 1$ and $R^1 = R$. If R is transitive, $R^m \subseteq R$, $m \geqslant 1$; hence $R = \hat{R}$ where $\hat{R}$ is the transitive closure of R, defined as $\hat{R} = R \cup R^2 \cup \cdots \cup R^m \cup \cdots$.

Generally, when R is not transitive but reflexive, $\hat{R}$ still exists because the sequence $\mu_{R^m}(x, y)$ is increasing with m and bounded by 1 (Tamura *et al.*, 1971).

Proof:

$$\mu_{R^m}(x, y) = \sup_{x_1, x_2, \ldots, x_{m-1}} \min(\mu_R(x, x_1), \mu_R(x_1, x_2), \ldots, \mu_R(x_{m-1}, y)).$$

Hence

$$\mu_{R^m}(x, y) \geqslant \sup_{x_1, \ldots, x_{m-2}} \min(\mu_R(x, x_1), \ldots, \mu_R(x_{m-2}, y), \mu_R(y, y)).$$

Because R is reflexive, the left-hand side of the inequality is equal to $\mu_{R^{m-1}}(x, y)$.   Q.E.D.

It is easy to show that (Tamura *et al.*, 1971)

$$\forall (x, y, z) \in X^3, \quad \mu_{R^{m+n}}(x, z) \geqslant \min(\mu_{R^m}(x, y), \mu_{R^n}(y, z)).$$

When $m \to +\infty$ and $n \to +\infty$, we obtain

$$\mu_{\hat{R}}(x, z) \geqslant \min(\mu_{\hat{R}}(x, y), \mu_{\hat{R}}(y, z)).$$

So the transitive closure $\hat{R}$ of R is max–min transitive.

When X is finite and $|X| = n$, $\exists k < n$, $R^k = \hat{R}$ because chains involving more than n elements must necessarily have cycles that do not alter the strength of the chains.

N.B.:   Note that if R models short-range interactions between elements, its transitive closure models long-range interaction.

β.   *Other Transitivities*

Other transitivities, associated with other kinds of composition of fuzzy relations, can be defined. Generally, R is said to be max-* transitive iff $R \boxdot R \subseteq R$.

Zadeh (1971) considered max–product transitivity. Bezdek and Harris (1978) introduced several other transitivities; max · * where $a * b$ is given by:

(1) $a \wedge b = \max(0, a + b - 1)$ (bold intersection);
(2) $a \square b = \frac{1}{2}(a + b)$ (arithmetic mean);
(3) $a \vee b = \max(a, b)$ (union);
(4) $a \,\hat{+}\, b = a + b - ab$ (probabilistic sum).

The appealing features of some of these transitivities will be discussed later.

C.   SIMILARITY RELATIONS AND RELATED TOPICS

a.   Definitions

"The concept of similarity relation is essentially a generalization of the concept of an equivalence relation" (Zadeh, 1971). More specifically, a similarity relation is a fuzzy relation in a universe X, denoted S, which is reflexive, symmetrical, and max–min transitive.

The complement of S, say $D = \bar{S}$, is called a dissimilarity relation ($\mu_D(x, y) = 1 - \mu_S(x, y)$). D is *antireflexive* (i.e., $\mu_D(x,x) = 0$, $\forall x \in X$), symmetrical, and min–max transitive (i.e., $\mu_D(x,z) \leqslant \max(\mu_D(x, y)$, $\mu_D(y,z))$ $\forall(x, y,z) \in X^3$).

$\mu_D(x, y)$ can be interpreted as a distance function, which is an ultrametric owing to the above inequality.

Let S_α be the α-cut of the similarity relation S. Zadeh (1971) proved the following proposition. If S is a similarity relation in X, then $\forall \alpha \in]0, 1]$, each S_α is an equivalence relation in X. Conversely, if the S_α, $0 < \alpha \leqslant 1$, are a nested sequence of distinct equivalence relations in X, with $\alpha_1 > \alpha_2$ iff $S_{\alpha_1} \subset S_{\alpha_2}$, S_1 nonempty, and $\mathrm{dom}(S_\alpha) = S_1$ $\forall \alpha$, then for any choice of α's in $]0, 1]$ which includes $\alpha = 1$, $S = \bigcup_\alpha \alpha S_\alpha$ is a similarity relation in X (Zadeh, 1971)

$$\left(\mu_S(x, y) = \sup_\alpha \min(\alpha, \mu_{S_\alpha}(x, y)) = \sup_\alpha \alpha\mu_{S_\alpha}(x, y) \right).$$

N.B.:   If $S_{\tilde{\alpha}}$ is a fuzzy α-cut of a max–min transitive fuzzy relation S, then $S_{\tilde{\alpha}}$ is also max–min transitive (Zadeh, Reference from IV.6, 1976).

b.   Partition Tree   (Zadeh, 1971)

Let Π_α denote the partition induced on X by S_α (α-cut of a similarity relation S). Clearly, $\Pi_{\alpha'}$ is a refinement of Π_α if $\alpha' \geqslant \alpha$. A nested sequence of partitions $\Pi_{\alpha_1}, \Pi_{\alpha_2}, \ldots, \Pi_{\alpha_k}$ may be represented diagrammatically in the form of a *partition tree*, as shown in Fig. 3. (The example is from Zadeh, 1971).

The concept of a partition tree plays the same role with respect to a similarity relation as the concept of a quotient does with respect to an equivalence relation.

$$\mu_S = \begin{bmatrix} 1 & 0.2 & 1 & 0.6 & 0.2 & 0.6 \\ 0.2 & 1 & 0.2 & 0.2 & 0.8 & 0.2 \\ 1 & 0.2 & 1 & 0.6 & 0.2 & 0.6 \\ 0.6 & 0.2 & 0.6 & 1 & 0.2 & 0.8 \\ 0.2 & 0.8 & 0.2 & 0.2 & 1 & 0.2 \\ 0.6 & 0.2 & 0.6 & 0.8 & 0.2 & 1 \end{bmatrix}$$

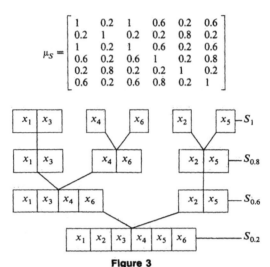

Figure 3

N.B.:   A similarity relation can be interpreted in terms of fuzzy similarity classes $S(x_j)$, one per element of the universe: $\mu_{S(x_j)}(x_i) = \mu_S(x_i, x_j)$, the grade of membership of x_i in the fuzzy class $S(x_j)$.

c.   Weaker Similarity Relations

Several authors (Zadeh, 1971; Bezdek and Harris, 1978) pointed out that max–min transitivity was too strong a property to impose on a fuzzy relation. For instance (Zadeh, 1971), suppose that X is a closed real interval $[a, b]$, and we want to model a proximity relation between elements of $[a, b]$ using a similarity relation S. A reasonable assumption is that $\mu_S(x, y)$ is continuous at $x = y$; then using max–min transitivity we can prove $\mu_S(x, y) = 1 \ \forall(x, y) \in [a, b]^2$. The paradox may be resolved by making S only max–product transitive (for example, $\mu_S(x, y) = e^{-k^2|x-y|}$) or max-∧ transitive (for instance, $\mu_S(x, y) = 1 - (|x - y|/|b - a|)$).

Let us compare the strength of the above introduced transitivities. Denoting by $\mathcal{R}_*$ the set of reflexive, symmetrical, max–∗ transitive fuzzy relations and by $\mathcal{R}$ the set of nonfuzzy equivalence relations, Bezdek and Harris (1978) showed that, since

$$\forall(a, b) \in [0, 1]^2,$$

$$\max(0, a + b - 1) \leqslant ab \leqslant \min(a, b) \leqslant \tfrac{1}{2}(a + b) \leqslant \max(a, b) \leqslant a + b - ab,$$

then,

$$\mathcal{R} \subseteq \mathcal{R}_+ \subseteq \mathcal{R}_\vee \subseteq \mathcal{R}_\square \subseteq \mathcal{R}_{\min} \subseteq \mathcal{R}. \subseteq \mathcal{R}_\wedge \qquad (8)$$

We see that the max-∧ and the max–product transitivities are the weakest ones and hence intuitively the most appealing. Transitivity max–min is too rigid; max–arithmetic mean, max–max, and max–probabilistic sum are a fortiori such.

A reflexive, symmetrical, max-∧ transitive fuzzy relation is called a *likeness* relation (Ruspini, Reference from IV.6, 1977).

If S is a likeness relation, then $1 - \mu_S$ is a pseudometric. Conversely, if d is a pseudometric valued in $[0, 1]$, then $1 - d$ is a characteristic function of a likeness relation (max-∧ transitivity is equivalent to the triangle inequality: $\mu_S(x, z) \geq \max(0, \mu_S(x, y) + \mu_S(y, z) - 1)$ is equivalent to $d(x, z) \leq \min(1, d(x, y) + d(y, z)) \leq d(x, y) + d(y, z)$).

d. Proximity Relation

A *proximity* relation (also called a *tolerance* relation) is a reflexive, symmetrical fuzzy relation.

To get a similarity relation from a proximity relation P, we must build the transitive closure $\hat{P}$ of the latter. Let P_α be the α-cut of P and $\widehat{(P_\alpha)}$ the transitive closure of the α-cut. Tamura *et al.* (1971) have shown that generally $\widehat{(P_\alpha)}$ refines $(\hat{P})_\alpha$, that is, $\forall (x, y) \in X^2$, if $x \widehat{(P_\alpha)} y$, then $x (\hat{P})_\alpha y$. However, when X is finite, $\widehat{(P_\alpha)} = (\hat{P})_\alpha$.

Some algorithms have been proposed to accelerate the computation of $\hat{P}$ when X is finite. Kandel and Yelowitz (1974) used a method much related to the Floyd (NF 1962) algorithm for shortest paths in a graph. Dunn (1974) noticed that a finite fuzzy proximity relation could be interpreted as a nonfuzzy capacitive graph where $\mu_P(x, y)$ is the capacity of the link x-y. The transitive closure of the relation is nothing but the maximal spanning tree of the capacitive graph. Hence, Prim's (NF 1957) algorithm can be used for computing $\hat{P}$. This algorithm is very fast.

e. Convex Hull of Equivalence Relations

Let conv($\mathcal{R}$) denote the convex hull of the nonfuzzy equivalence relations in X (finite). conv($\mathcal{R}$) is made of all the convex combinations of elements of $\mathcal{R}$. Bezdek and Harris (1978) very recently exhibited a relationship between conv($\mathcal{R}$) and max-∧ transitivity: conv($\mathcal{R}$) $\subset \mathcal{R}_\wedge$ for $|X| > 3$.

The convex decomposition $\sum_i c_i R_i$, where $R_i \in \mathcal{R}$ and $\sum_i c_i = 1$, of an element in conv($\mathcal{R}$) provides an alternative to the partition tree decomposition. Each R_i is equivalent to a nonfuzzy partition of X, and c_i expresses the "percentage" of R_i needed to build the fuzzy relation $\sum_i c_i R_i$. Note that the partitions so generated are not nested hierarchically. Unfortunately, given a likeness relation S, there is as yet no efficient algorithm for deciding whether S belongs to conv($\mathcal{R}$) or not and a fortiori for computing the c_i when they exist.

f. A Connection between Fuzzy Partitions and Likeness Relations

Given a fuzzy partition, it is possible to induce a likeness relation. Let $A_1, \ldots, A_p$ be a fuzzy partition of X (1.B.e). An associated likeness relation is defined by (Bezdek and Harris, 1978).

$$\mu_S(x, y) = \sum_{i=1}^{p} \min(\mu_{A_i}(x), \mu_{A_i}(y)).$$

Note that $1 - \mu_S$ is a pseudometric because

$$\mu_S(x, y) = 1 - \frac{1}{2} \sum_{i=1}^{p} |\mu_{A_i}(x) - \mu_{A_i}(y)|$$

owing to $\min(a, b) = \frac{1}{2}(a + b - |a - b|)$.

The converse transformation is unfortunately generally not possible.

g. Comments

The most patent conclusion of this section is that a similarity relation is a very restricted notion because it is equivalent to an ultrametric. A likeness relation seems more promising, although it is equivalent to a well-known nonfuzzy concept, a pseudometric. However, the concept of a fuzzy relation renews the semantics of pseudometrics, possibly adapting them to situations in which the classes involved do not have sharply defined boundaries.

D. FUZZY ORDERINGS

As equivalences can be generalized into similarities and likenesses, classical orderings can also be fuzzified. In this section we consider reflexive and max–min transitive fuzzy relations.

a. Antisymmetries

For binary classical relations R, antisymmetry is defined by $\forall (x, y) \in X^2$, if $x\,R\,y$ and $y\,R\,x$, then $x = y$, which is equivalent to

$$\forall(x, y) \in X^2, \quad \text{if} \quad x \neq y, \quad \text{then} \quad \mu_R(x, y) = 0 \quad \text{or} \quad \mu_R(y, z) = 0.$$

Two definitions of antisymmetry can be found in the literature. They coincide with the above definition for nonfuzzy relations:

Perfect antisymmetry (Zadeh, 1971): a fuzzy relation R is perfectly antisymmetric iff

$$\forall (x, y) \in X^2, \quad \text{if} \quad x \neq y \quad \text{and} \quad \mu_R(x, y) > 0, \quad \text{then} \quad \mu_R(y, x) = 0;$$

Antisymmetry (Kaufmann, 1975): a fuzzy relation R is antisymmetric iff

$$\forall (x, y) \in X^2, \quad \text{if} \quad x \neq y, \quad \text{either} \quad \mu_R(x, y) \neq \mu_R(y, x)$$
$$\text{or} \quad \mu_R(x, y) = \mu_R(y, x) = 0.$$

Note that perfect antisymmetry implies antisymmetry.

b. Fuzzy Partial Orderings (Zadeh, 1971)

A fuzzy relation P in X is a *fuzzy partial ordering* iff it is reflexive, max–min transitive, and perfectly antisymmetric.

When X is finite, it is possible to represent P as a triangular matrix or a Hasse diagram. A fuzzy Hasse diagram is a valued, oriented graph whose nodes are the elements of X. The link $x \rightarrow y$ exists iff $\mu_P(x, y) > 0$. Each link is valued by $\mu_P(x, y)$. Owing to perfect antisymmetry and transitivity, the graph has no cycle. An example (Zadeh, 1971) is provided in Fig. 4, where

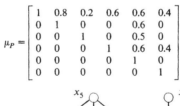

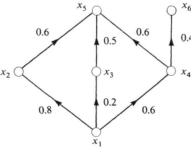

Figure 4

With each $x \in X$, we associate two fuzzy sets: the dominating class denoted by $P_{\geqslant}(x)$ and defined by $\mu_{P_{\geqslant}(x)}(y) = \mu_P(y, x)$ and the dominated class denoted by $P_<(x)$ and defined by $\mu_{P_<(x)}(y) = \mu_P(x, y)$. And x is said to be undominated iff $P_>(x) = \emptyset$, i.e., $\forall y$, $\mu_P(y, x) = 0$; x is said to be undominating iff $P_<(x) = \emptyset$, i.e., $\forall y$, $\mu_P(x, y) = 0$. It is evident that the sets of undominated and undominating elements of any fuzzy partial ordering are nonempty when X is a finite set $\{x_1, \ldots, x_n\}$. Assume that X is ordered in such a way that $\forall i$, $\forall j$ if $\mu_P(x_i, x_j) > 0$, then $i < j$, i.e., the corresponding matrix is triangular. It is obvious that x_1 is undominated and x_n undominating.

A related concept is that of a *fuzzy upper bound* of a nonfuzzy subset of X. Specifically, let A be a nonfuzzy subset of X. The fuzzy upper bound of A, denoted $U(A)$, is a fuzzy set defined by (Zadeh, 1971)

$$U(A) = \bigcap_{x \in A} P_>(x).$$

For a nonfuzzy partial ordering, this reduces to the conventional definition of an upper bound.

N.B.: An α-cut of a fuzzy partial ordering in X is a nonfuzzy partial ordering. The converse also holds in the same sense as for similarity (see C.a, see Zadeh, 1971).

c. Linear Ordering

A fuzzy linear ordering L is a fuzzy partial ordering such that $\forall x, \forall y$ if $x \neq y$, either $\mu_L(x, y) > 0$ or $\mu_L(y, x) > 0$.

Any α-cut of a fuzzy linear ordering is a nonfuzzy linear ordering.

Spilrajn's theorem: Let P be a fuzzy partial ordering in X. Then there exists a fuzzy linear ordering L in a set Y of the same finite cardinality as X and a one-to-one mapping σ from X onto Y such that if $\mu_P(x, y) > 0$, then $\mu_L(\sigma(x), \sigma(y)) = \mu_P(x, y)$.

Zadeh (1971) gives a proof of this "fuzzy extension" of a very well-known result. Informally, this theorem states that any fuzzy partial ordering can be mapped onto a fuzzy linear ordering that is consistent with it. The construction of L may be visualized as a projection of the Hasse diagram of P on an "inclined" line. See Fig. 5 (Zadeh, 1971). Specifically σ is such that $\forall x_i$, $\forall x_j$, $x_i \neq x_j$,

$$\mu_L(\sigma(x_i), \sigma(x_j)) = \begin{cases} \mu_P(x_i, x_j) & \text{if } \mu_P(x_i, x_j) > 0, \\ 0 & \text{if } \mu_P(x_i, x_j) = 0 \text{ and } \mu_P(x_j, x_i) > 0, \\ \epsilon & \text{if } \mu_P(x_i, x_j) = \mu_P(x_j, x_i) = 0 \text{ and } i < j, \\ 0 & \text{if } \mu_P(x_i, x_j) = \mu_P(x_j, x_i) = 0 \text{ and } j < i, \end{cases}$$

where P is triangular, $\sigma(x_i) = y_i \in Y$, and ϵ is any positive constant that is smaller than or equal to the smallest positive entry in the matrix P. Generally, σ and L are not unique.

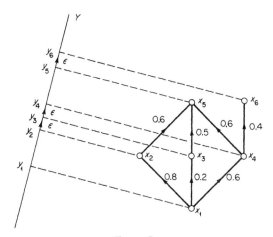

Figure 5

d. Fuzzy Preorder

A *fuzzy preorder* is a reflexive and transitive fuzzy relation that is not assumed to be perfectly antisymmetric.

Let P be a fuzzy preorder. If there exists an ordinary subset A of X such that

$$\forall(x, y) \in A^2, \quad \mu_P(x, y) = \mu_P(y, x) \neq 0,$$

the restriction of P in A is a similarity called a similarity subrelation of P. A similarity subrelation is maximal iff A is maximal. A maximal A is called a similarity class of the preorder P. Each x in X belongs to a similarity class, at least $\{x\}$. Hence, the set of similarity classes of P is a cover of X. A fuzzy preorder is said to be *reducible* (Kaufmann, 1975) iff the set of similarity classes is a partition of X.

N.B.: A nonfuzzy preorder is always reducible.

When P is reducible, elements in the same similarity class need not be distinguished and we get a fuzzy preorder between similarity classes.

Whether the preorder is reducible or not, Orlovsky (1978) proved the following proposition: any fuzzy preorder P on a finite or compact universe X has undominated elements, i.e.,

$$\exists x \in X, \quad \forall y \in X, \quad \mu_P(x, y) \geqslant \mu_P(y, x).$$

Let P^* be the antisymmetrized P, i.e.,

$$\mu_{P^*}(x, y) = \max(0, \mu_P(x, y) - \mu_P(y, x)).$$

P^* is a fuzzy partial order. Obviously, the undominated elements of P^* are the same as those of P. When X is finite, it is thus easy to find the undominated (undominating) elements of P.

e. Comment

Orlovsky's result—the existence of undominated elements for any fuzzy preorder on a compact set—is very important from a philosophical point of view. The assumption of max–min transitivity, which some authors considered as unnatural in a fuzzy situation, is equivalent to the existence of nonfuzzy preferred elements in the sense of the preorder, which looks paradoxical in such a fuzzy situation. The main contribution of the notion of fuzzy preorder is to propose grades of preference, without blurring the choice itself.

E. EQUATIONS OF FUZZY RELATIONS

As in the three previous sections, we consider here only binary relations and study the equation

$$Q \circ R = S \tag{9}$$

where Q is a fuzzy relation on $X \times Y$, R a fuzzy relation on $Y \times Z$, and S a fuzzy relation on $X \times Z$.

Knowing Q and R in (9), it is easy to find S. The converse problem, i.e., find Q (resp. R) knowing S and R (resp. Q), is as interesting but may seem more difficult. Most of the published works concerning this problem were authored by E. Sanchez.

a. The General Problem   (Sanchez, 1976)

The involved fuzzy relations are supposed to be valued only on a Brouwerian lattice L (1.G.a). Recall the operation α on L defined by $a \alpha b = \sup\{x \in L, \ \inf(a, x) \leqslant b\}$, $\forall (a, b) \in L^2$. The following properties obviously hold:

$$\forall (a, b, c) \in L^3, \qquad a \alpha (\sup(b, c)) \geqslant a \alpha b \quad (\text{or } a \alpha c), \tag{10}$$

$$\forall (a, b) \in L^2, \qquad a \alpha (\inf(a, b)) \geqslant b. \tag{11}$$

Sanchez introduced the operator @ to compose fuzzy relations:

$$\mu_{Q @ R}(x,z) = \inf_{y} (\mu_{Q}(x,y) \alpha \, \mu_{R}(y,z)).$$

The following propositions give the main properties of @:

for every pair of fuzzy relations Q on $X \times Y$ and R on $Y \times Z$, we have

$$R \subseteq Q^{-1} @ (Q \circ R) \tag{12}$$

and

$$Q \subseteq \left(R @ (Q \circ R)^{-1} \right)^{-1}; \tag{13}$$

for every pair of fuzzy relations Q on $X \times Y$ and S on $X \times Z$, we have

$$Q \circ (Q^{-1} @ S) \subseteq S \tag{14}$$

and

$$(R @ S^{-1})^{-1} \circ R \subseteq S. \tag{15}$$

Proof: Let $U = Q^{-1} @ (Q \circ R)$.

$$\mu_{U}(y,z) = \inf_{x \in X} \left[\mu_{Q^{-1}}(y,x) \alpha \left[\sup_{t \in Y} \inf(\mu_{Q}(x,t) \mu_{R}(t,z)) \right] \right]$$

or

$$\mu_{U}(y,z) = \inf_{x \in X} \left[\mu_{Q}(x,y) \alpha \left[\sup \left\{ \inf(\mu_{Q}(x,y), \mu_{R}(y,z)), \right. \right. \right.$$
$$\left. \left. \left. \sup_{t \neq y} \inf(\mu_{Q}(x,t), \mu_{R}(t,z)) \right\} \right] \right];$$

hence, using (10),

$$\mu_{U}(y,z) \geqslant \inf_{x \in X} \left[\mu_{Q}(x,y) \alpha \inf(\mu_{Q}(x,y), \mu_{R}(y,z)) \right];$$

hence, using (11),

$$\mu_{U}(y,z) \geqslant \mu_{R}(y,z), \quad \text{Q.E.D.}$$

The other inclusions are proved in the same fashion (Sanchez, 1976).

We can now state two fundamental results that give the greatest solutions of (9) (Sanchez, 1976).

(1). Let Q be a fuzzy relation on $X \times Y$, S a fuzzy relation on $X \times Z$, and $\mathcal{R}$ the set of fuzzy relations R on $Y \times Z$ such that $Q \circ R = S$. Then, either $\mathcal{R} = \emptyset$ or $Q^{-1} @ S \in \mathcal{R}$. If $\mathcal{R} \neq \emptyset$, $Q^{-1} @ S$ is the greatest element in $\mathcal{R}$.

(2).   Let R be a fuzzy relation on $Y \times Z$ and S a fuzzy relation on $X \times Z$, and $\mathcal{Q}$ the set of fuzzy relations on $X \times Y$ such that $Q \circ R = S$, then either $\mathcal{Q} = \emptyset$ or $(R \, @ \, S^{-1})^{-1} \in \mathcal{Q}$. If $\mathcal{Q} \neq \emptyset$, $(R \, @ \, S^{-1})^{-1}$ is the greatest element in $\mathcal{Q}$.

Proof: We prove only the first proposition. Assume $\mathcal{R} \neq \emptyset$ and $R \in \mathcal{R}$. From (12) we have $R \subseteq Q^{-1} @ \, S$. Hence, $S = Q \circ R \subseteq Q \circ (Q^{-1} @ \, S)$ (see B.b.α). But from (14) $Q \circ (Q^{-1} @ \, S) \subseteq S$, hence $Q \circ (Q^{-1} \alpha S) = S$, i.e., $Q^{-1} @ \, S \in \mathcal{R}$.   Q.E.D.

When the Brouwerian lattice L is just $[0, 1]$, recall that $a \, \alpha \, b = 1$ iff $a \leqslant b$ and $a \, \alpha \, b = b$ iff $a > b$: so the greatest solutions in (9) can be easily computed.

N.B.:   1.   Inf–max fuzzy relations equations $(Q \,\overline{\circ}\, R = S)$ can be solved on a dually Brouwerian lattice (1.6.a). The operator ϵ such that $\forall (a, b) \in L^2$, $a \, \epsilon \, b = \inf\{x \in L, \sup(a, x) \geqslant b\}$ replaces α. The associated $\text{\textcircled{$\epsilon$}}$-composition is defined by

$$\mu_{Q \,\text{\textcircled{ϵ}}\, R}(x, z) = \sup_{y \in Y} (\mu_Q(x, y) \epsilon \, \mu_R(y, z)).$$

Then $Q^{-1} \text{\textcircled{ϵ}} S((R \, \text{\textcircled{ϵ}} \, S^{-1})^{-1}$ resp.) are the least R (Q resp.) such that $Q \,\overline{\circ}\, R = S$ when solutions exists.

2.   The above results are still valid when we relax (9) into $Q \circ R \subseteq S$ (Sanchez, 1977a), but now the inequality obviously always has solutions.

b.   **Particular Case 1**   (Sanchez, 1977a)

We consider the following problem: find R such that $A \circ R = B$ where A is a fuzzy set on X, B a fuzzy set on Y, and R an unknown fuzzy relation in $X \times Y$, valued in $[0, 1]$. X and Y are assumed to be finite.

Sanchez defines the operator σ in $[0, 1]$ such that $a \sigma b = 0$ iff $a < b$, $a \sigma b = b$ iff $a \geqslant b$. It is easy to check that $a \sigma b \leqslant \min(a, b)$.

Let A and B be two fuzzy sets on X and Y, respectively. The fuzzy relation $A \, \text{\textcircled{$\sigma$}} \, B$ in $X \times Y$ has membership function

$$\mu_{A \,\text{\textcircled{σ}}\, B}(x, y) = \mu_A(x) \sigma \, \mu_B(y).$$

Let $\mathcal{R} = \{R, A \circ R = B\}$; if $\mathcal{R} \neq \emptyset$, then $A \, \text{\textcircled{$\sigma$}} \, B \in \mathcal{R}$.

Proof:

$$\mu_{A \,\circ\, (A \,\text{\textcircled{σ}}\, B)}(y) = \sup_{x \in X} \min(\mu_A(x), \mu_A(x) \sigma \, \mu_B(y))$$

$$= \sup_{x} (\mu_A(x) \sigma \, \mu_B(y)).$$

Hence,

$$\mu_{A \circ (A \textcircled{0} B)}(y) = \sup_{\substack{x \\ \mu_A(x) \geqslant \mu_B(y)}} \mu_B(y) = \mu_B(y)$$

because if $\mathfrak{R} \neq \varnothing$, $\forall y$, $\exists x$, $\mu_A(x) \geqslant \mu_B(y)$ (obvious since $\mu_B(y) = \sup_x \min(\mu_A(x), \mu_R(x, y))$). Q.E.D.

Moreover, Sanchez (1977a) showed the following results:

$\mathfrak{R}$ has a least element iff $\forall y$, $(\exists! x, \mu_A(x) \geqslant \mu_B(y)$ or $\mu_B(y) = 0)$; when it exists, it is $A \textcircled{0} B$;

if $\mathfrak{R} \neq \varnothing$, $\forall R$ such that $A \textcircled{0} B \subseteq R \subseteq A \textcircled{@} B$, then $A \circ R = B$.

Note that owing to the result of the general case $A \textcircled{@} B$ is the greatest element in $\mathfrak{R} \neq \varnothing$.

c. Particular Case 2

Now we turn to the following problem: find A such that $A \circ R = B$ where A is an unknown fuzzy set on X (finite), B a fuzzy set on Y (finite), and R a fuzzy relation in $X \times Y$.

We have $\mu_B(y) = \sup_x \min(\mu_A(x), \mu_R(x, y))$. Note first that the problem has no solution as soon as $\exists y$, $\forall x$, $\mu_R(x, y) < \mu_B(y)$.

The following proposition characterizes the solution μ_A when it exists: A is a solution iff:

(1). $\forall y$, $(\exists x, \mu_A(x) \geqslant \mu_R(x, y) = \mu_B(y))$
or $(\exists x, \mu_A(x) = \mu_B(y) < \mu_R(x, y))$;
(2). $\forall x \forall y$, $\mu_A(x) \leqslant \mu_B(y)$ if $\mu_R(x, y) > \mu_B(y)$.

Proof: Let $K(y) = \{x \in X, \mu_B(y) = \min(\mu_A(x), \mu_R(x, y))\}$.

(i) $\forall x$, $\min(\mu_A(x), \mu_R(x, y)) \leqslant \mu_B(y)$ since A is a solution. Hence, if $\mu_R(x, y) > \mu_B(y)$, then $\mu_A(x) \leqslant \mu_B(y)$. This proves 2. Moreover, $\forall x \in K(y) \neq \varnothing$ (since A is a solution)

if $\mu_R(x, y) = \mu_B(y)$, then $\mu_A(x) \geqslant \mu_B(y)$;
if $\mu_R(x, y) > \mu_B(y)$, then $\mu_A(x) = \mu_B(y)$.

(ii) y is supposed fixed. Assume A satisfies 1 and 2. It is a solution because

$$\mu_B(y) = \max\left[\sup_{\substack{x \\ \mu_R(x, y) = \mu_B(y)}} \min(\mu_B(y), \mu_A(x)) \right.$$

$$\left. \sup_{\substack{x \\ \mu_R(x, y) > \mu_B(y)}} \min(\mu_R(x, y), \mu_A(x)) \right]. \quad \text{Q.E.D.}$$

For each y, the feasible domain of $\mu_A(x)$ (such that A is a solution) is defined by the following:

Let

$$X^0(y) = \{x, \mu_R(x, y) = \mu_B(y)\}$$

and

$$X^+(y) = \{x, \mu_R(x, y) > \mu_B(y)\}.$$

(a) For $x \in X^-(y) = \{x, \mu_R(x, y) < \mu_B(y)\}$: $\mu_A(x)$ is unconstrained.
(b) $\forall x \in X^+(y)$, $\mu_A(x) \in [0, \mu_B(y)]$.
(c) $\exists x \in X^0(y)$, $\mu_A(x) \in [\mu_B(y), 1]$ or $\exists x \in X^+(y)$, $\mu_A(x) = \mu_B(y)$.

For a given y, let $p(y) = |X^0(y)|$ and $q(y) = |X^+(y)|$. Let $I^k(y)$ be an n-tuple of intervals $I^k(x, y)$ ($n = |X|$) satisfying the above three requirements. When $X^-(y) \neq X$, we have: if ($\forall x \in X$, $\mu_A(x) \in I^k(x, y)$), then $\mu_B(y) = \max_{x \in X} \min(\mu_A(x), \mu_R(x, y))$.

Generally, several $I^k(y)$ exist.

When $X^0(y) = \emptyset$, the number of $I^k(y)$ is $q(y)$ ($k = 1, q(y)$). They are obtained by forcing $\mu_A(x(k)) = \mu_B(y)$ for an arbitrarily chosen $x(k) \in X^+(y)$ and setting $I^k(x(k), y) = \mu_B(y), I^k(x, y) = [0, \mu_B(y)]$, $\forall x \in X^+(y) - \{x(k)\}$ and $I^k(x, y) = [0, 1]$ for $x \in X^-(y)$.

When $X^0(y) \neq \emptyset$, the number of $I^k(y)$ is $p(y) + q(y)$. The first $q(y)$ ones are obtained as above with $I^k(x, y) = [0, 1]$ for $x \in X^0(y)$. For $k > q(y)$, $I^k(y)$ is defined by $I^k(x(k), y) = [\mu_B(y), 1]$ for an arbitrarily chosen $x(k) \in X^0(y)$, $I^k(x, y) = [0, 1]$ for $x \in X^0(y) - \{x(k)\}$, $I^k(x, y) = [0, 1]$, for $x \in X^-(y)$, $I^k(x, y) = [0, \mu_B(y)]$ for $x \in X^+(y)$.

A possible set of admissible intervals $\{\varphi_i(x), x \in X\}$ for $\mu_A(x)$ such that $B = A \circ R$ is obtained as follows. For each $y \in Y$, choose one of the $I^k(y)$ (k is not necessarily the same for all y), denoted $I_i(y)$. If the $I_i(y)$ are such that $\forall x$, $\bigcap_{y \in Y} I_i(x, y) \neq \emptyset$, then

$$\forall x \in X, \quad \varphi_i(x) = \bigcap_{y \in Y} I_i(x, y) = \mu_{\Phi_i}(x).$$

Note that Φ_i is a Φ-fuzzy set (see 1.G.d, 2.C.b). Usually several Φ_i can be built. Moreover, the greatest feasible solution can be found at once when it exists, namely $R^{-1} @ B$ ($\mu_{R^{-1} @ B}(x) = \inf_y(\mu_R(y, x) \alpha \mu_B(y))$) owing to the result of the general case. Thus, the $\varphi_i(x)$ are of the form $[\alpha_i(x), \beta(x)]$ $\forall x \in X$, where $\beta(x)$ does not depend on i. However, several incomparable least solutions $\int_X \alpha_i(x)/x$ may exist.

Example

$$|X| = |Y| = 4; \quad X = \{x_1, x_2, x_3, x_4\}; \quad Y = \{y_1, y_2, y_3, y_4\};$$

$$R = \begin{bmatrix} 1 & 0.1 & 0.9 & 0.2 \\ 0.2 & 1 & 0.4 & 1 \\ 0.1 & 0.4 & 0.3 & 0.5 \\ 0.1 & 0.2 & 0.9 & 0.8 \end{bmatrix} \quad B = [0.9 \quad 0.5 \quad 0.5 \quad 0.8];$$

$$X^-(y_1) = \{x_2, x_4\}; \quad X^0(y_1) = \{x_3\}; \quad X^+(y_1) = \{x_1\};$$
$$X^-(y_2) = \{x_1, x_3\}; \quad X^0(y_2) = \varnothing; \quad X^+(y_2) = \{x_2, x_4\};$$
$$X^-(y_3) = \{x_1, x_2, x_3\}; \quad X^0(y_3) = \{x_4\}; \quad X^+(y_3) = \varnothing;$$
$$X^-(y_4) = \{x_1, x_2\}; \quad X^0(y_4) = \{x_4\}; \quad X^+(y_4) = \{x_3\}.$$

The possible choices are

$$I^1(y_1) = (0.9, [0,1], [0,1], [0,1]); \quad I^2(y_1) = ([0,0.9], [0,1], [0.9,1], [0,1]);$$
$$I^1(y_2) = ([0,1], 0.5, [0,1], [0,0.5]); \quad I^2(y_2) = ([0,1], [0,0.5], [0,1], 0.5);$$
$$I^1(y_3) = ([0,1], [0,1], [0,1], [0.5,1]);$$
$$I^1(y_4) = ([0,1], [0,1], 0.8, [0,1]); \quad I^2(y_4) = ([0,1], [0,1], [0,0.8], [0.8,1]).$$

$I^2(y_1)$ is consistent with neither $I^1(y_4)$ nor $I^2(y_4)$, and $I^2(y_4)$ is consistent with neither $I^1(y_2)$ nor $I^2(y_2)$. Both can be rejected.

Hence there are two possible solutions

$$\mu_{A_1} = I^1(y_1) \cap I^1(y_2) \cap I^1(y_3) \cap I^1(y_4) = (0.9, 0.5, 0.8, 0.5),$$

$$\mu_{A_2} \in I^1(y_1) \cap I^2(y_2) \cap I^1(y_3) \cap I^1(y_4) = (0.9, [0,0.5], 0.8, 0.5).$$

Note that $A_1 = R^{-1} @ B$. Knowledge of the greatest solution can accelerate the a priori cancellation of some $I^k(y)$. The final range of the possible values of μ_A is $\mu_A(x_1) = 0.9$; $\mu_A(x_2) \in [0, 0.5]$; $\mu_A(x_3) = 0.8$; $\mu_A(x_4) = 0.5$.

An algorithm for the determination of the possible values of μ_A can be found in Tsukamoto and Terano (1977). Their approach is very similar to the one outlined here. Tashiro (1977) extended Tsukamoto and Terano's method to the case when B and R are interval-valued. This extension is possible because of the following remark. Write $\mu_B(y_j) = [b_j^-, b_j^+]$, $\mu_R(x_i, y_i) = [r_{ij}^-, r_{ij}^+]$, and $\mu_A(x_i) = [a_i^-, a_i^+]$, then

$$\mu_B(y_j) = \widetilde{\max_i} \widetilde{\min}([a_i^-, a_i^+], [r_{ij}^-, r_{ij}^+]),$$

$$\mu_B(y_j) = \widetilde{\max_i}([\min(a_i^-, r_{ij}^-), \min(a_i^+, r_{ij}^+)]).$$

Hence $[b_j^-, b_j^+] = [\max_i \min(a_i^-, r_{ij}^-), \max_i \min(a_i^+, r_{ij}^+)]$. We see that the Φ-fuzzy equation is equivalent to two ordinary fuzzy ones.

d. Eigen Fuzzy Sets   (Sanchez, 1977b, 1978)

An *eigen* fuzzy set A of a fuzzy relation R in $X \times X$ is a fuzzy set on X such that $A \circ R = A$. Sanchez has proved the following results which characterize the greatest eigen fuzzy set of R:

(a)   Let A_0 be the fuzzy set such that

$$\forall x, \quad \mu_{A_0}(x) = \inf_{x' \in X} \sup_{x \in X} \mu_R(x, x').$$

This constant fuzzy set is an eigen fuzzy set of R.

(b)   Let A_1 be the fuzzy set such that

$$\forall x, \quad \mu_{A_1}(x) = \sup_{x' \in X} \mu_R(x', x).$$

The sequence (A_m) defined by $A_m = A_{m-1} \circ R$, $m \geqslant 2$, is decreasing and bounded by A_0 and A_1:

$$A_0 \subseteq \cdots \subseteq A_{m+1} \subseteq A_m \subseteq \cdots \subseteq A_2 \subseteq A_1.$$

(c) $\exists k \leqslant |X|$ such that $A_k = A_{k+m}, m > 0$, and A_k is the greatest eigen fuzzy set of R and also of the transitive closure $\hat{R}$.

In Sanchez (1978) some algorithms for the determination of A_k are provided.

e. Comment

Let us quote Sanchez (1977a): "The composition of a fuzzy relation R with a fuzzy set A corresponds to the concept of a conditio: ed fuzzy set and can be interpreted in terms of a fuzzy metaimplication: if A then B by R." See [III.1.E]. "One can infer diagnosis and prognosis from observed symptoms by means of a specific knowledge." The determination of R in $A \circ R = B$ models the acquisition of knowledge from experiments, the determination of A in $A \circ R = B$ models the search of a fuzzy cause (see IV.7).

F. GENERALIZED FUZZY RELATIONS

Until now we have focused our attention upon fuzzy relations in the sense of fuzzy sets on a Cartesian product of universes, which express a relationship between elements. Obviously other kinds of relations may involve fuzziness. In this section we give only some definitions and suggestions for the setting of several generalized fuzzy relations.

a. Nonfuzzy relation between Fuzzy Sets

Zadeh (Reference from IV.2, 1976) introduced *tableaus* of fuzzy sets whose columns refer to the universes and rows contain $(n + 1)$-tuples of labels of fuzzy sets. The $(n + 1)$th fuzzy set is considered as the image of the n others through a nonfuzzy mapping. In fact, the tableaus play a basic role in the description and the execution of fuzzy algorithms.

b. Interval-valued Fuzzy Relations

Ponsard (1977) has extended some results of sections C and D to interval-valued binary fuzzy relations using the operators W and M (see 2.C.b). The reflexivity of a Φ-fuzzy relation R_Φ in X^2 is defined by

$$\forall x \in X \quad \mu_{R_\Phi}(x,x) = [1,1] = 1.$$

The transitivity of R is defined by

$$\forall (x,y,z) \in X^3, \quad \mu_{R_\Phi}(x,z) \geqslant (\mu_{R_\Phi}(x,y) \, M \, \mu_{R_\Phi}(y,z))$$

($\geqslant$ in the sense of 2.C.b). The symmetry of R_Φ is defined by

$$\forall (x,y) \in X^2, \quad \mu_{R_\Phi}(x,y) = \mu_{R_\Phi}(y,x).$$

Using these definitions, Ponsard (1977) develops Φ-fuzzy preorders and Φ-fuzzy similarities.

c. Fuzzy-Valued Fuzzy Relations

A fuzzy-valued fuzzy relation in $X \times Y$ is a type 2 fuzzy set on $X \times Y$. The composition of such relations Q in $X \times Y$ and R in $X \times Z$ can be performed using $\widetilde{\max}$ and $\widetilde{\min}$:

$$\mu_{Q \circ R}(x,z) = \widetilde{\max_y} \, \widetilde{\min} \, (\mu_Q(x,y), \mu_R(y,z)).$$

This definition holds for Y finite. Note that the $\widetilde{\max}$–$\widetilde{\min}$ composition of interval-valued fuzzy relations (a particular case of fuzzy-valued fuzzy relations) is different from the W–M Ponsard composition (see b above). A direct extension of definitions of properties specific to fuzzy relations to fuzzy-valued fuzzy relations may appear too strict; for instance symmetry would mean

$$\forall (x,y), \quad \mu_R(x,y) = \mu_R(y,x) \quad \text{in} \quad \tilde{\mathscr{P}}[0,1];$$

weaker symmetry could be stated using approximate equality in the sense of 1.E.c.

d. Fuzzy Relation between (Non)Fuzzy Sets

C. L. Chang (Reference from III.3) has proposed a way of inducing a fuzzy relation $\tilde{R}$ in $\tilde{\mathcal{P}}(X) \times \tilde{\mathcal{P}}(Y)$ from a fuzzy relation R in $X \times Y$:

$$\forall(A, B) \in \tilde{\mathcal{P}}(X) \times \tilde{\mathcal{P}}(Y), \quad \mu_{\tilde{R}}(A, B) = \sup_{x, y} \min(\mu_A(x), \mu_B(y), \mu_R(x, y)).$$

$\mu_{\tilde{R}}(A, B)$ is nothing but the degree of consistency of $A \circ R$ and B (or A and $B \circ R^{-1}$): $\mathrm{hgt}((A \circ R) \cap B)$. Note that one must not confuse $\tilde{R}$ and the extension of μ_R, a mapping from $X \times Y$ to $[0, 1]$, by means of the extension principle. The latter would be a fuzzy-valued fuzzy relation between fuzzy sets.

Sanchez (1977c) has studied $\tilde{R}$ for A and B ordinary sets. He defines two kinds of inverses for $\tilde{R}$:

a lower inverse $\tilde{R}_*$, characterized by $\mu_{\tilde{R}_*}(B, A) = \sup_{C \subseteq B} \mu_{\tilde{R}}(A, C)$;
an upper inverse $\tilde{R}^*$, characterized by $\mu_{\tilde{R}}^*(B, A) = \sup_{C, C \cap B \neq \varnothing} \mu_{\tilde{R}}(A, C)$.

For an extensive treatment of these inverse relations, see Sanchez (1977c).

e. Tolerance Classes of Fuzzy Sets

In order to deal with the fact that membership functions are always partially out of reach, higher order fuzzy sets were defined (type 2 fuzzy sets 1.G.d, 2.C.b, probabilistic sets 1.G.e, level 2 fuzzy sets 2.C.a), an alternative approach can be to use a proximity relation in $\tilde{\mathcal{P}}(X) \times \tilde{\mathcal{P}}(X)$ to sketch "fuzzy tolerance classes" for the admissible membership functions of a given ill-known fuzzy set. Denote this proximity relation by $\sim$. For the sake of the consistency, $\sim$ must be compatible with most of the operations $*$ on $\tilde{\mathcal{P}}(X)$. Specifically, if $A \sim A'$ and $B \sim B'$, then $(A * B) \sim (A' * B')$, where $*$ may be $\cap$, $\cup$, ... or even $\oplus$, ... and A, A', B, B' are ordinary fuzzy sets.

This is interpreted as "if A looks like A' and B like B', then $A * B$ must look like $A' * B'$." There are several possible choices for $\sim$.

The consistency condition can be expressed as follows. There exists an increasing (in the sense of 2.B) operation $\perp$ in $[0, 1]$ such that

$$\mu_\sim(A * B, A' * B') \geqslant \mu_\sim(A, A') \perp \mu_\sim(B, B').$$

For instance, Nowakowska (Reference from IV.1) showed that the above condition holds for $\sim = S_4$ (see 1.E.c.β), $* = \cap$ or $* = \cup$, and $\perp = \min$.

REFERENCES

Bezdek, J. C., and Harris, J. D. (1978). Fuzzy partitions and relations: An axiomatic basis for clustering. *Int. J. Fuzzy Sets Syst.* **1**, No. 2, 111–127.

Chen, C. (1974). Realizability of communication nets: An application of the Zadeh criterion. *IEEE Trans. Circuits Syst.* **21**, No. 1, 150–151.

Dunn, J. C. (1974). A graph-theoretic analysis of pattern classification via Tamura's fuzzy relation. *IEEE Trans. Syst., Man Cybern.* **4**, 310–313.

Kandel, A. (1975). Properties of fuzzy matrices and their applications to hierarchical structures. *Asilomar Conf. Circuits, Syst. Comput., 9th*, pp. 531–538.

Kandel, A., and Yelowitz, L. (1974). Fuzzy chains. *IEEE Trans. Syst., Man Cybern.* **4**, 472–475.

Kaufmann, A. (1975). "Introduction to the Theory of Fuzzy Subsets. Vol. 1: Fundamental Theoretical Elements," Academic Press, New York. (Reference from I.)

Leenders, J. H. (1977). Some remarks on an article by Raymond T. Yeh and S. Y. Bang dealing with fuzzy relations. *Simon Stevin, Wis- Natuurk. Tijdschr.* **51**, 93–100.

Orlovsky, S. A. (1978). Decision-making with a fuzzy preference relation. *Int. J. Fuzzy Sets Syst.* **1**, No. 3, 155–168.

Pappis, C. P., and Sugeno, M. (1976). Fuzzy relational equations and the inverse problem. In "Discrete Systems and Fuzzy Reasoning" (E. H. Mamdani, B. R. Gaines, eds.) Workshop Proc. Queen Mary College, Univ. of London.

Ponsard, C. (1977). Hiérarchie des places centrales et graphes Φ-flous. *Environ. Plann. A* **9**, 1233–1252. (Reference from V, 1977a.)

Sanchez, E. (1976). Resolution of composite fuzzy relation equations. *Inf. Control* **30**, 38–48.

Sanchez, E. (1977a). Solutions in composite fuzzy relation equations: Application to medical diagnosis in Brouwerian logic. In "Fuzzy Automata and Decision Processes" (M. M. Gupta, G. N. Saridis, and B. R. Gaines, eds.), pp. 221–234. North-Holland Publ., Amsterdam.

Sanchez, E. (1977b). "Eigen Fuzzy Sets and Fuzzy Relations," Memo UCB/ERL. M77-20 Univ. of California, Berkeley.

Sanchez, E. (1977c). Inverses of fuzzy relations. Application to possibility distributions and medical diagnosis. *Proc. IEEE Conf. Decision Control, New Orleans* **2**, 1384–1389. [Also in *Int. J. Fuzzy Sets Syst.* **2**, No. 1, 75–86 (1979).]

Sanchez, E. (1978). Resolution of eigen fuzzy sets equations. *Int. J. Fuzzy Sets Syst.* **1**, No. 1, 69–74.

Tamura, S., Higuchi, S., and Tanaka, K. (1971). Pattern classification based on fuzzy relations. *IEEE Trans. Syst., Man Cybern.* **1**, 61–66. (Reference from IV.6.)

Tashiro, T. (1977). Method of solution to inverse problem of fuzzy correspondence model. In "Summary of Papers on General Fuzzy Problems." No. 3, pp. 70–79. Working Group Fuzzy Syst., Tokyo.

Thomason, M. G. (1977). Convergence of powers of a fuzzy matrix. *J. Math. Anal. Appl.* **57**, 476–480.

Tsukamoto, Y., and Terano, T. (1977). Failure diagnosis by using fuzzy logic. *Proc. IEEE Conf. Decision Control, New Orleans* **2**, 1390–1395. (Reference from IV.7.)

Yeh, R. T. (1973). Toward an algebraic theory of fuzzy relational systems. *Proc. Int. Congr. Cybern. Namur*, pp. 205–223.

Yeh, R. T., and Bang, S. Y. (1975). Fuzzy relations, fuzzy graphs, and their applications to clustering analysis. In "Fuzzy Sets and Their Applications to Cognitive and Decision Processes" (L. A. Zadeh, K. S. Fu, K. Tanaka, and M. Shimura, eds.), pp. 125–149. Academic Press, New York. (Reference from III.4.)

Zadeh, L. A. (1971). Similarity relations and fuzzy orderings. *Inf. Sci.* **3**, 177–200.

Zadeh, L. A. (1975a). Calculus of fuzzy restrictions. In "Fuzzy Sets and Their Applications to Cognitive and Decision Processes" (L. A. Zadeh, K. S. Fu, K. Tanaka, and M. Shimura, eds.), pp. 1–39. Academic Press, New York.

Zadeh, L. A. (1975b). The concept of a linguistic variable and its application to approximate reasoning. Parts 1, 2, and 3. *Inf. Sci.* **8**, 199–249; **8**, 301–357; **9**, 43–80. (Reference from II.2.)

Chapter *4*

FUZZY FUNCTIONS

Under the name *fuzzy functions* are gathered various kinds of mappings between sets generalizing ordinary mappings in some sense. They are described in the first section of this chapter and interpreted. Strangely enough, most of them have received little attention in the literature, except from specific points of view (fuzzy topology).

The problem of maximizing a function over a fuzzy domain or a fuzzy function over a nonfuzzy domain is investigated in the second section.

The two following parts are devoted to the integration and differentiation of a special kind of fuzzy functions—closely related to some fuzzy relations on $\mathbb{R}^2$. The results that are presented here are a first attempt to extend elementary notions in real analysis.

Lastly, fuzzy topology and categories of fuzzy sets are briefly surveyed. Because of very specific and abstract features, neither is detailed here. The interested reader is referred to the extensive bibliography of these topics at the end of the chapter.

A. VARIOUS KINDS OF FUZZY FUNCTIONS

A *fuzzy function* can be understood in several ways according to where fuzziness occurs. Roughly there are three basic kinds of fuzzy functions, from an interpretive point of view:

ordinary functions having fuzzy properties or satisfying fuzzy constraints;

functions that just "carry" the fuzziness of their argument(s) without generating extra fuzziness themselves: the image of a nonfuzzy element is a nonfuzzy element;

ill-known functions of nonfuzzy arguments: the image of an element is blurred by the jiggling of the function.

Of course, hybrid types may be considered. Moreover, we have the abstract concept of an ordinary function between sets of fuzzy sets.

a. Fuzzily Constrained Functions

α. *Fuzzy Domain—Fuzzy Range* (Negoita and Ralescu, 1975)

Let X and Y be two universes and f be an ordinary function from X to Y: $x \in X \mapsto f(x) \in Y$. Let A and B be two fuzzy sets on X and Y, respectively. f is said to have a fuzzy domain A and a fuzzy range B iff

$$\forall x \in X, \quad \mu_B(f(x)) \geqslant \mu_A(x). \tag{1}$$

Example 1 "Big trucks must go slowly": X is a set of trucks, Y is a scale of speeds, f assigns a speed limit $f(x)$ to each truck x. A is the fuzzy set of big trucks; B is the fuzzy set of low speeds. The constraint (1) means, "The bigger the truck, the lower its speed limit."

Example 2 Many proverbs as well as regulations can be modeled by a function with a fuzzy domain and a fuzzy range. For instance: "The smaller the drink, the cooler the blood, the clearer the head." "The more thy years, the nearer thy grave."

Now, consider a function g from Y to Z with a fuzzy domain B and a fuzzy range C. $g \circ f$ is a function from X to Z with a fuzzy domain A and a fuzzy range C since $\mu_B(f(x)) \geqslant \mu_A(x)$, $\mu_C(g(y)) \geqslant \mu_B(y)$, and $y = f(x)$ imply $\mu_C(g(f(x))) \geqslant \mu_A(x)$.

N.B.: This kind of fuzzification is similar to the one that defines fuzzy groups (1.F.b, 2.A.d).

β. *Fuzzy Injection, Fuzzy Continuity, Fuzzy Surjection*

Let f be an ordinary function from X to Y. f is said to be injective iff $\forall(x_1, x_2) \in X^2$, $f(x_1) = f(x_2)$ implies $x_1 = x_2$. Let P be a fuzzy proximity relation (3.C.d) in X^2. f is said to be ϵ-fuzzily injective iff $\forall(x_1, x_2) \in X^2$, $f(x_1) = f(x_2)$ implies $\mu_P(x_1, x_2) \geqslant \epsilon$.

A more general definition is: f is fuzzily injective iff

$$\forall(x_1, x_2) \in X^2, \quad \mu_P(x_1, x_2) \geqslant \mu_Q(f(x_1), f(x_2)) \tag{2}$$

where Q is a fuzzy proximity relation in Y^2. This constraint, very similar to (1), means "the closer the images, the closer their antecedents."

The composition of fuzzily injective functions is still fuzzily injective.

f is said to be fuzzily continuous iff

$$\forall(x_1, x_2) \in X^2, \quad \mu_Q(f(x_1), f(x_2)) \geqslant \mu_P(x_1, x_2). \tag{3}$$

(3) obviously means "the closer the elements, the closer their images," which may appear consistent with our intuition of continuity. Note that fuzzy continuity and fuzzy injection are here dual concepts.

Note that the usual definition of continuity is

$$\forall \epsilon, \quad \exists \eta, \quad d(x_1, x_2) \leqslant \eta \quad \text{implies} \quad d'(f(x_1), f(x_2)) \leqslant \epsilon$$

where d and d' are distances on X and Y, respectively. A relaxation of this definition is:

$$\forall \epsilon \in [0, 1], \quad \exists \eta \in [0, 1], \quad \mu_P(x_1, x_2) \geqslant \eta \quad \text{implies}$$
$$\mu_Q(f(x_1), f(x_2)) \geqslant \epsilon. \tag{4}$$

Both definitions are equivalent when P and Q are likeness relations (3.C.c) such that $\mu_p(x_1, x_2)$ (resp.: $\mu_Q(y_1, y_2)$) $= 1$ implies $x_1 = x_2$ (resp.: $y_1 = y_2$) and d and d' are metrics valued on $[0, 1]$. Note that (3) implies (4) ($\eta = \epsilon$). Conversely, the dual of (4) provides a less strict definition of fuzzy injection:

$$\forall \epsilon \in [0, 1], \quad \exists \eta \in [0, 1], \quad \mu_Q(f(x_1), f(x_2)) \geqslant \eta \quad \text{implies}$$
$$\mu_P(x_1, x_2) \geqslant \epsilon.$$

The composition of fuzzily continuous functions (in the sense of (3) or (4)) is still fuzzily continuous.

Recall that f is said to be onto (surjective) iff

$$\forall y \in Y, \quad \exists x \in X, \quad y = f(x).$$

Given a proximity relation Q in Y, f is said to be ϵ-fuzzily onto iff

$$\forall y \in Y, \quad \exists x \in X, \quad \mu_Q(y, f(x)) \geqslant \epsilon. \tag{5}$$

More generally, f is said to be fuzzily surjective on the fuzzy set B iff

$$\forall y \in Y, \quad \exists x \in X, \quad \mu_Q(y, f(x)) \geqslant \mu_B(y). \tag{6}$$

(6) means the more y belongs to B, the closer is a neighbor of y having an antecedent.

Remarks 1 Definitions (1), (2), (3), and (6) implicitly assume that membership grades in different fuzzy sets can be compared. In fact, we tacitly use relative membership.

2 The same definitions are related to the implication $\Rightarrow$ (see III.1.B.b. β). The truth value of a consequence is at least equal to the truth value of the premise.

b. Fuzzy Extension of a Nonfuzzy Function (Zadeh, Reference from II.1, 1965)

Let f be a nonfuzzy function from X to Y; the image of a fuzzy set $\tilde{x}$ on X is defined by means of the extension principle. It is $f(\tilde{x})$ defined as

$$\mu_{f(\tilde{x})}(y) = \sup_{x \in f^{-1}(y)} \mu_{\tilde{x}}(x)$$

$$= 0 \quad \text{if} \quad f^{-1}(y) = \emptyset$$

where $f^{-1}(y)$ is the set of antecedents of y. A function of a fuzzy variable from $\tilde{\mathcal{P}}(X)$ to $\tilde{\mathcal{P}}(Y)$ is thus constructed; its restriction to X is nonfuzzy. Moreover, note that the image of a fuzzy singleton λ/x is $\lambda/f(x)$. In that sense f carries fuzziness without altering it.

Examples 1 $\tilde{y} = (a \odot \tilde{x}) \oplus b$, $(a, b) \in \mathbb{R}^2$, $X = Y = \mathbb{R}$.
2 $y = e^{\lambda \tilde{x}}$, $\lambda \in \mathbb{R}$.

It is easy to see that the composition of two extended functions from $\tilde{\mathcal{P}}(X)$ to $\tilde{\mathcal{P}}(Y)$ and from $\tilde{\mathcal{P}}(Y)$ to $\tilde{\mathcal{P}}(Z)$, respectively, is the extension of the composition of the original functions. This composition is associative. Note also that $f(\tilde{x}) = \tilde{x} \circ R$ where R is defined by

$$\mu_R(x, y) = \begin{cases} 1 & \text{iff} \quad y = f(x), \\ 0 & \text{otherwise.} \end{cases}$$

c. Fuzzy Function of a Nonfuzzy Variable

Two points of view can be developed depending on whether the image of $x \in X$ is a fuzzy set $\tilde{f}(x)$ on Y or x is mapped to Y through a fuzzy set of functions.

α. Fuzzifying Function

A fuzzifying function from X to Y is an ordinary function from X to $\tilde{\mathcal{P}}(Y)$, $\tilde{f} : x \mapsto \tilde{f}(x)$.

The concept of a fuzzifying function and that of a fuzzy relation are mathematically equivalent: $\tilde{f}$ is associated with a fuzzy relation R such that

$$\forall (x, y) \in X \times Y, \quad \mu_{\tilde{f}(x)}(y) = \mu_R(x, y).$$

$\tilde{f}(x)$ is a section of R (see 3.A.a).

Example $\tilde{y} = (\tilde{a} \odot x) \oplus \tilde{b}$ where $(\tilde{a}, \tilde{b}) \in [\tilde{\mathcal{P}}(\mathbb{R})]^2$, and more generally any function with fuzzy parameters.

The composition of fuzzifying functions is defined by

$$\mu_{\tilde{g} \circ \tilde{f}(x)}(z) = \sup_{y \in Y} \min\left(\mu_{\tilde{f}(x)}(y), \mu_{\tilde{g}(y)}(z)\right)$$

where $\tilde{g}$ is a fuzzifying function from Y to Z. The interpretation is: given an intermediary point y, the membership of an element z in $g \circ f(x)$ is bounded by the membership of y in $\tilde{f}(x)$ and by that of z in $\tilde{g}(y)$. The final membership of z in $\tilde{g} \circ \tilde{f}(x)$ is given by the best intermediary point. Note that the composition of fuzzifying functions is nothing but the sup–min composition of their associated fuzzy relations. This composition is thus associative.

N.B.: 1. Fuzzifying functions have been studied by Sugeno (1977) under the name of fuzzy correspondences.

2. Fuzzifying functions (resp. fuzzy relations) may have fuzzy domain and fuzzy range in the sense of a $\cdot \alpha$ (Negoita and Ralescu, 1975):

$$\mu_R(x, y) = \mu_{\tilde{f}(x)}(y) \leqslant \min(\mu_A(x), \mu_B(y))$$

where A and B are respectively the fuzzy domain and the fuzzy range. Such fuzzy functions can also be composed.

β. *Fuzzy Bunch of Functions*

A fuzzy bunch F of functions from X to Y is a fuzzy set on Y^X, that is, each function f from X to Y has a membership value $\mu_F(f)$ in F.

This definition is not equivalent to that of a fuzzifying function. A fuzzifying function $\tilde{f}$ is a fuzzy bunch F in the following sense: $\forall \alpha \in [0, 1]$, the equation $\mu_{\tilde{f}(x)}(y) = \alpha$ defines one or several univalued functions f_α^i from X to Y and the fuzzy bunch is $F = \bigcup_i F^i$ where $F^i = \int_{\alpha \in]0, 1]} \alpha / f_\alpha^i$.

Conversely, a fuzzy bunch is not reducible to a fuzzifying function since there may be two functions f and g from X to Y such that $\exists x, f(x) = g(x) = y$ and $\mu_F(f) \neq \mu_F(g)$. This can never happen for a fuzzifying function because to each pair (x, y) is assigned a unique membership value $\mu_{\tilde{f}(x)}(y) = \mu_R(x, y)$. In a fuzzy bunch each pair (x, y) has several possible membership values. In that sense a fuzzy bunch is a multivalued fuzzy relation. However, if we want to reduce the bunch to a fuzzifying function, we can suppress the ambiguity of the membership value by choosing a combination rule r (sup, inf, . . .) according to the situation:

$$\mu_R(x, y) = \sup_{\substack{f \\ y = f(x)}} \mu_F(f) \qquad (\text{if } r = \sup).$$

Let F and G be two fuzzy bunches from X to Y and from Y to Z, respectively. The composition $H = G \circ F$ of two fuzzy bunches is a fuzzy bunch from X to Z defined by

$$\forall h, \quad \mu_H(h) = \sup_{\substack{f, g \\ h = g \circ f}} \min(\mu_F(f), \mu_G(g)).$$

This composition is associative.

Remark Fuzzy functions of a nonfuzzy variable may have two semantic interpretations:

we do not know the precise image y of x; we know only a distribution of possibility, of probability, of belief (see Chapter 5) of the value of y;

the image of x is actually blurred; it is a fuzzy point (or spot) $\tilde{f}(x)$ in Y.

d. Nonfuzzy Function of a Fuzzy Variable

Let f be a function from $\tilde{\mathcal{P}}(X)$ to $\tilde{\mathcal{P}}(Y)$. An example of such an f is the extension of an ordinary function from X to Y. Another example is a fuzzy relation R using sup–min composition: $\tilde{x} \mapsto \tilde{x} \circ R = \tilde{y}$. Note that in terms of fuzzifying function $(\tilde{f})$ we can define $\tilde{f}(\tilde{x}) = \tilde{x} \circ R$, which naturally extends the domain of $\tilde{f}$ to $\tilde{\mathcal{P}}(X)$. The composition of such extended fuzzifying functions is obviously consistent with that of fuzzy relations: $\tilde{g}[\tilde{f}(\tilde{x})] = (\tilde{x} \circ R) \circ Q = (\tilde{g} \circ \tilde{f})(\tilde{x})$ where Q is the fuzzy relation associated with $\tilde{g}$.

An ordinary function from $\tilde{\mathcal{P}}(X)$ to $\tilde{\mathcal{P}}(Y)$ is more general than an extended fuzzifying function. For instance, an extended fuzzifying function $\tilde{f}$ is such that

$$\forall(\tilde{x}, \tilde{x}') \in [\tilde{\mathcal{P}}(X)]^2, \quad \begin{cases} \tilde{x} \subseteq \tilde{x}' \text{ implies } \tilde{f}(\tilde{x}) \subseteq \tilde{f}(\tilde{x}'), \\ \tilde{f}(\tilde{x} \cup \tilde{x}') = \tilde{f}(\tilde{x}) \cup \tilde{f}(\tilde{x}'). \end{cases}$$

On the contrary, consider the complementation function φ from $\tilde{\mathcal{P}}(X)$ to $\tilde{\mathcal{P}}(X)$: $\varphi(\tilde{x}) = \overline{(\tilde{x})}$. Obviously, we have $\tilde{x} \subseteq \tilde{x}'$ implies $\varphi(\tilde{x}') \subseteq \varphi(\tilde{x})$ and $\varphi(\tilde{x} \cup \tilde{x}') = \varphi(\tilde{x}) \cap \varphi(\tilde{x}')$. Hence, φ is not an extended fuzzifying function and there is no fuzzy relation associated with φ.

An extended fuzzifying function is entirely characterized by its restriction to the ordinary singletons of its domain, i.e., by its associated fuzzy relation, and does not carry more information. This is not true for any function from $\tilde{\mathcal{P}}(X)$ to $\tilde{\mathcal{P}}(Y)$.

N.B.: When the fuzzy relation R associated with a fuzzifying function $\tilde{f}$ from X to $\tilde{\mathcal{P}}(X)$ is reflexive, then

$$\forall \tilde{x} \in \tilde{\mathcal{P}}(X), \quad \tilde{x} \subseteq \tilde{f}(\tilde{x}).$$

The converse holds.

B. FUZZY EXTREMUM

Usually, the maximum (or minimum) of a function f over a given domain D is attained at a precise point x_0. However, we may be interested in the behavior of the function in a neighborhood of x_0; the concept of a maximizing set (minimizing set) provides a tool for modeling this situation. The notion of an extremum also must be generalized to deal with problems such as an extremum of a function over a fuzzy domain or an extremum of a fuzzy function over a domain.

a. Maximizing and Minimizing Set   (Zadeh, 1972)

Let f be a real-valued function whose domain is a set X. f is assumed to be bounded from below by $\inf(f)$ and from above by $\sup(f)$. The maximizing set is a fuzzy set M in X such that:

$$\forall x \in X; \quad \mu_M(x) = \frac{f(x) - \inf(f)}{\sup(f) - \inf(f)} .$$

We always have $\mu_M(x_0) = 1 \; \forall x_0$ such that $f(x_0) = \sup(f)$, and $\mu_M(x) = 0$ $\forall x$ such that $f(x) = \inf(f)$.

Clearly, the maximizing set provides essential information about the effect on the value of the objective function f of choosing values of x other than x_0.

Remark   Another possible membership function for M is the nth power of the normalized f, for instance,

$$\mu_M(x) = \left[\frac{f(x) - \inf(f)}{\sup(f) - \inf(f)} \right]^n .$$

The maximizing set is invariant under linear scaling, that is, M does not change when f is replaced by kf, $k \in \mathbb{R}$.

The fuzzy maximum of f, i.e., a fuzzy set of Y, the range of f ($Y \subseteq \mathbb{R}$), is the image under f of the maximizing set, i.e., $f(M)$:

$$\forall y \in Y, \quad \mu_{f(M)}(y) = \sup_{x \in f^{-1}(y)} \mu_M(x).$$

N.B.:   The minimizing set of f is defined as the maximizing set of $-f$.

b. Maximum of a Nonfuzzy Function over a Fuzzy Domain

Two approaches exist for this problem according to whether a nonfuzzy maximum or a fuzzy one is sought.

α. Nonfuzzy Maximum

Let D be the domain over which we want to maximize f, a function from X to $\mathbb{R}$. Let M be the maximizing set of f. When D is nonfuzzy, an element x_0 that maximizes f in D is such that

$$\mu_M(x_0) = \sup_{x \in D} \mu_M(x) = \sup_{x \in X} \min(\mu_M(x), \mu_D(x)).$$

When D is fuzzy, the maximization problem can be understood as: find an element of X that belongs as much as possible both to the maximizing set and to the fuzzy domain D. The corresponding membership grade is the consistency of M and D, i.e.,

$$\text{hgt}(M \cap D) = \sup_{x \in X} \min(\mu_M(x), \mu_D(x)) = \mu(x_0).$$

Zadeh (Reference from III.2, 1965) first used the product $f(x) \cdot \mu_D(x)$ instead of min and the maximizing set.

An analysis of the search of a maximum for $\mu(x) = \min(\mu_M(x), \mu_D(x))$ was carried out by Tanaka *et al.* (1973). They used the resolution of D into its α-cuts D_α, noting that

$$\mu(x_0) = \sup_{x \in X} \min(\mu_M(x), \mu_D(x)) = \sup_{\alpha \in [0,1]} \min\left(\alpha, \sup_{x \in D_\alpha} \mu_M(x)\right).$$

The function $g(\alpha) = \sup_{x \in D_\alpha} \mu_M(x)$ is nonincreasing ($\alpha_1 > \alpha_2$ implies $D_{\alpha_1} \subseteq D_{\alpha_2}$); hence if g is continuous, the maximum is attained for α^* such that $\alpha^* = \sup_{x \in D_{\alpha^*}} \mu_M(x) = \mu_M(x_0)$; hence $x_0 \in D_{\alpha^*}$ and $\mu_D(x_0) \geqslant \mu_M(x_0)$. The initial maximizing problem is thus equivalent to the maximization of μ_M over the nonfuzzy domain $T = \{x, \mu_D(x) \geqslant \mu_M(x)\}$, provided that $g(\alpha)$ is continuous. A sufficient condition is given by Tanaka *et al.* (1973): if D is a strictly convex fuzzy set on $\mathbb{R}^n$ ($\forall(x, y) \in \text{supp}\,D$, $x \neq y$, $\forall \lambda \in \,]0, 1[$, $\mu_D(\lambda x + (1 - \lambda)y) > \min(\mu_D(x), \mu_D(y))$), then g is continuous.

The main drawback of this approach is that when $\mu_D(x_0)$ is small, the solution is not very satisfactory because x_0 does not belong "enough" to D; we may prefer a solution that belongs more to D although it will entail a shift of x_0 toward smaller values of μ_M. The second approach copes with this difficulty.

β. Fuzzy Maximum

Let $N(\alpha) = \{x_0 \in X, f(x_0) = \sup_{x \in D_\alpha} f(x)\}$ and $R = \bigcup_\alpha N(\alpha)$. $N(\alpha)$ is the set of elements maximizing f on D_α. The fuzzy set of maximizing elements is $N = D \cap R$. The following proposition holds:

$$\forall x \in R, \quad \mu_N(x) = \sup_{x \in N(\alpha)} \alpha.$$

(Note that $N(\alpha)$ is *not* the α-cut of N.)

Proof: Let $x_0 \in R$. Obviously, $x_0 \in D_\lambda \cap R$ for $\lambda \leqslant \mu_D(x_0)$ only. Note that

$$D_\lambda \cap R = D_\lambda \cap \left(\bigcup_\alpha N(\alpha) \right) = D_\lambda \cap \left(\bigcup_{\alpha \geqslant \lambda} N(\alpha) \right)$$

because $\forall \alpha < \lambda$, $N(\alpha) - N(\lambda) = \emptyset$ or $N(\alpha) - N(\lambda) \not\subset D_\lambda$. Hence, $x_0 \in \bigcup_{\alpha \geqslant \mu_D(x_0)} N(\alpha)$.

Moreover, if $\exists \alpha > \mu_D(x_0)$ such that $x_0 \in N(\alpha)$, then $x_0 \in D_\alpha$, which is contradictory when $\alpha > \mu_D(x_0)$. Hence, $x_0 \in N(\alpha)$ for some $\alpha \leqslant \mu_D(x_0)$. On the whole $x_0 \in N(\mu_D(x_0))$ and $x_0 \notin N(\alpha)$ for $\alpha > \mu_D(x_0)$.

Hence, $\sup_{x_0 \in N(\alpha)} \alpha = \mu_D(x_0) = \mu_N(x_0)$ since $x_0 \in R$. Q.E.D.
Conventionally, $\forall x \notin R$, $\sup_{x \in N(\alpha)} \alpha = 0$.

The fuzzy value of the maximum of f over D is the fuzzy set of $\mathbb{R}$ induced from N through f, i.e., $f(N)$ such that

$$\mu_{f(N)}(y) = \sup_{x \in f^{-1}(y)} \mu_N(x).$$

Note that $\mu_{f(N)}$ is a nonincreasing function on supp $f(N)$. If we want to improve the value y of the maximum, we must broaden the maximization domain D_α, i.e., diminish α. In Zadeh's notation

$$f(N) = \int_\mathbb{R} \alpha \Big/ \sup_{x \in D_\alpha} f(x).$$

This approach was developed by Orlovsky (1977). He also developed another definition of N, by considering the maximal elements over X, in the sense of Pareto, of the set of pairs $(f(x), \mu_D(x))$; $(f(x_0), \mu_D(x_0))$ is said to be *Pareto maximal* iff

$$\{(f(x), \mu_D(x)), f(x) \geqslant f(x_0) \text{ and } \mu_D(x) \geqslant \mu_D(x_0)\} = \{(f(x_0), \mu_D(x_0))\}.$$

Let P be the set of elements x such that $(f(x), \mu_D(x))$ is Pareto maximal. When $X = \mathbb{R}^n$ and the functions f and μ_D are continuous, Orlovsky has proved that if $N' = D \cap P$, then $f(N') = f(N)$, which expresses the equivalence between both of Orlovsky's approaches.

N.B.: The approach developed in β is consistent with the extension principle. The function to be extended is from $\mathcal{P}(X)$ to $\mathbb{R}$: $D \mapsto \sup_{x \in D} f(x)$, the membership function of the extension, when D is fuzzy, is

$$\mu(y) = \sup_{\substack{A: \\ y = \sup_{x \in A} f(x)}} \mu_D(A) \quad \text{where} \quad \mu_D(A) = \begin{cases} \alpha & \text{if } A = D_\alpha, \\ 0 & \text{otherwise.} \end{cases}$$

D is viewed as a fuzzy set on $\mathcal{P}(X)$ whose support is the set of its α-cuts.

Then $\mu(y) = \mu_{f(N)}(y)$. With a less drastic definition of μ_D on $\mathcal{P}(X)$, consistency with the extension principle may fail.

Example Fig. 1 illustrates both approaches α and β. $\forall \alpha > \mu_D(b)$, the element that maximizes f over D_α is a. $\forall \alpha \in]\mu_D(b), \mu_D(c)]$, the element that maximizes f over D_α is on the right edge of D_α. $\forall \alpha < \mu_D(c)$, this element is always c. Hence supp $N = \{a\} \cup [b,c]$, supp $f(N) = [f(b), f(c)]$ $= [f(a), f(c)]$. Note that $\mu_{f(N)}(f(a)) = 1$ and $\mu_{f(N)}(y) < \mu_D(b)$ for $y > f(a)$. The maximizing set of f is sketched with a dashed line; the method of Tanaka et al gives (x^*, α^*). Generally $x^* \in $ supp N as in Fig. 1; that is to say, approach α is included in Orlovsky's.

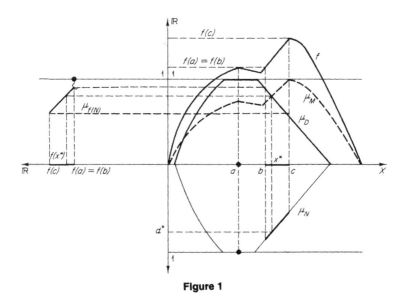

Figure 1

c. Fuzzy Maximum of a Fuzzy Function on a Nonfuzzy Domain

Let $\tilde{f}$ be a fuzzifying function from X to $\mathbb{R}$ and D a nonfuzzy domain of X over which we want to maximize $\tilde{f}$. For simplicity, assume X is finite.

Since $\tilde{f}(x)$ is a fuzzy set on $\mathbb{R}$, a first idea for defining the value $\tilde{m}$ of the maximum of f over D is to use the operator $\widetilde{\max}$ and state $\tilde{m} = \widetilde{\max}_{x \in D} \tilde{f}(x)$. This quantity exists because D is finite. It is a fuzzy set on $\mathbb{R}$. There is ambiguity for the choice of an element in D realizing $\tilde{m}$ because $\tilde{m}$ is generally not one of the $\tilde{f}(x)$'s (see 2.B.d.ϵ). Hence, we must keep track of the x's that actually contribute to the membership function of $\tilde{m}$.

Note that if $|D| = n$:

$$\mu_{\tilde{m}}(y) = \sup_{\substack{y_1, \ldots, y_n \\ y = \max(y_1, \ldots, y_n)}} \min\left(\mu_{\tilde{f}(x_1)}(y_1), \ldots, \mu_{\tilde{f}(x_n)}(y_n)\right)$$

$$= \sup_{\substack{f \in \mathbb{R}^D \\ y = \max_{i=1,n}(f(x_i))}} \min_{j=1,n} \mu_{\tilde{f}(x_j)}(f(x_j))$$

or in Zadeh's notation:

$$m = \int_{f \in \mathbb{R}^D} \min_{j=1,n} \mu_{\tilde{f}(x_j)}(f(x_j)) \bigg/ \sup_{x \in D} f(x).$$

An alternative method considers equations $\mu_{\tilde{f}(x)}(y) = \alpha$, $\alpha \in]0, 1]$. When these equations define univalued functions $y = f_\alpha(x)$, we may think of maximizing each f_α over D and state

$$\tilde{m}' = \int_{\alpha \in]0, 1]} \alpha \bigg/ \sup_{x \in D} f_\alpha(x)$$

N.B.: There are possibly several f_α such that $\mu_{\tilde{f}(x)}(f_\alpha(x)) = \alpha$, $\forall x \in X$. When $\mathbb{R}^D$ can be replaced by the set $\{f_\alpha, \alpha \in]0, 1]\}$, then $\tilde{m}' = \tilde{m}$.

Example Let $\tilde{f}(x)$ be a fuzzifying function from $\mathbb{R}$ to $\mathbb{R}$ such that $\tilde{f}(x)$ is a triangular fuzzy number for any x. $D = \{x_1, x_2, x_3, x_4, x_5\}$. On the left-hand part of Fig. 2 are represented the elements of D and the curves $f_1, f_\alpha^+, f_\alpha^-$ that satisfy, $\forall x \in X$,

$$\mu_{\tilde{f}(x)}(f_1(x)) = 1, \qquad \mu_{\tilde{f}(x)}(f_\alpha^-(x)) = \mu_{\tilde{f}(x)}(f_\alpha^+(x)) = \alpha.$$

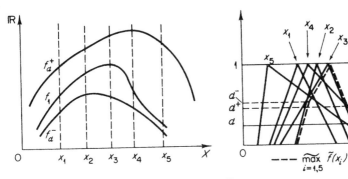

Figure 2

The right-hand part of Fig. 2 pictures the five fuzzy numbers $\tilde{f}(x_i)$, $i = 1, 5$. $\tilde{m}$ is the dashed line. Only x_2, x_3, x_4 contribute to building $\mu_{\tilde{m}}$. Moreover,

$$\max_{x \in D} f_\alpha^+(x) = f_\alpha^+(x_4), \qquad \max_{x \in D} f_1(x) = f_1(x_3), \qquad \max_{x \in D} f_\alpha^-(x) = f_\alpha^-(x_2).$$

$\tilde{m}$ and $\tilde{m}'$ are thus consistent for these points. From Fig. 2, right-hand side, we see that

$$\tilde{m} = \int_{\alpha \in]0,\,\alpha^-]} \alpha/f_\alpha^-(x_2) + \int_{\alpha \in [\alpha^-,\,1]} \alpha/f_\alpha^-(x_3) + \int_{\alpha \in [\alpha^+,\,1]} \alpha/f_\alpha^+(x_3)$$
$$+ \int_{\alpha \in]0,\,\alpha^+]} \alpha/f_\alpha^+(x_4) = \tilde{m}'$$

in Zadeh's notation. α^- and α^+ are such that $f_{\alpha^-}(x_2) = f_{\alpha^-}(x_3)$ and $f_{\alpha^+}(x_4) = f_{\alpha^+}(x_3)$, respectively. More specifically, for

$$\alpha \in [0, \alpha^-], \qquad f_\alpha^-(x_2) \geqslant f_\alpha^-(x_i) \qquad \forall i,$$
$$\alpha \in [\alpha^-, 1], \qquad f_\alpha^-(x_3) \geqslant f_\alpha^-(x_i) \qquad \forall i,$$
$$\alpha \in [\alpha^+, 1] \qquad f_\alpha^+(x_3) \geqslant f_\alpha^+(x_i) \qquad \forall i,$$
$$\alpha \in [0, \alpha^+] \qquad f_\alpha^+(x_4) \geqslant f_\alpha^+(x_i) \qquad \forall i.$$

The fuzzy set on D maximizing f is here $N = \alpha^-/x_2 + 1/x_3 + \alpha^+/x_4$. More generally, if

$$\tilde{m}' = \int_{\alpha \in]0,\,1]} \alpha/ \sup_{x \in D} f_\alpha(x) = \int_{\alpha \in]0,\,1]} \alpha/f_\alpha(x(\alpha))$$

then

$$N = \int_{\alpha \in]0,\,1]} \alpha/x(\alpha) \qquad \text{(symbolically)}.$$

C. INTEGRATION OF FUZZY FUNCTIONS OVER (NON)FUZZY INTERVALS

This section is concerned with the possibility of extending elementary results in the analysis of real-valued ordinary functions to fuzzifying functions from $\mathbb{R}$ to $\mathbb{R}$. This attempt is very similar to that of Chapter 2, section B where results in real algebra have been extended to fuzzy numbers. Unsurprisingly, the most remarkable properties will be obtained for fuzzifying functions that map into the set of real fuzzy numbers. Integration over a nonfuzzy and a fuzzy interval are investigated. The main reference for this material is Dubois and Prade (1978).

a. Integral of a Fuzzifying Function over a Nonfuzzy Interval

α. Definition

Let f be a fuzzifying function from $[a, b] \subseteq \mathbb{R}$ to $\mathbb{R}$ such that $\forall x \in [a, b]$, $\tilde{f}(x)$ is a fuzzy number, i.e., a piecewise continuous convex normalized fuzzy set on $\mathbb{R}$. $\forall \alpha \in]0, 1]$, the equation $\mu_{\tilde{f}(x)}(y) = \alpha$ with x and α as parameters is assumed to have two and only two continuous solutions $y = f_\alpha^+(x)$ and $y = f_\alpha^-(x)$ for $\alpha \neq 1$ and only one, $y = f(x)$, for $\alpha = 1$, which is also continuous. f_α^+ and f_α^- are defined such that

$$f_{\alpha'}^+(x) \geqslant f_\alpha^+(x) \geqslant f(x) \geqslant f_\alpha^-(x) \geqslant f_{\alpha'}^-(x) \qquad \forall \alpha, \alpha', \text{ with } \alpha' \leqslant \alpha.$$

These functions will be called α-level curves of $\tilde{f}$ (see Fig. 3). The integral of any continuous α-level curve of $\tilde{f}$ over $[a,b]$ always exists. Unless specified, $\tilde{f}$ always satisfies these assumptions.

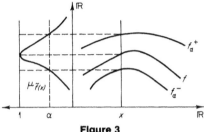

Figure 3

An intuitive way of defining the integral $\tilde{I}(a,b)$ of $\tilde{f}$ over $[a,b]$ is to assign the membership value α to the integral of any α-level curve of $\tilde{f}$ over $[a,b]$. Using Zadeh's notation, $\tilde{I}(a,b)$ is the fuzzy set on $\mathbb{R}$

$$\tilde{I}(a,b) = \int_{\alpha \in]0,\,1]} \alpha \Big/ \int_a^b f_\alpha^-(x)\,dx + \int_{\alpha \in]0,\,1]} \alpha \Big/ \int_a^b f_\alpha^+(x)\,dx. \qquad (7)$$

This definition is consistent with the extension principle. Let us show this for a particular case.

Let L be the set of functions l from $\mathbb{R}$ to $\mathbb{R}$ such that $\int_a^b l(x)\,dx$ exists and l is made of a denumerable union of pieces of level curves (see Fig. 4). Hence $l = \bigcup_{i \in N} l_i$ where l_i is continuous. The curve l_i delimits an area A_i whose surface is T_i. The fuzzifying function is viewed as a fuzzy set on L such that $\mu_{\tilde{f}}(l) = \inf_{i \in N} \mu_{\tilde{f}}(l_i)$ with $\mu_{\tilde{f}}(l_i) = \alpha_i$ iff l_i is part of an α_i-level curve of $\tilde{f}$.

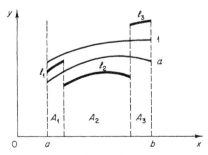

Figure 4

According to the extension principle, we have

$$\mu_{\tilde{I}(a,b)}(T) = \sup_{l \in L \,:\, T = \int_a^b l(x)\,dx} \inf_i \mu_{\tilde{f}}(l_i). \tag{8}$$

The following proposition holds when $\mu_{\tilde{f}(x)}(y)$ is continuous with respect to y and x and $\forall x$, $\tilde{f}(x)$ is a fuzzy number without constant-membership intervals: $\forall T \in \mathbb{R}$, $\exists$ an α-level curve f_α, delimiting an area A whose surface is T, such that

$$\mu_{\tilde{I}(a,b)}(T) = \alpha \qquad \text{and} \qquad T = \int_a^b f_\alpha(x)\,dx.$$

Proof: The double continuity of $\mu_{\tilde{f}(x)}(y)$ with respect to y and x implies that of α-level curves with respect to α. Denote by f_α a generic α-level curve of $\tilde{f}$. $\forall T$, $\exists \alpha$ and f_α such that $\int_a^b f_\alpha(x)\,dx = T$ and f_α is unique. Let $L_T = \{ l \in L \mid \int_a^b l(x)\,dx = T \}$. $\forall l \in L_T - \{ f_\alpha \}$, l is made of pieces l_i on each side of f_α because were they on the same side of f_α, the so-delimited area could not have a surface T (see Fig. 4). Owing to the bell shape of $f(x)$, $\inf_i \mu_{\tilde{f}}(l_i) \leqslant \alpha$. Hence, $\mu_{\tilde{I}(a,b)}(T) = \alpha$. Formulas (7) and (8) are consistent.   Q.E.D.

N.B.:   When $\mu_{\tilde{f}(x)}(y)$ is only piecewise continuous with respect to y, the mapping $T \mapsto f_\alpha$, $\int_a^b f_\alpha(x)\,dx = T$ can be multivalued because some level curves may overlap. However, the level curves of any fuzzifying function can never cross each other since they are defined by the equations $\alpha = \mu_{\tilde{f}(x)}(y), \alpha \in \,]0, 1]$.

When $\tilde{f}(x)$ has constant-membership intervals, the level curves may degenerate into "level areas."

Formula (8) could be extended naturally by replacing L with the set G of functions g such that $\int_a^b g(x)\,dx$ exists:

$$\mu_{\tilde{I}(a,b)}(T) = \sup_{g \in G \,:\, T = \int_a^b g(x)\,dx} \mu_{\tilde{f}}(g) \tag{9}$$

with $\mu_{\tilde{f}}(g) = \inf_{x \in [a,b]} \mu_{\tilde{f}(x)}(g(x))$. Consistency of (8) and (9) can be conjectured, but the proof requires some precise mathematical tools and is beyond the scope of this book.

N.B.:   Since interval-valued functions are particular cases of fuzzy-valued functions, the above approach may be viewed as an attempt to generalize integrals of ordinary set-valued functions (see Aumann, NF 1965). The latter have arisen in connection with economics problems.

β. Calculations of $\tilde{I}(a,b)$ when f in a L-R type Fuzzifying Function

A fuzzifying function is said to be an L-R type fuzzifying function iff it satisfies the requirements of α and $f(x) = (f(x), s(x), t(x))_{LR}$ is an L-R type fuzzy number $\forall x \in [a,b]$. f, s, and t are positive integrable functions on $[a,b]$. Note that the 1-level curve of $\tilde{f}(x)$ is $f(x)$, i.e., the mean value of

$\tilde{f}(x)$ $\forall x$. Obviously, the two α-level curves of $f(x)$ are

$$f_\alpha^-(x) = f(x) - s(x)L^{-1}(\alpha), \qquad f_\alpha^+(x) = f(x) + t(x)R^{-1}(\alpha).$$

Integrating $f_\alpha^-(x)$ over $[a,b]$ gives

$$\int_a^b f_\alpha^-(x)\,dx = \int_a^b f(x)\,dx - L^{-1}(\alpha)\int_a^b s(x)\,dx = Z.$$

Denoting by F, S, T, antiderivatives of f, s, t, respectively, we get

$$Z = F(b) - F(a) - L^{-1}(\alpha)(S(b) - S(a))$$

or

$$L\left(\frac{F(b) - F(a) - Z}{S(b) - S(a)}\right) = \alpha \qquad \forall Z \leqslant F(b) - F(a).$$

Note that $S(b) - S(a) \geqslant 0$ since $b \geqslant a$. The same reasoning holds for f_α^+, and we get

$$\tilde{I}(a,b) = \left(\int_a^b f(x)\,dx, \int_a^b s(x)\,dx, \int_a^b t(x)\,dx\right)_{LR}, \tag{10}$$

which is the result of (8) when $\tilde{f}$ is an L-R fuzzifying function.

To integrate an L-R fuzzifying function over a nonfuzzy interval $[a,b]$, it is sufficient to integrate mean value and spread functions over $[a,b]$. The result is an L-R fuzzy number.

γ. *Relationship with Riemann Sums*

Let $(x_1, \ldots, x_n) \in [a,b]^n$ be made up of n real numbers such that

$$a = x_1 < x_2 < \cdots < x_{n-1} < x_n = b$$

and $\tilde{\Sigma}_n$ be the fuzzy sum $(x_2 - x_1)\tilde{f}(x_2) \oplus (x_3 - x_2)\tilde{f}(x_3) \cdots \oplus (x_n - x_{n-1})$ $f(x_n)$. When f is an L-R type fuzzifying function, the fuzzy Riemann sum $\tilde{\Sigma}_n$ can be written.

$$\tilde{\Sigma}_n = \left(\sum_{i=2}^n (x_i - x_{i-1})f(x_i), \sum_{i=2}^n (x_i - x_{i-1})s(x_i), \sum_{i=2}^n (x_i - x_{i-1})t(x_i)\right)_{LR}.$$

Owing to the continuity of L and R and to the existence of the integrals over $[a,b]$ for f,s,t, the limit of $\tilde{\Sigma}_n$ exists and is

$$\lim_{n \to +\infty} \tilde{\Sigma}_n =$$

$$\left(\lim_{n \to +\infty} \sum_{i=2}^n (x_i - x_{i-1})f(x_i), \lim_{n \to +\infty} \sum_{i=2}^n (x_i - x_{i-1})s(x_i), \lim_{n \to +\infty} \sum_{i=2}^n (x_i - x_{i-1})t(x_i)\right)_{LR}$$

$$= \left(\int_a^b f(x)\,dx, \int_a^b s(x)\,dx, \int_a^b t(x)\,dx\right)_{LR} = \tilde{I}(a,b).$$

So, when f is an L-R type fuzzifying function, the extension principle does generalize Riemann sums, and hence the integration defined in this section generalizes Riemann's integration.

N.B.:   Denoting by $\tilde{F}(x)$ the L-R type fuzzifying function $(F(x), S(x), T(x))_{LR}$, which we may call an "antiderivative" of $\tilde{f}$, the formula

$$\tilde{I}(a,b) = \tilde{F}(b) \ominus \tilde{F}(a)$$

does not hold any longer because $\tilde{I}(a,b) = (F(b) - F(a), S(b) - S(a), T(b) - T(a))_{LR}$, which differs from

$$\tilde{F}(b) \ominus \tilde{F}(a) = (F(b) - F(a), S(b) + S(a), T(b) + T(a))_{LR}.$$

b.   Integral of a Nonfuzzy Function over a Fuzzy Interval

α.   Definition

Let A and B be two fuzzy sets on $\mathbb{R}$. The extension principle allows defining the integral of a real-valued ordinary function f over the fuzzy interval (A, B) bounded by A and B, say $I(A, B)$:

$$\mu_{I(A, B)}(Z) = \sup_{x, y \,:\, Z = \int_x^y f(u)\,du} \min(\mu_A(x), \mu_B(y)). \tag{11}$$

β.   One of the Bounds is Not Fuzzy

We consider the integral of f over $[a, B)$:

$$\mu_{I(a, B)}(Z) = \sup_{y \,:\, Z = \int_a^y f(u)\,du} \mu_B(y) = \sup_{y \,:\, Z = F(y) - F(a)} \mu_B(y)$$

where F is an antiderivative of f. We see that $I(a, B) = F(B) \ominus F(a)$ is the value of the extended $F(x) - F(a)$, when $x = B$.

γ.   Both Bounds are Fuzzy

(11) can be changed into

$$\mu_{I(A, B)}(Z) = \sup_{Z = F(y) - F(x)} \min(\mu_A(x), \mu_B(y))$$

$$= \sup_{x \in \mathbb{R}} \min\left(\mu_A(x), \sup_{y \,:\, Z = F(y) - F(x)} \mu_B(y)\right)$$

$$= \sup_{x \in \mathbb{R}} \min(\mu_A(x), \mu_{I(x, B)}(Z)),$$

that is, $I(A, B) = A \circ I(\cdot, B) = A \circ (F(B) \ominus F(\cdot))$. $I(A, B)$ is the fuzzy value of the extended fuzzifying function $y = F(B) \ominus F(x)$ for $x = A$, using the results of A.d. Hence, $I(A, B) = F(B) \ominus F(A)$, which can be denoted $\int_A^B f(x)\,dx$.

N.B.: When A and B are fuzzy numbers of L-R type, the calculation of $I(A,B)$ is not especially simplified. For instance, $A = (a, \underline{a}, \bar{a})_{LR}$, then

$$\mu_{F(A)}(Z) = L\left(\frac{F^{-1}(Z) - a}{\underline{a}}\right) \qquad \text{for} \quad F^{-1}(Z) \leqslant a \text{ and } F \text{ injective.}$$

$I(A,B)$ will not generally be an L-R type fuzzy number when A and B are.

Remark An alternative approach to the integral of a nonfuzzy function over a fuzzy interval could be the following.

Let C be a fuzzy interval modeled by a flat fuzzy number (see 2.B.e.η.). The integral over C of the function f can be $i(C) = \int_\mathbb{R} \mu_C(x) \cdot f(x) dx$. What is obtained is a median value between the crisp integrals $i(C_1)$ and $i(\text{supp} C)$ where C_1 is the 1-cut of C. This point of view departs from the fuzzy evaluation of the fuzzy surface bounded by A, B, f and the abscissa axis.

c. Integral of a Fuzzifying Function over a Fuzzy Interval

Zadeh's extension principle gives now

$$\mu_{\tilde{I}(A, B)}(Z) = \sup_{\substack{l \in L, (x, y) \in \mathbb{R}^2, x \leqslant y \\ \int_x^y l(t) dt = Z}} \min(\mu_A(x), \mu_B(y), \mu_{\tilde{f}}(l)). \qquad (12)$$

Where $\tilde{f}$ is a fuzzifying function satisfying the assumptions of a and A, B are fuzzy numbers that delimit a fuzzy interval, $\tilde{I}(A,B)$ is the fuzzy integral. (12) can be written

$$\mu_{\tilde{I}(A, B)}(Z) = \sup_{x \leqslant y} \min\left[\mu_A(x), \mu_B(y), \sup_{\substack{l \in L \\ Z = \int_x^y l(t) dt}} \mu_{\tilde{f}}(l) \right]$$

$$= \sup_{x \leqslant y} \min\left[\mu_A(x), \mu_B(y), \mu_{\tilde{I}(x, y)}(Z) \right].$$

Note that since $\tilde{I}(x, y) = -\tilde{I}(y, x)$, the condition $x \leqslant y$, which was a priori imposed in (12) for the sake of consistency with (8), can be dropped. Thus, $\tilde{I}(A, B)$ is the value of the extended $\tilde{I}(x, y)$ for $x = A$, $y = B$.

N.B.: 1. We could have considered changing (12) into

$$\mu_{\tilde{I}(A, B)}(Z) = \sup_{l \in L} \min\left(\mu_{\tilde{f}}(l), \sup_{\substack{x, y, x \leqslant y \\ Z = \int_x^y l(x) dx}} \min(\mu_A(x), \mu_B(y)) \right). \qquad (13)$$

The calculation carried out in this way can be shown to be very unwieldy (see Dubois and Prade, 1978). Even if we assume that the upper bound is attained for an α-level curve of $\tilde{f}$, the determination of α can be very difficult because the already-mentioned upper bound is not necessarily attained for equal values of the arguments of the min.

N.B.: 2. Since from a, $\tilde{I}(x, y) \neq \tilde{F}(y) \ominus \tilde{F}(x)$ when $\tilde{F}$ is an anti-derivative of $\tilde{f}$, we do not have $\tilde{I}(A, B) = \tilde{F}(B) \ominus \tilde{F}(A)$ either. We remark that, denoting by $\mathfrak{L}$ an antiderivative of l, (13) can be written:

$$\mu_{\tilde{I}(A, B)}(Z) = \sup_{l \in L} \min(\mu_{\tilde{f}}(l),\ \mu_{\mathfrak{L}(B) \ominus \mathfrak{L}(A)}(Z)),$$

using the results of b. The intractability of (13) is thus related to the fact that $\tilde{I}(A, B) \neq \tilde{F}(B) \ominus \tilde{F}(A)$.

d. Some Properties of the Integral Operator

α. Linearity

It is easy to see using definition (8) that

$$\int_a^b (\tilde{f}(x) \oplus \tilde{g}(x))\, dx = \int_a^b \tilde{f}(x)\, dx \oplus \int_a^b \tilde{g}(x)\, dx.$$

$\int_a^b \tilde{f}(x)\, dx$, $\int_a^b \tilde{g}(x)\, dx$ denote the integrals of the fuzzifying functions $\tilde{f}$ and $\tilde{g}$, respectively. When $\tilde{f}$ and $\tilde{g}$ are L-R type fuzzifying functions, the result is also easily obtained by considering Riemann sums.

We also have

$$\int_{\tilde{a}}^{\tilde{b}} (f(x) + g(x))\, dx = \int_{\tilde{a}}^{\tilde{b}} f(x)\, dx \oplus \int_{\tilde{a}}^{\tilde{b}} g(x)\, dx \qquad (14)$$

which is easy to see reasoning with α-cuts.

On the contrary,

$$\int_{\tilde{a}}^{\tilde{b}} f(x)\, dx \oplus \int_{\tilde{b}}^{\tilde{c}} f(x)\, dx \supseteq \int_{\tilde{a}}^{\tilde{c}} f(x)\, dx. \qquad (15)$$

The equality does not hold except if $\tilde{b}$ is a real number.

Proof: Let $[\underline{a}(\alpha), \bar{a}(\alpha)]$, $[\underline{b}(\alpha), \bar{b}(\alpha)]$, $[\underline{c}(\alpha), \bar{c}(\alpha)]$ be the α-cuts of $\tilde{a}$, $\tilde{b}$, $\tilde{c}$ respectively.

The α-cut of the left-hand side of (14) is

$$\left[\int_{\underline{a}}^{\underline{b}} (f(x) + g(x))\, dx, \int_{a}^{\bar{b}} (f(x) + g(x))\, dx \right].$$

It obviously equals the α-cut of the right-hand side of (14).

The α-cut of the left-hand side of (15) is

$$\left[\int_{\underline{a}}^{b} f(x)\,dx + \int_{\overline{b}}^{c} f(x)\,dx, \int_{a}^{\overline{b}} f(x)\,dx + \int_{b}^{\overline{c}} f(x)\,dx \right]$$

which contains

$$\left[\int_{\overline{a}}^{c} f(x)\,dx, \int_{a}^{\overline{c}} f(x)\,dx \right] \quad \text{Q.E.D.}$$

β. Relationship with $\cup$, $\cap$

Denote by $\int_a^B f(x)\,dx$ the integral of the nonfuzzy function f over the fuzzy interval $[a, B)$, $a \in \mathbb{R}$, $B \in \tilde{\mathscr{P}}(\mathbb{R})$; then:

$$\int_a^{B \cup C} f(x)\,dx = \left(\int_a^B f(x)\,dx \right) \cup \left(\int_a^C f(x)\,dx \right), \qquad C \in \tilde{\mathscr{P}}(\mathbb{R});$$

when f is positive,

$$\int_a^{B \cap C} f(x)\,dx = \left(\int_a^B f(x)\,dx \right) \cap \left(\int_a^C f(x)\,dx \right).$$

These properties are particular cases of $h(B \cup C) = h(B) \cup h(C)$; and, when h is injective, $h(B \cap C) = h(B) \cap h(C)$ where h is any extended function from $\tilde{\mathscr{P}}(X)$ to $\tilde{\mathscr{P}}(Y)$ $\forall B, C \in \tilde{\mathscr{P}}(X)$.

e. Example

Consider the function $\tilde{a}x \oplus \tilde{b} = \tilde{y}$ where $\tilde{a} = (a, \underline{a}, \overline{a})_{LL}$ $\tilde{b} = (b, \underline{b}, \overline{b})_{LL}$ fuzzy numbers of the L-L type. The expressions for $\mu_{\tilde{y}}(t)$ are

$$x \geqslant 0, \quad t \leqslant ax + b: \quad \mu_{\tilde{y}}(t) = L\left[\frac{ax + b - t}{\underline{a}x + \underline{b}} \right];$$

$$x \geqslant 0, \quad t \geqslant ax + b: \quad \mu_{\tilde{y}}(t) = L\left[\frac{t - (ax + b)}{\overline{a}x + \overline{b}} \right];$$

$$x \leqslant 0, \quad ax + b \geqslant t: \quad \mu_{\tilde{y}}(t) = L\left[\frac{ax + b - t}{-\overline{a}x + \underline{b}} \right];$$

$$x \leqslant 0, \quad ax + b \leqslant t: \quad \mu_{\tilde{y}}(t) = L\left[\frac{t - (ax + b)}{-\underline{a}x + \overline{b}} \right].$$

Assuming $w = L^{-1}(\alpha)$, the α-level curves are respectively in each case

$$\text{for } x \geq 0 \quad \begin{cases} g_\alpha^-(x) = x(a - \underline{a}w) + b - \underline{b}w, \\ g_\alpha^+(x) = x(a + \bar{a}w) + b + \bar{b}w, \end{cases}$$

$$\text{for } x \leq 0 \quad \begin{cases} g_\alpha^-(x) = x(a + \bar{a}w) + b - \underline{b}w, \\ g_\alpha^+(x) = x(a - \underline{a}w) + b + \bar{b}w, \end{cases}$$

and are pictured on Fig. 5. Note that $g_\alpha^+(x)$, $x \geq 0$, has the same slope as $g_\alpha^-(x)$, $x \leq 0$, and $g_\alpha^-(x)$, $x \geq 0$, has the same slope as $g_\alpha^+(x)$, $x \leq 0$. Using the results about integration, $\int_0^x (\tilde{a}s \oplus \tilde{b}) ds$ has membership function

$$x \geq 0, \quad \frac{a}{2} x^2 + bx \geq t, \quad \mu(t) = L\left[\frac{(a/2)x^2 + bx - t}{(\underline{a}/2)x^2 + \underline{b}x} \right];$$

$$x \geq 0, \quad \frac{a}{2} x^2 + bx \leq t, \quad \mu(t) = L\left[\frac{t - [(a/2)x^2 + bx]}{(\bar{a}/2)x^2 + \bar{b}x} \right];$$

and $\int_x^0 (\tilde{a}s \oplus \tilde{b}) ds$ has membership function:

$$x \leq 0, \quad -\frac{a}{2} x^2 - bx \geq t, \quad \mu(t) = L\left[\frac{-(a/2)x^2 - bx - t}{(\bar{a}/2)x^2 - \underline{b}x} \right];$$

$$x \leq 0, \quad -\frac{a}{2} x^2 - bx \leq t, \quad \mu(t) = L\left[\frac{(a/2)x^2 + bx + t}{(\underline{a}/2)x^2 - \bar{b}x} \right];$$

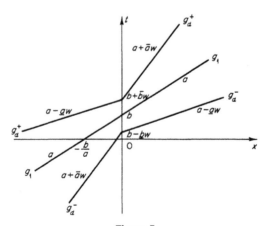

Figure 5

and

$$\int_0^x (\tilde{a}s \oplus \tilde{b})\, ds = \left[\frac{a}{2}\, x^2 + bx,\, \underline{a}\,\frac{x^2}{2} + \underline{b}x,\, \bar{a}\,\frac{x^2}{2} + \bar{b}x \right]_{LL}, \qquad x \geq 0,$$

$$\int_x^0 (\tilde{a}s \oplus \tilde{b})\, ds = \left[-\frac{a}{2}\, x^2 - bx,\, \bar{a}\,\frac{x^2}{2} - \underline{b}x,\, \underline{a}\,\frac{x^2}{2} - \bar{b}x \right]_{LL}, \qquad x \leq 0,$$

It is nothing but the fuzzy function $\tilde{I}(0,x) = \tilde{y} = \tilde{a}(x^2/2) \oplus \tilde{b}x$, an extended primitive of $y = ax + b$, for $x \geq 0$, and $\tilde{I}(x,0) = -(\tilde{a}(x^2/2) \oplus \tilde{b}x)$ for $x \leq 0$.

The integral $\int_0^{\tilde{c}}(ax + b)\, dx$ where a and b are no longer fuzzy is, from b . γ, $(a\tilde{c}^2/2) \oplus b\tilde{c}$, which, in this example, can be computed easily through approximate sum and product formulas (see 2.B.e). Lastly, for $\tilde{c} > 0$,

$$\int_0^{\tilde{c}} (\tilde{a}x \oplus \tilde{b})\, dx = \left[\frac{\tilde{a}x^2}{2} \oplus \tilde{b}x \right]_0^{\tilde{c}} = \tilde{a} \odot \frac{(\tilde{c})^2}{2} \oplus \tilde{b} \odot \tilde{c}.$$

Now consider $\int_{\tilde{u}}^{\tilde{v}}(\tilde{a}s \oplus \tilde{b})\, ds$, where $\tilde{u}$ and $\tilde{v}$ are positive fuzzy numbers. First,

$$\tilde{I}(u,v) = \left[\frac{a}{2}\,(v^2 - u^2) + b(v - u),\, \underline{a}\,\frac{(v^2 - u^2)}{2} + \underline{b}(v - u), \right.$$

$$\left. \bar{a}\,\frac{(v^2 - u^2)}{2} + \bar{b}(v - u) \right]_{LL}$$

with $v > u$, using the main result of a.β.
Spreads are positive since $v > u$. Thus,

$$\tilde{I}(u,v) = \tilde{a}\left[\frac{v^2 - u^2}{2} \right] \oplus \tilde{b}(v - u)$$

and from b.γ,

$$\tilde{I}(\tilde{u},\tilde{v}) = \tilde{a} \odot \left[\frac{(\tilde{v})^2 \ominus (\tilde{u})^2}{2} \right] \oplus \tilde{b} \odot (\tilde{v} \ominus \tilde{u})$$

which can be easily calculated using the methods developed in 2.B.e provided that $\tilde{u}$ and $\tilde{v}$ are fuzzy numbers of L-R and R-L type, respectively. Note that $\tilde{I}(\tilde{u},\tilde{v}) \neq \tilde{I}(0,\tilde{v}) \ominus \tilde{I}(\tilde{u},0)$, which is

$$\left[\frac{\tilde{a} \odot (\tilde{v})^2}{2} \oplus \tilde{b} \odot \tilde{v} \right] \ominus \left[\frac{\tilde{a} \odot (\tilde{u})^2}{2} \oplus \tilde{b} \odot \tilde{u} \right]$$

because of the nondistributivity of $\odot$ over $\ominus$ (see 2.B.d.β).

D. DIFFERENTIATION   (Dubois and Prade, 1978)

In the preceding section we have extended the concept of integration to a real fuzzifying function (i.e., a mapping from $\mathbb{R}$ to $\tilde{\mathcal{P}}(\mathbb{R})$). Conversely, differentiation is introduced here.

a. Definition of the Extended Derivative

Let $\tilde{f}$ be a fuzzifying function from $\mathbb{R}$ to $\mathbb{R}$. The image of any $x \in D \subseteq \mathbb{R}$ is assumed to be a fuzzy number (i.e., a convex and normalized fuzzy set in $\mathbb{R}$). Moreover, each α-level curve f_α of $\tilde{f}$ is assumed to have a derivative at any $x_0 \in D$. Then, the derivative of $\tilde{f}$ at x_0, denoted $(d\tilde{f}/dx)(x_0)$ is defined by its membership function

$$\mu_{(d\tilde{f}/dx)(x_0)}(P) = \sup_{f_\alpha \,:\, (df_\alpha/dx)(x_0)=P} \mu(f_\alpha) \qquad (14)$$

where $\mu(f_\alpha) = \alpha$ by definition. ($\mu_{(d\tilde{f}/dx)(x)}(P) = 0$ if $\nexists \alpha, (df_\alpha/dx)(x_0) = P$).

Thus the membership value of P to $(d\tilde{f}/dx)(x_0)$ is the greatest level of all the α-level curves whose slope at x_0 is P. $(d\tilde{f}/dx)(x_0)$ is an estimate of the parallelism of the bundle of level curves at x_0. The less fuzzy $(d\tilde{f}/dx)(x_0)$, the more parallel the level curves.

b. L-R Type Fuzzifying Functions

Let $\tilde{f}$ be such that $\forall x \in D$, $\tilde{f}(x) = (f(x), s(x), t(x))_{LR}$. If $\alpha \neq 1$, there are two α-level curves f_α^- and f_α^+, whose equations are

$$f_\alpha^-(x) = f(x) - L^{-1}(\alpha) \cdot s(x), \qquad f_\alpha^+(x) = f(x) + R^{-1}(\alpha) \cdot t(x),$$

and $f_1(x) = f(x)$.

For the sake of simplicity, $\tilde{f}(x)$ is assumed to be a strictly convex fuzzy number (i.e., L and R are continuous and strictly decreasing on $[0, +\infty)$, and thus $\forall \alpha \neq 0$, $\nexists(a, b)$ with $a \neq b$, such that $\forall u \in [a, b]$, $\mu_{\tilde{f}(x)}(u) = \alpha$). Moreover, f, s, and t are assumed to have derivatives at any $x \in D$. Hence

$$\frac{df_\alpha^-}{dx}(x_0) = \frac{df}{dx}(x_0) - L^{-1}(\alpha)\frac{ds}{dx}(x_0),$$

$$\frac{df_\alpha^+}{dx}(x_0) = \frac{df}{dx}(x_0) + R^{-1}(\alpha)\frac{dt}{dx}(x_0).$$

According to the sign of $(ds/dx)(x_0)$ and $(dt/dx)(x_0)$, the bundle of α-level curves may have different features.

(i) $(ds/dx)(x_0) > 0$, $(dt/dx)(x_0) > 0$. If $(ds/dx)(x_0) > 0$ and $(dt/dx)(x_0) > 0$ (see Fig. 6), $s(x)$ and $t(x)$ are increasing functions

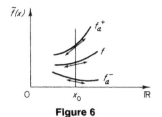

Figure 6

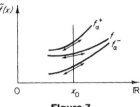

Figure 7

in a neighborhood of x_0; so when x increases, the α-level curves get away from $f(x)$ and we have, if $\beta < \alpha < 1$,

$$\frac{df_\beta^+}{dx}(x_0) > \frac{df_\alpha^+}{dx}(x_0) > \frac{df}{dx}(x_0) > \frac{df_\alpha^-}{dx}(x_0) > \frac{df_\beta^-}{dx}(x_0).$$

Given a slope P, there exists at most one level curve whose derivative in x_0 is equal to P. Hence, (14) gives

$$\text{if } \quad P \leqslant \frac{df}{dx}(x_0), \qquad \mu_{(d\tilde{f}/dx)(x_0)}(P) = L\left[\frac{(df/dx)(x_0) - P}{(ds/dx)(x_0)}\right];$$

$$\text{if } \quad P \geqslant \frac{df}{dx}(x_0), \qquad \mu_{(d\tilde{f}/dx)(x_0)}(P) = R\left[\frac{P - (df/dx)(x_0)}{(dt/dx)(x_0)}\right];$$

and

$$\frac{d\tilde{f}}{dx}(x_0) = \left(\frac{df}{dx}(x_0), \frac{ds}{dx}(x_0), \frac{dt}{dx}(x_0)\right)_{LR}.$$

(ii) $(ds/dx)(x_0) < 0$, $(dt/dx)(x_0) < 0$. This is the opposite case. $s(x)$ and $t(x)$ are decreasing functions in a neighborhood of x_0, and if $\beta < \alpha < 1$ $((ds/dx)(x_0) < 0, (dt/dx)(x_0) < 0)$,

$$\frac{df_\beta^-}{dx}(x_0) > \frac{df_\alpha^-}{dx}(x_0) > \frac{df}{dx}(x_0) > \frac{df_\alpha^+}{dx}(x_0) > \frac{df_\beta^+}{dx}(x_0).$$

Similarly to (i), we get

$$\frac{d\tilde{f}}{dx}(x_0) = \left(\frac{df}{dx}(x_0), -\frac{dt}{dx}(x_0), -\frac{ds}{dx}(x_0)\right)_{RL}.$$

(iii) $(ds/dx)(x_0) < 0, (dt/dx)(x_0) > 0$. Figure 7 sketches the shape of the bundle in a neighborhood of x_0. Here,

$$\forall \alpha, \quad \min\left[\frac{df_\alpha^+}{dx}(x_0), \frac{df_\alpha^-}{dx}(x_0)\right] \geqslant \frac{df}{dx}(x_0).$$

Thus, if $P < (df/dx)(x_0)$, then $\mu_{(d\tilde{f}/dx)(x_0)}(P) = 0$. If $P \geqslant (df/dx)(x_0)$, there may be two level curves f_α^-, f_β^+ whose derivatives at x_0 are equal to P. Hence,

$$\frac{d\tilde{f}}{dx}(x_0) = \left(\frac{df}{dx}(x_0), 0, -\frac{ds}{dx}(x_0) \right)_L \cup \left(\frac{df}{dx}(x_0), 0, \frac{dt}{dx}(x_0) \right)_R.$$

Usually, $(d\tilde{f}/dx)(x_0)$ has no particular type (i.e., L or R).

(iv) $(ds/dx)(x_0) > 0; (dt/dx)(x_0) < 0$.   A similar discussion would lead to

$$\frac{d\tilde{f}}{dx}(x_0) = \left(\frac{df}{dx}(x_0), \frac{ds}{dx}(x_0), 0 \right)_L \cup \left(\frac{df}{dx}(x_0), -\frac{dt}{dx}(x_0), 0 \right)_R.$$

Remarks 1   It is clear that

$$\frac{d\tilde{f}}{dx}(x_0) \neq \lim_{h \to 0} \frac{\tilde{f}(x_0 + h) \ominus \tilde{f}(x)}{h} = \left(\frac{df}{dx}(x_0), +\infty, +\infty \right)$$

because even in an extended subtraction spreads must be added.

2   If $\tilde{f}(x) = (f(x), s(x), t(x))_{LR}$, $\tilde{g}(x) = (g(x), u(x), v(x))_{LR}$, then

$$\frac{d}{dx}(\tilde{f} \oplus \tilde{g})(x_0) = \frac{d\tilde{f}}{dx}(x_0) \oplus \frac{d\tilde{g}}{dx}(x_0)$$

if $ds/dx, dt/dx, du/dx, dv/dx$ have the same sign at x_0. It would be possible to show that in some cases the usual formula for differentiation of a product still holds for L-R type fuzzifying functions.

c. Example

Let $\tilde{f}(x) = e^{\tilde{\lambda} \odot x}$ where $\tilde{\lambda}$ is a strictly convex continuous positive L-R type fuzzy number. Then, $(d\tilde{f}/dx)(x) = \tilde{\lambda} \odot e^{\tilde{\lambda} \odot x}$ since it is possible to differentiate along the α-level curves: $(df_\alpha/dx)(x) = \lambda_\alpha e^{\lambda_\alpha \cdot x}$ and to apply the theorem of 2.B.a because $ue^{v \cdot x}$ is increasing with (u, v). But it is not always so easy!

Remark   *Integration and differentiation of a fuzzy bunch of functions.*   S. S. L. Chang and Zadeh (1972) have defined the derivative and the integral of a function with a fuzzy parameter (viewed as a fuzzy bunch of ordinary functions) in the following way: Let $x \mapsto f(x, a)$ be a function from $\mathbb{R}$ to $\mathbb{R}$ depending on a real parameter a. Its extension $\tilde{f}$ when the parameter is a fuzzy set A on $\mathbb{R}$ is defined by

$$\mu_{\tilde{f}(x, A)}(y) = \sup_{a, \, y = f(x, a)} \mu_A(a).$$

This fuzzifying function $x \mapsto \tilde{f}(x, A)$ can also be viewed as a fuzzy bunch $F = \int \mu_A(a)/f(\cdot, a)$; in this latter approach the membership function of the derivative and the integral are respectively

$$\mu_{\frac{df}{dx}(x, A)}(y) = \sup_{a, \, y = \frac{df}{dx}(x, a)} \mu_A(a);$$

$$\mu_{\int_{x_0}^x f(t, A)\, dt}(y) = \sup_{a, \, y = \int_{x_0}^x f(t, a)\, dt} \mu_A(a).$$

In the above example (c) the α-level curves are precisely the elements of the support of the associated fuzzy bunch; thus the two approaches give the same results.

E. FUZZY TOPOLOGY

The notion of a fuzzy topological space was introduced by C. L. Chang (1968). It is a straightforward extension of the concept of ordinary topological space (i.e., a pair $(X, \mathfrak{T})$ such that: (1) $X, \emptyset \in \mathfrak{T}$; (2) $A, B \in \mathfrak{T}$ implies $A \cap B \in \mathfrak{T}$; (3) $A_i \in \mathfrak{T}, \forall i \in I$ implies $\bigcup_{i \in I} A_i \in \mathfrak{T}$ where $\mathfrak{T} \subseteq \mathcal{P}(X)$). It is beyond the scope of this book to present all the notions and results that have been developed in fuzzy topology. Only some of the basic definitions and propositions are given here. A rather extensive bibliography is listed at the end of this chapter.

A *fuzzy topology* is a family $\tilde{\mathfrak{T}}$ of fuzzy sets on X satisfying the following conditions: (1) $X, \emptyset \in \tilde{\mathfrak{T}}$; (2) if $A, B \in \tilde{\mathfrak{T}}$, then $A \cap B \in \tilde{\mathfrak{T}}$; (3) if $\forall i \in I$, $A_i \in \tilde{\mathfrak{T}}$, then $\bigcup_{i \in I} A_i \in \tilde{\mathfrak{T}}$. (Chang, 1968).

$(X, \tilde{\mathfrak{T}})$ is said to be a *fuzzy topological space*. Each member of $\tilde{\mathfrak{T}}$ is called a $\tilde{\mathfrak{T}}$-*open fuzzy set*. A fuzzy set is $\tilde{\mathfrak{T}}$-*closed* iff its complement is $\tilde{\mathfrak{T}}$-open. For instance, $(X, \tilde{\mathcal{P}}(X))$ is a fuzzy topological space, namely the *discrete* fuzzy topology of X.

A fuzzy set $N \in \tilde{\mathfrak{T}}$ is a *neighborhood* of A iff $\exists O \in \tilde{\mathfrak{T}}$ such that $A \subseteq O \subseteq N$. A is open iff for each fuzzy set B contained in A, A is a neighborhood of B. The above definition is somewhat different from the ordinary one in that we do not consider here the neighborhood of a point but of a fuzzy set.

Let A and B be fuzzy sets of $\tilde{\mathfrak{T}}$ such that $A \supseteq B$. Then B is said *interior* to A iff A is a neighborhood of B. The interior of A, denoted $A°$, is the union of all interior fuzzy sets A. $A°$ is the largest open fuzzy set contained in A. A is open iff $A = A°$.

Let f be a function from X to Y. Let $\tilde{\mathcal{U}}$ be a fuzzy topology on Y. The inverse, denoted $f^{-1}(B)$, of a fuzzy set B in Y is a fuzzy set in X whose membership function is

$$\forall x \in X, \quad \mu_{f^{-1}(B)}(x) = \mu_B(f(x)).$$

f is said to be *F-continuous* iff the inverse of each $\tilde{\mathcal{U}}$-open set is $\tilde{\mathcal{T}}$-open. Then, for each fuzzy set A in X, the inverse of every neighborhood of $f(A)$ is a neighborhood of A.

In the literature, several definitions of fuzzy compactness have been proposed and investigated (see C. L. Chang, 1968; Christoph, 1977; Goguen, 1973; Lowen, 1976, 1977; Wong, 1973, 1974a; Weiss, 1975; Takahashi, 1978). Not all these definitions are equivalent. The interested reader should consult the two comparative studies (in the sense of the existence of a Tychonoff theorem) by Gantner *et al.* (1978) and by Lowen (1978).

Local properties (Wong, 1974b) and normality (Hutton, 1975) have also been studied. Katsaras and Liu (1977)'s fuzzy vector spaces are particularly worth considering.

Questions such as, What does fuzzy continuity mean for an ordinary function? or What is continuity for a fuzzifying function (A.c.α)? also seem worth considering in the framework of topology.

F.   CATEGORIES OF FUZZY SETS

Category theory is a very general theory whose aim is "to lay bare some of the underlying principles common to diverse fields in the mathematical sciences." (Arbib and Manes, NF 1975). It has been used by several authors (Goguen, 1969, 1974; Eytan, 1977; Negoita and Ralescu, 1975) who tried to provide an abstract foundation to fuzzy set theory, independent of ordinary set theory. This approach contrasts with Chapin's (1.B.d).

A category K is a collection of *objects* denoted obj(K) together with, for each pair (A, B) of objects, a collection of entities called *morphisms*. The set of morphisms f between A and B is denoted $K(A, B)$. We write $f: A \rightarrow B$. Morphisms f and g respectively in $K(A, B)$ and $K(B, C)$ can be composed to make a unique morphism $g \odot f$ in $K(A, C)$. Symbolically, we write

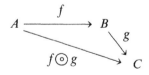

and say that "the diagram commutes." For K to be a category, the following properties must be satisfied:

the composition law of morphisms is associative;
$\forall A \in \mathrm{obj}(k)$, there is a unique morphism $1_A \in K(A, A)$ such that

$$\forall f : A \to B, \quad f \odot 1_A = 1_A \odot f = f;$$

1_A is called the identity morphism.

Example An example is the category SET of sets. Morphisms are mappings between sets.

A functor F is an assignment between two categories K and K', mapping objects onto objects and morphisms onto morphisms, and such that

$$\forall f \in K(A, B), \quad F(f) \in K'(F(A), F(B));$$
$$\forall f : A \to B, \quad g : B \to C, \quad F(g \odot f) = F(g) \odot F(f);$$
$$\forall A \in \mathrm{obj}(K), \quad F(1_A) = 1_{F(A)}.$$

Thus, a functor is a kind of homomorphism between categories.

Goguen (1969) introduced the first category of fuzzy sets, denoted Set(L) where L is a complete lattice. Objects of Set(L) are L-fuzzy sets, i.e., pairs (X, μ) where X is an ordinary set and μ a function from X to L. Morphisms are ordinary functions $f : X \to Y$ such that $\mu \circ f \geqslant \chi$ where $(X, \mu), (Y, \chi)$ are objects of set(L). Hence, morphisms are fuzzy functions in the sense of A.a. $\geqslant$ is the order relation induced from that in L. This category is also used in Eytan (1977). An extensive presentation of Set(L), its properties, and its ability to represent concepts can be found in Goguen's (1974) paper.

Other categories of fuzzy sets include: $\mathrm{Set}_f(L)$, the category whose objects are pairs (X, μ), $\mu : X \to L$ and whose morphisms are fuzzy relations $X \times Y \to L$ such that

$$\forall x \in X, \quad \forall y \in Y, \quad \mu_R(x, y) \leqslant \inf(\mu(x), \chi(y))$$

$$\text{with} \quad (X, \mu), (Y, \chi) \in \mathrm{obj}(\mathrm{Set}_f(L)).$$

$\mathrm{Set}_f(L)$ is thus the category of fuzzy sets and fuzzy relations in the sense of N.B.2 in A.c.α.

There exists also $\mathrm{Set}_g(L)$, which has the same definition as $\mathrm{Set}_f(L)$ except the condition for fuzzy relations; for $\mathrm{Set}_g(L)$,

$$\sup_{x \in X} \inf(\mu_R(x, y), \mu(x)) \leqslant \chi(y) \qquad \forall y \in Y.$$

A discussion of $\mathrm{Set}_f(L)$ and $\mathrm{Set}_g(L)$ can be found in Negoita and Ralescu (1975); two functors between Set(L) and $\mathrm{Set}_f(L)$, and between $\mathrm{Set}_f(L)$ and $\mathrm{Set}_g(L)$ respectively are constructed.

Different from the categories of fuzzy sets are the fuzzy categories that are extended categories as fuzzy sets are extended sets. Two points of view exist. Given a category K, we build $\tilde{K}$ over K such that

$$\text{obj}(\tilde{K}) = \text{obj}(K).$$

$\forall A, B \in \text{obj}(\tilde{K})$, a morphism of $\tilde{K}$ is an L-fuzzy set of $K(A, B)$ (see Negoita and Ralescu, 1975). It can be proved that $\tilde{K}$ is a category. When $K = \text{SET}$, the morphisms of $\tilde{K}$ are nothing but fuzzy bunches (A.c.β).

A *fuzzy theory* (Arbib and Manes, Reference from III.2, 1975b) is a triple $\mathcal{F} = \{F, o, i\}$ where F is a function from $\text{obj}(K)$ to $\text{obj}(K)$, o is a function $K(A, F(B)) \times K(B, F(C)) \to (K(A, F(C))$, and i a collection of morphisms $A \to F(A)$, $A \in \text{obj}(K)$. It is possible to equip $\mathcal{F}$ with properties such that there is a category $\mathcal{F}(K)$ with $\text{obj}(\mathcal{F}(K)) = \text{obj}(K)$, $\mathcal{F}(K)(A, B) = K(A, F(B))$ $\forall A, B$ in $\text{obj}(K)$. When $K = \text{SET}$ and $F(X) = \tilde{\mathscr{P}}(X)$, the morphisms of $\mathcal{F}(K)$ are fuzzifying functions.

The role played by the interval $[0, 1]$ in the definition of fuzzy concepts is discussed under the name "fuzzy characters" in a categorical framework by Negoita and Ralescu (1975). The latter author also recently studied a fuzzy generalization of the notion of subobject in a category (Ralescu, 1978).

Lastly, we must mention a completely different group of works which use category theory terminology but are not related to Goguen's approach.

Poston (1971) defines a category called "Fuz" whose objects are sets equipped with a nonfuzzy proximity relation. Dodson (1974, 1975) generalizes Fuz by considering sets X with a nonfuzzy proximity relation on $X \times \mathscr{P}(X)$, called *hazy spaces*. Both authors are motivated by an extension of the usual topology (not in the sense of E) and differential geometry to spaces were usual distances no longer exist. According to Dodson (1974) "the situation in Fuz [and in hazy spaces] resembles that in real experiments: making measurements with limited precision." Dodson (1975) indicates with an example how hazy spaces are a good tool for modeling the notion of an elementary particle, in accordance with an uncertainty principle similar to Heisenberg's.

REFERENCES

Chang, S. S. L., and Zadeh, L. A. (1972). On fuzzy mapping and control. *IEEE Trans. Syst., Man Cybern.* **2**, 30–34. (Reference from III.2.)

Dubois, D., and Prade, H. (1978). Toward fuzzy analysis: Integration and differentiation of fuzzy functions. In "Fuzzy Algebra, Analysis, Logics," Tech. Rep. TR-EE 78/13. Purdue Univ., Lafayette, Indiana.

Negoita, C. V., and Ralescu, D. A. (1975). "Applications of Fuzzy Sets to Systems Analysis." Chap. 1, pp. 18–24. Birkhaeuser, Basel. (Reference from I.)

Orlovsky, S. A. (1977). On programming with fuzzy constraint sets. *Kybernetes* **6**, No. 3, 197–201.

Sugeno, M. (1977). "Fuzzy Systems with Underlying Deterministic Systems. (Towards Fuzzy Descriptions of Complex Systems)," Res. Rep., L.A.A.S., Toulouse, France. (Also in "Summary of Papers on General Fuzzy Problems," No. 3, pp. 25–52. Working Group Fuzzy Syst., Tokyo, (1977.)

Tanaka, H., Okuda, T., and Asai, K. (1973). On fuzzy mathematical programming. *J. Cybern.* **3**, No. 4, 37–46.

Zadeh, L. A. (1972). "On Fuzzy Algorithms," Memo UCB/ERL M-325. Univ. of California, Berkeley. (Reference from III.3, 1972a.)

Fuzzy Topology

Auray, J. P. and Duru, G. (1976). Introduction à la théorie des espaces multiflous. Working paper No. 16, IME University of Dijon, Dijon, France.

Chang, C. L. (1968). Fuzzy topological spaces. *J. Math. Anal. Appl.* **24**, 182–190.

Christoph, F. T. (1977). Quotient fuzzy topology and local compactness. *J. Math. Anal. Appl.* **57**, 497–504.

Gantner, T. E., Steinlage, R. C., and Warren, R. H. (1978). Compactness in fuzzy topological spaces. *J. Math. Anal. Appl.* **62**, 547–562.

Goguen, J. A. (1973). The fuzzy Tychonoff theorem. *J. Math. Anal. Appl.* **43**, 734–742.

Höhle, U. (1978). Probalistische Topologien. *Manuscripta Math.* **26**, 223–245.

Hutton, B. (1975). Normality in fuzzy topological spaces. *J. Math. Anal. Appl.* **50**, 74–79.

Hutton, B. (1977). Uniformities in fuzzy topological spaces. *J. Math. Anal. Appl.* **58**, 559–571.

Katsaras, A. K., and Liu, D. B. (1977). Fuzzy vector spaces and fuzzy topological vector spaces. *J. Math. Anal. Appl.* **58**, 135–146.

Kaufmann, A. (1977). Topologie plane. In "Introduction à la Théorie des Sous-Ensembles flous. Vol. 4: Compléments et Nouvelles Applications," Masson, Paris. (Reference from I.)

Kramosil, I., and Michálek, J. (1975). Fuzzy metrics and statistical metric spaces. *Kybernetika* (Prague), **11**, 336–344.

Lowen, R. (1974). Topologies floues. *C. R. Acad. Sci., Ser. A* **278**, 925–928.

Lowen, R. (1975). Convergence floue. *C. R. Acad. Sci., Ser. A* **280**, 1181–1183.

Lowen, R. (1976). Fuzzy topological spaces and fuzzy compactness. *J. Math. Anal. Appl.* **56**, 621–633.

Lowen, R. (1977). Initial and final fuzzy topologies and the fuzzy Tychonoff theorem. *J. Math. Anal. Appl.* **58**, 11–21.

Lowen, R. (1978). A comparison of different compactness notions in fuzzy topological spaces. *J. Math. Anal. Appl.* **64**, 446–454.

Ludescher, H., and Roventa, E. (1976). Sur les topologies floues définies à l'aide de voisinages. *C. R. Acad. Sci.* Paris, Sér. A, **283**, 575–577.

Michálek, J. (1975). Fuzzy topologies. *Kybernetika* (Prague), **11**, 345–354.

Nazaroff, G. J. (1973). Fuzzy topological polysystems. *J. Math. Anal. Appl.* **41**, 478–485. (Reference from III.2.)

Takahashi, W. (1978). Minimax theorems for fuzzy sets. Tokyo Institute of Technology, Tokyo.

Warren, R. H. (1976). Optimality in fuzzy topological polysystems. *J. Math. Anal. Appl.* **54**, 309–315. (Reference from III.2.)

Warren, R. H. (1977). Boundary of a fuzzy set. *Indiana Univ. Math. J.* **26**, No. 2, 191–197.

Warren, R. H. (1978). Neighborhoods, bases and continuity in fuzzy topological spaces. *Rocky Mount. J. Math.* **8**, No. 2, 459–470.

Weiss, M. D. (1975). Fixed points, separation, and induced topologies for fuzzy sets. *J. Math. Anal. Appl.* **50**, 142–150.

Wong, C. K. (1973). Covering properties of fuzzy topological spaces. *J. Math. Anal. Appl.* **43**, 697–704.

Wong, C. K. (1974a). Fuzzy topology: Product and quotient theorems. *J. Math. Anal. Appl.* **45**, 512–521.

Wong, C. K. (1974b). Fuzzy points and local properties of fuzzy topology. *J. Math. Anal. Appl.* **46**, 316–328.

Wong, C. K. (1975). Fuzzy topology. In "Fuzzy Sets and Their Applications to Cognitive and Decision Processes" (L. A. Zadeh, K. S. Fu, K. Tanaka, and M. Shimura, eds.), pp. 171–190. Academic Press, New York.

Wong, C. K. (1976). Categories of fuzzy sets and fuzzy topological spaces. *J. Math. Anal. Appl.* **53**, 704–714.

Categories of Fuzzy Sets

Dodson, C. T. J. (1974). Hazy spaces and fuzzy spaces. *Bull. London Math. Soc.* **6**, 191–197.

Dodson, C. T. J. (1975). Tangent structures for hazy spaces. *J. London Math. Soc.* **2**, No. 11, 465–473.

Eytan, M. (1977). Sémantique préordonnée des ensembles flous. *Cong. AFCET Model. Maitrise Syst., Versailles*, pp. 601–608.

Goguen, J. A. (1969). Categories of V-sets. *Bull Am. Math. Soc.* **75**, 622–624.

Goguen, J. A. (1974). Concept representation in natural and artificial languages: Axioms, extension, and applications for fuzzy sets. *Int. J. Man-Mach. Stud.* **6**, 513–561.

Goguen, J. A. (1975). Objects. *Int. J. Gen. Syst.* **1**, 237–243.

Negoita, C. V., and Ralescu, D. A. (1975). "Applications of Fuzzy Sets to Systems Analysis," Chaps. 1 and 2. Birkhaeuser, Basel. (Reference from I.)

Poston, T. (1971). "Fuzzy Geometry." Ph.D. Thesis, Univ. of Warwick, Warwick, England.

Ralescu, D. A. (1978). Fuzzy subobjects in a category and the theory of C-sets. *Int. J. Fuzzy Sets Syst.* **1**, No. 3, 193–202.

Chapter **5**

FUZZY MEASURES.
PROBABILITIES/
POSSIBILITIES

Whereas the four previous chapters were devoted to sets, we are concerned here with measures of sets. Instead of considering membership grades, we now deal with degrees of belief, possibility, probability that a given unlocated element belongs to a (fuzzy or nonfuzzy) set.

The first section presents Sugeno's fuzzy measures and integral. Fuzzy measures assume only monotonicity and thus are very general. Probability and possibility measures, Shafer's belief functions, and Shackle's consonant belief functions are shown to be particular cases of fuzzy measures. Fuzzy integrals in the sense of Sugeno are analogous to Lebesgue integrals. A result on conditional fuzzy measures is reminiscent of Bayes' theorem.

In Section B basic notions of a theory of possibility, following Zadeh, are provided. Similarities between possibility and probability theory are emphasized. A possibility distribution can be induced from a fuzzy set and does not underlie the idea of a replicated experiment, nor does a possibility measure satisfy the additivity property.

The next section deals with fuzzy events modeled as fuzzy sets and with their fuzzy or nonfuzzy probability and possibility. Lastly, fuzzy distributions of probability and possibility are briefly investigated.

A. FUZZY MEASURES AND SUGENO'S INTEGRALS

In the four preceding chapters we were interested in the grade of membership $\mu_A(x)$ of a known element $x \in X$ in a set A without precise boundary.

On the contrary, we are now concerned with guessing (most often subjectively) whether an a priori nonlocated element in X belongs to a subset A of X. A is fuzzy or not here. Such an uncertainty is sometimes conveniently expressed using probabilities.

Sugeno's approach (see Sugeno, 1974, 1977; Terano and Sugeno, 1975), which is the topic of this section, generalizes probability measures by dropping the additivity property and replacing it by a weaker one, i.e., monotonicity.

a. Fuzzy Measures

Note: In this section a we consider only *non*fuzzy subsets.

Let g be a function from $\mathcal{P}(X)$ to $[0, 1]$. g is said to be a *fuzzy measure* iff:

(1) $g(\emptyset) = 0$; $g(X) = 1$;
(2) $\forall A, B \in \mathcal{P}(X)$, if $A \subseteq B$, then $g(A) \leqslant g(B)$ (monotonicity).
(3) if $\forall i \in \mathbb{N}, A_i \in \mathcal{P}(X)$ and $(A_i)_i$ is monotonic $(A_1 \subseteq A_2 \subseteq \cdots \subseteq A_n \subseteq \cdots$ or $A_1 \supseteq A_2 \supseteq \cdots \supseteq A_n \cdots)$, then

$$\lim_{i \to \infty} g(A_i) = g\left(\lim_{i \to \infty} A_i \right) \quad \text{(continuity)}.$$

N.B.:   More generally, a fuzzy measure can be defined on a Borel field $\mathcal{B} \subset \mathcal{P}(X)$, i.e., (1) $\emptyset \in \mathcal{B}$; (2) if $A \in \mathcal{B}$, then $\bar{A} \in \mathcal{B}$; (3) if $\forall i \in \mathbb{N}$, $A_i \in \mathcal{B}$, then $\bigcup_{i \in \mathbb{N}} A_i \in \mathcal{B}$.

g is associated with a nonlocated element x of X. $g(A)$ is called by Sugeno a "grade of fuzziness" of A. It expresses an evaluation of the statement "x belongs to A" in a situation in which one subjectively guesses whether x is within A. The monotonicity of g means that "$x \in A$" is less certain than "$x \in B$" when $A \subseteq B$. It is easy to check that

$$\forall A, B \in \mathcal{P}(X), \quad g(A \cup B) \geqslant \max(g(A), g(B))$$

and

$$\forall A, B \in \mathcal{P}(X), \quad g(A \cap B) \leqslant \min(g(A), g(B)).$$

Several examples of fuzzy measures are provided.

α. Probability Measures

P is a probability measure iff:

(1) $\forall A, P(A) \in [0, 1]$; $P(X) = 1$;

(2) if $\forall i \in \mathbb{N}$, $A_i \in \mathcal{P}(X)$ and $\forall i \neq j$ $A_i \cap A_j = \emptyset$, then

$$P\left(\bigcup_{i \in \mathbb{N}} A_i\right) = \sum_{i=1}^{\infty} P(A_i).$$

P is obviously a fuzzy measure.

β. Dirac Measures

A Dirac measure is a fuzzy measure μ defined by

$$\forall A \in \mathcal{P}(X), \mu(A) = \begin{cases} 1 & \text{iff} \quad x_0 \in A, \\ 0 & \text{otherwise}, \end{cases}$$

where x_0 is a given element in X. μ is nothing but the membership value of x_0 in any subset of X.

γ. λ-Fuzzy Measures

g_λ fuzzy measures were proposed by Sugeno (1974) by relaxing the additivity property of probabilities into: $\forall A, B \in \mathcal{P}(X)$, such that $A \cap B = \emptyset$,

$$g_\lambda(A \cup B) = g_\lambda(A) + g_\lambda(B) + \lambda g_\lambda(A) g_\lambda(B), \qquad -1 < \lambda. \qquad (1)$$

where $g_\lambda(X) = 1$ and g_λ satisfies the continuity property of fuzzy measures.

λ-fuzzy measures are indeed fuzzy measures for $\lambda > -1$.

Proof: From (1) we have

$$g_\lambda(X) = g_\lambda(X) + g_\lambda(\emptyset)(1 + \lambda g_\lambda(X))$$

Hence, since $\lambda \neq -1$, $g_\lambda(\emptyset) = 0$. If $A \subseteq B$, then $\exists C, B = A \cup C$ and $A \cap C = \emptyset$. We have

$$g_\lambda(B) = g_\lambda(A) + g_\lambda(C)(1 + \lambda g_\lambda(A)) \geqslant g_\lambda(A)$$

since $\lambda > -1$.   Q.E.D.

N.B.:   For $\lambda = 0$, λ-fuzzy measures are probability measures. Taking $B = \bar{A}$, we get from (1)

$$g_\lambda(\bar{A}) = \frac{1 - g_\lambda(A)}{1 + \lambda g_\lambda(A)}$$

This expression is exactly the same as the λ-complement formula in 1.B.b.

More generally, when A and B are any subsets of X, the following formula holds:

$$g_\lambda(A \cup B) = \frac{g_\lambda(A) + g_\lambda(B) - g_\lambda(A \cap B) + \lambda \cdot g_\lambda(A) \cdot g_\lambda(B)}{1 + \lambda \cdot g_\lambda(A \cap B)},$$

which is easy to prove, expressing $g_\lambda(A \cup B)$ in terms of $g_\lambda(A)$ and $g_\lambda(\overline{A} \cap B)$, expressing $g_\lambda(B)$ in terms of $g_\lambda(\overline{A} \cap B)$ and $g_\lambda(A \cap B)$, and eliminating $g_\lambda(\overline{A} \cap B)$ from the two expressions.

If $X = \mathbb{R}$, a λ-fuzzy measure is easily obtained from a function h such that (1) if $x \leqslant y$, then $h(x) \leqslant h(y)$; (2) h is continuous; (3) $\lim_{x \to -\infty} h(x) = 0$; (4) $\lim_{x \to +\infty} h(x) = 1$. h is very similar to a probability cumulative distribution function and we have (Sugeno, 1977):

$$\forall [a,b] \subset \mathbb{R}, \quad g_\lambda([a,b]) = \frac{h(b) - h(a)}{1 + \lambda h(a)}.$$

If we iterate (1) using a family of disjoint subsets A_i, we get

$$g_\lambda\left(\bigcup_{i \in \mathbb{N}} A_i\right) = \sum_{i=1}^{\infty} g_\lambda(A_i), \quad \lambda = 0$$

$$= \frac{1}{\lambda}\left[\prod_{i=1}^{\infty} (1 + \lambda g_\lambda(A_i)) - 1\right], \quad \lambda \neq 0.$$

When X is a finite set $\{x_1, \ldots, x_n\}$, a fuzzy measure g_λ is obtained from the values $g_i = g_\lambda(\{x_i\}) \in [0, 1]$ using the above formula, provided that the g_i satisfy the normalization constraint

$$g_\lambda(X) = \frac{1}{\lambda}\left[\prod_{i=1}^{n} (1 + \lambda g_i) - 1\right] = 1.$$

δ. *Belief Functions*   (Shafer, NF 1976)

A belief function b is a measure on X finite, such that

(1) $b(\emptyset) = 0$, $b(X) = 1$; $\forall A \in \mathcal{P}(X)$, $0 \leqslant b(A) \leqslant 1$;
(2) $\forall A_1, A_2, \ldots, A_n \in \mathcal{P}(X)$,

$$b(A_1 \cup A_2 \cup \cdots \cup A_n) \geqslant \sum_{i=1}^{n} b(A_i) - \sum_{i<j} b(A_i \cap A_j) + \cdots$$

$$+ (-1)^{n+1} b(A_1 \cap A_2 \cap \cdots \cap A_n)$$

$b(A)$ is interpreted as a grade of belief that a given element of X belongs to A. Note that $b(A) + b(\overline{A}) \leqslant 1$, which means that a lack of belief in $x \in A$ does not imply a strong belief in $x \in \overline{A}$. Particularly, a *total ignorance* is modeled by the belief function b_i such that $b_i(A) = 0$ if $A \neq X$ and

$b_i(A) = 1$ if $A = X$. A probability measure is a special case of belief functions. Belief functions are fuzzy measures. Let $B \subseteq A$, hence $\exists C$, $A = B \cup C$ and $B \cap C = \emptyset$. (2) becomes $b(A) = b(B \cup C) \geqslant b(B) + b(C) \geqslant b(B)$.

Belief functions can be defined by a so-called basic probability function m from $\mathcal{P}(X)$ to $[0, 1]$ such that: (1) $m(\emptyset) = 0$; (2) $\sum_{A \in \mathcal{P}(X)} m(A) = 1$ (the total belief has a measure 1). It is easy to show that $\sum_{A \subseteq B} m(A)$ is a belief function $b(B)$. Conversely, for any belief function b, $\sum_{B \subseteq A} (-1)^{|A - B|} b(B)$ is a basic probability function $m(A)$. The subsets A of X such $m(A) > 0$ are called focal elements of b. "$m(A)$ measures the belief that one commits *exactly* to A, *not* the *total* belief that one commits to A" (Shafer, NF 1976).

A λ-fuzzy measure is a belief function iff $\lambda \geqslant 0$.

Proof: (Banon 1978)
Let A be a subset of X finite. Developing the expression of $g_\lambda(A)$ in terms of g_i's yields

$$g_\lambda(A) = \sum_{B \subseteq A} \lambda^{|B| - 1} \cdot \prod_{x_i \in B} g_i.$$

We may state $m(B) = \lambda^{|B| - 1} \cdot \prod_{x_i \in B} g_i$ iff $\lambda \geqslant 0$.
Moreover, $\sum_{B \subseteq X} m(B) = 1$ due to the normalization constraint on the g_i's.   Q.E.D.
However, belief functions are more general than λ-fuzzy measures for $\lambda \geqslant 0$; knowledge of $b(A)$ and $b(B)$ is not always sufficient to calculate $b(A \cup B)$. Moreover, b_i is not a λ-fuzzy measure.

ϵ.   *Consonant Belief Functions*   (Shackle, NF 1961, Cohen NF 1973)
A consonant belief function is a belief function whose focal elements $A_1, \ldots, A_n$ are nested: $A_1 \subset A_2 \subset \cdots \subset A_n$. In particular, b_i is a consonant belief function.

A consonant belief function is a λ-fuzzy measure iff it is a Dirac measure (Banon, 1978).

Shafer (NF 1976) showed that the above definition of a consonant belief function was equivalent to:

(1) $b(\emptyset) = 0$; $b(X) = 1$;
(2) $b(A \cap B) = \min(b(A), b(B))$.

This latter definition was independently introduced by Shackle (NF 1961).

Other properties of consonant belief functions are: $\forall A, \min(b(A), b(\overline{A})) = 0$; $\forall b, \exists A, B, b(A \cup B) > \max(b(A), b(B))$, except if b is a Dirac measure. The first equality means that a positive grade of membership is never granted to both sides of a dichotomy at the same time. $b(\overline{A})$ is interpreted by Shackle as a potential grade of surprise.

Particular cases of consonant belief functions are *certainty measures* such that

$$\exists Y_0 \subseteq X, \quad c(A) = 1 \text{ if } A \supseteq Y_0, \text{ and } 0 \text{ otherwise.}$$

ζ.   Plausibility Measures

The plausibility of a subset A of X (finite) has been defined by Shafer (NF 1976) as

$$Pl(A) = 1 - b(\overline{A}) \tag{2}$$

where b is a belief function.

A plausibility measure satisfies the following axioms:

(1) $Pl(\emptyset) = 0$; $Pl(X) = 1$.
(2) $\forall A_1, \ldots, A_n \subseteq X$,

$$Pl(A_1 \cap \cdots \cap A_n) \leqslant \sum_{i=1,n} Pl(A_i) - \sum_{i<j} Pl(A_i \cup A_j) + \cdots$$
$$+ (-1)^{n+1} Pl(A_1 \cup \cdots \cup A_n).$$

Plausibility measures are particular cases of fuzzy measures.

Proof: Noticing that $\forall A \subseteq X$, $\forall B \subseteq X$, $Pl(A \cup B) \leqslant Pl(A) + Pl(B) - Pl(A \cap B)$, let $C \subseteq A$ and $B = C \cup \overline{A}$, hence $A \cup B = X$, $A \cap B = C$, and $1 \leqslant Pl(A) + Pl(B) - Pl(C)$; since $Pl(B) = 1$, $Pl(A) \geqslant Pl(C)$. Q.E.D.

Moreover, $Pl(A) + Pl(\overline{A}) \geqslant 1$.

N.B.:   Plausibility measures and belief functions have been introduced by Dempster (NF 1967) under the names *upper* and *lower probabilities*, induced from a probability measure by a multivalued mapping.

A λ-fuzzy measure is a plausibility measure iff $-1 < \lambda \leqslant 0$.

Proof: Let g_λ be a λ-fuzzy measure with $-1 < \lambda \leqslant 0$. Denote $f(A) = 1 - g_\lambda(\overline{A})$. Expressing for any A and $B f(A \cup B) = 1 - g_\lambda(\overline{A \cup B})$ in terms of $g_\lambda(\overline{A})$, $g_\lambda(\overline{B})$ and $g_\lambda(\overline{A \cap B})$, owing to the λ-complementation formula, shows that f is nothing but g_μ with $\mu = -\lambda/(1 + \lambda)$.

Note that the function $\lambda \mapsto -\lambda/(1 + \lambda)$ is an involutive bijection from $]-1, 0]$ to $[0, +\infty)$.

Thus, due to the definition of the plausibility measures in terms of belief functions and to the fact that g_λ is a belief function iff $\lambda \geqslant 0$, the proposition holds.   Q.E.D.

In terms of basic probability functions, we have

$$\forall A \subseteq X, \quad Pl(A) = \sum_{A \cap B \neq \emptyset} m(B).$$

η. *Possibility Measures* (Zadeh, 1978)

A *possibility measure* Π is a function from $\mathcal{P}(X)$ to $[0, 1]$ such that

(1) $\Pi(\emptyset) = 0; \Pi(X) = 1$;
(2) For any collection $\{A_i\}$ of subsets of X, $\Pi(\bigcup_i A_i) = \sup_i \Pi(A_i)$.

A possibility measure can be built from a possibility distribution, i.e., a function π from X to $[0, 1]$ such that $\sup_{x \in X} \pi(x) = 1$ (normalization condition). More specifically, we have

$$\forall A, \quad \Pi(A) = \sup_{x \in A} \pi(x). \tag{3}$$

Finding the associated possibility distribution from the knowledge of Π can be achieved by stating $\pi(x) = \Pi(\{x\})$, at least for denumerable universes X.

The following propositions are due to Banon (1978):
· A possibility measure is a belief function iff it is a Dirac measure.
· A possibility measure is a λ-fuzzy measure iff it is a Dirac measure.

Note also that $\forall \Pi, \exists A, B, \Pi(A \cap B) < \min(\Pi(A), \Pi(B))$ except if Π is a Dirac measure.

Lastly, it is easy to check that a possibility measure on X finite is a plausibility measure.

N.B.: 1. Some authors prefer in some contexts non-normalized possibility measures, i.e. $\Pi(X) \leqslant 1$. Viewing π as a membership function, the interpretation of such measures is closely related to that of non-normalized fuzzy sets (see 2.B.f.).

N.B.: 2. Let g be a function from $\mathcal{P}(X)$ to $[0, 1]$ such that

(1) $g(\emptyset) = 0; g(X) = 1$;
(2) $\forall A \subseteq X, \forall B \subseteq X$, if $A \cap B = \emptyset$ then $g(A \cup B) = \max(g(A), g(B))$.

When X is finite, g is a possibility measure.

Proof: If $A \subseteq C, \exists B$, such that $A \cap B = \emptyset$ and $A \cup B = C$. Hence, $g(C) = \max(g(A), g(B)) \geqslant g(A)$, i.e., g is a fuzzy measure.

$$\forall A \subseteq X, \quad \forall B \subseteq X, \quad g(A \cup B) = \max\left(g(A), g(\overline{A} \cap B)\right)$$

$$g(B) = \max\left(g(A \cap B), g(\overline{A} \cap B)\right).$$

Hence,

$$\max(g(A \cup B), g(A \cap B)) = \max\Big(\max\big(g(A), g(\overline{A} \cap B)\big), g(A \cap B)\Big)$$

$$= \max(g(A), g(B)).$$

But since g is a fuzzy measure, $g(A \cap B) \leqslant g(A \cup B)$.   Q.E.D.

N.B.:   3. $\Pi_m(A) = 1 - b_i(A)$, (i.e., $\Pi_m(A) = 1$ iff $A \neq \emptyset$) is called *maximum possibility measure*.

N.B.:   4.   *Crisp possibility measures* are defined by

$$\exists Y_0 \subseteq X, \quad c\Pi(A) = 1 \text{ if } A \cap Y_0 \neq \emptyset, \text{ and } c\Pi(A) = 0 \text{ otherwise.}$$

Note that $c\Pi(A) = 1 - c(\overline{A})$, where c is a certainty measure.

Remark   A consonant belief function b can be built from a possibility measure and reciprocally by setting

$$\Pi(A) = 1 - b(\overline{A}).$$

Hence

$$b(A) = 1 - \sup_{x \notin A} \pi(x) = \inf_{x \notin A} (1 - \pi(x)) = \inf_{x \notin A} \nu(x). \tag{4}$$

By analogy with modal logic where "A is necessary" is equivalent to "non-A is not possible" ($\Box A \equiv \neg \Diamond \neg A$), we could interpret a consonant belief function as a *necessity* measure.

Fig. 1 pictures the inclusion relationships that exist between the various sets of fuzzy measures on finite sets.

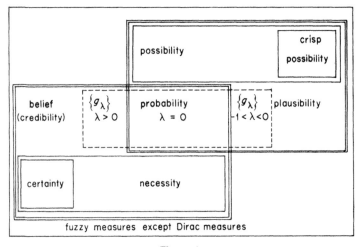

Figure 1

b. Sugeno's Fuzzy Integral

α. Nonfuzzy Domain

Using fuzzy measures Sugeno (1974) defined fuzzy integrals that are very similar to Lebesgue integrals. Let h be a function from X to $[0, 1]$. The fuzzy integral over the nonfuzzy set $A \subseteq X$ of the function h with respect to a fuzzy measure g is defined as

$$\oint_A h(x) \circ g(\cdot) = \sup_{\alpha \in [0, 1]} \min[\alpha, g(A \cap H_\alpha)] \qquad (5)$$

where $H_\alpha = \{x, h(x) \geqslant \alpha\}$.

N.B.: The analogy of (5) with a Lebesgue integral can be clearly exhibited as follows (Sugeno, 1974). Let $(E_1, \ldots, E_n)$ be an ordinary partition of X and assume

$$h(x) = \sum_{i=1}^{n} \alpha_i \mu_{E_i}(x)$$

where μ_{E_i} is the crisp characteristic function of E_i. Let l be a Lebesgue measure on $(X, \mathcal{P}(X))$. The Lebesgue integral of h over A is

$$\int_A h \, dl = \sum_{i=1}^{n} \alpha_i l(A \cap E_i).$$

Now assume $0 \leqslant \alpha_1 \leqslant \cdots \leqslant \alpha_n \leqslant 1$. Let $F_i = \bigcup_{j=i}^{n} E_j$. Then, defining $h(x) = \max_{i=1, n} \min(\alpha_i, \mu_{F_i}(x))$:

$$\oint_A h(x) \circ g(\cdot) = \max_{i=1, n} \min(\alpha_i, g(A \cap F_i)).$$

We give here some of the properties of the fuzzy integral:

· $\oint_X a \circ g(\cdot) = a \qquad \forall a \in [0, 1]$;

· if $\forall x, h(x) \leqslant h'(x)$, then $\oint_X h(x) \circ g(\cdot) \leqslant \oint_X h'(x) \circ g(\cdot)$ (monotonicity);

· $g(A) = \oint_X \mu_A(x) \circ g(\cdot)$ where $A \in \mathcal{P}(X)$;

· let $M = \oint_X h(x) \circ g(\cdot)$ and

$$h'(x) = \begin{cases} M, & \forall x \in H_M = \{x, h(x) \geqslant M\} \\ h(x) & \text{otherwise,} \end{cases}$$

then $\oint_A h'(x) \circ g(\cdot) = \oint_A h(x) \circ g(\cdot)$;

· if $A \subseteq B$, then $\oint_A h(x) \circ g(\cdot) \leqslant \oint_B h(x) \circ g(\cdot)$;

let P be a probability measure on X, then

$$\left| \oint_X h(x) \circ P(\cdot) - \int_X h(x)\, dP \right| \leqslant \frac{1}{4}.$$

These properties are proved in Sugeno (1974).

Kandel (1978a) proved the following result, when g is continuous

$$\oint_A h(x) \circ g(\cdot) = \inf_{\alpha \in [0,\,1]} \max[\alpha, g(A \cap H_\alpha)],$$

by noting that $g(A \cap H_\alpha)$ is a nonincreasing function of α. When X is finite, this result no longer holds.

Assume $X = \{x_1, \ldots, x_n\}$ and $h(x_1) \leqslant \cdots \leqslant h(x_n)$, then we have

$$\oint_X h(x) \circ g(\cdot) = \max_{i=1,\,n} \min(h(x_i), g(H_i))$$

where $H_i = \{x_i, x_{i+1}, \ldots, x_n\}$. Let i_0 be such that $\oint_X h(x) \circ g(\cdot) = \min(h(x_{i_0}), g(H_{i_0})) = M$. Note that the $n-1$ following terms are less than or equal to M: $\{g(H_i), i > i_0\}$ and $\{h(x_i), i < i_0\}$. There are $n-1$ other terms that are greater than or equal to M: $\{g(H_i), i < i_0\}$ and $\{h(x_i), i > i_0\}$. Moreover $g(H_1) = 1$. So M is obviously the median of the set of $2n-1$ terms $\{h(x_i), i = 1, n\} \cup \{g(H_i), i = 2, n\}$, once this set has been ordered. Thus, as indicated by Kandel (1978a, b), Sugeno's fuzzy integral can be interpreted as a "weighted median."

β. Fuzzy Domain

Let A be a fuzzy set on X, the fuzzy integral of a function h from X to $[0, 1]$ over A with respect to g is

$$\oint_A h(x) \circ g(\cdot) = \oint_X \min(\mu_A(x), h(x)) \circ g(\cdot). \tag{6}$$

The following properties hold:

$$\forall A, B, \quad \oint_{A \cup B} h(x) \circ g(\cdot) \geqslant \max\left(\oint_A h(x) \circ g(\cdot), \oint_B h(x) \circ g(\cdot) \right)$$

$$\oint_{A \cap B} h(x) \circ g(\cdot) \leqslant \min\left(\oint_A h(x) \circ g(\cdot), \oint_B h(x) \circ g(\cdot) \right);$$

it is easy to see that (6) gives (5) when A is nonfuzzy; the fuzzy measure of a fuzzy set A is

$$g(A) = \oint_X \mu_A(x) \circ g(\cdot) \quad \text{where} \quad A \in \tilde{\mathcal{P}}(X).$$

In particular, possibility measures of fuzzy sets can thus be defined.

c. Conditional Fuzzy Measure

One may think that the notion of conditional fuzzy measure is to that of fuzzy relation what the concept of fuzzy measure is to that of fuzzy set. Moreover, it generalizes conditional probabilities.

Let X and Y be two universes. A conditional fuzzy measure on Y with respect to X is a fuzzy measure $\sigma_Y(\cdot \,|\, x)$ on Y for any fixed $x \in X$. A fuzzy measure g_Y on Y is induced by $\sigma_Y(\cdot \,|\, x)$ and a fuzzy measure g_X as follows, for B nonfuzzy,

$$g_Y(B) = \fint_X \sigma_Y(B \,|\, x) \circ g_X(\cdot) \qquad \forall B \in \mathscr{P}(Y). \tag{7}$$

g_X corresponds to an a priori probability and $\sigma_Y(\cdot \,|\, x)$ to a conditional probability. For this reason, g_X may be called an a priori fuzzy measure. $\sigma_Y(B \,|\, x)$ measures the grade of fuzziness of the statement, "One of the elements of B results because of x." In some applications $\sigma_Y(\cdot \,|\, x)$ models subjectivity, which modifies the information g_X.

The fuzzy integral of a function h from Y to $[0, 1]$ with respect to g_Y will be (shown in Sugeno, 1974)

$$\fint_B h(y) \circ g_Y(\cdot) = \fint_B h(y) \circ \left[\fint_X \sigma_Y(\cdot \,|\, x) \circ g_X(\cdot) \right]$$

$$= \fint_X \left[\fint_B h(y) \circ \sigma_Y(\cdot \,|\, x) \right] \circ g_X(\cdot).$$

When h is interpreted as a membership function μ_B of a fuzzy set B in Y, $g_Y(B)$ is calculated using the above formula as

$$g_Y(B) = \fint_X \sigma_Y(B \,|\, x) \circ g_X(\cdot)$$

with

$$\sigma_Y(B \,|\, x) = \fint_Y \mu_B(y) \circ \sigma_Y(\cdot \,|\, x).$$

Similarly to (7) we can consider

$$g_X'(A) = \fint_Y \sigma_X(A \,|\, y) \circ g_Y(\cdot) \qquad \forall A \in \mathscr{P}(X).$$

If we can choose $\sigma_X(\cdot \,|\, y)$ such that $g_X' = g_X$, then Sugeno (1974) has shown that $\sigma_X(\cdot \,|\, y)$ and $\sigma_Y(\cdot \,|\, x)$ were linked by

$$\fint_B \sigma_X(A \,|\, y) \circ g_Y(\cdot) = \fint_A \sigma_Y(B \,|\, x) \circ g_X(\cdot). \tag{8}$$

This result is similar to Bayes's theorem. Using this identity and knowing for instance g_X and $\sigma_Y(\cdot \,|\, x)$, hence g_Y, we can infer $\sigma_X(\cdot \,|\, y)$, with

$B = \{y\}$. $\sigma_X(\cdot \mid y)$ is not always uniquely determined. A possible interpretation of this model is that subjective incomplete information g_X may be improved to σ_X by extra information σ_Y.

Remark An extension of Sugeno's fuzzy integrals to evaluate fuzzy measures of L-fuzzy sets or Φ-fuzzy sets could be carried out.

B. POSSIBILITY AND PROBABILITY

This section is devoted to a comparison between possibility and probability. Similar quantities can be evaluated for each kind of measure. An analogue of Bayesian inference for possibility exists. Possibility distributions and probability distributions are loosely related through a consistency principle.

a. Possibility and Fuzzy Sets   (Zadeh, 1978)

Let A be a nonfuzzy set of X and v a variable on X. To say that v takes its value in A indicates that any element in A could possibly be a value of v and that any element not in A cannot be a value of v. The statement "v takes its value in A" can be viewed as inducing a possibility distribution π over X associating with each element x the possibility that x is a value of v:

$$\Pi(v = x) = \pi(x) = \begin{cases} 1 & \text{if } x \in A, \\ 0 & \text{otherwise.} \end{cases}$$

Next, assume A is a fuzzy set that acts as a fuzzy restriction on the possible value of v see 3.A.a). An extension of our above interpretation is that A induces a possibility distribution that is equal to μ_A on the values of v:

$$\Pi(v = x) = \pi(x) = \mu_A(x).$$

Since the expression of a possibility distribution can be viewed as a fuzzy set, possibility distributions may be manipulated by the combination rules of fuzzy sets, and more particular of fuzzy restrictions.

N.B.: 1. Note that although a fuzzy set and a possibility distribution have a common mathematical expression, the underlying concepts are different. A fuzzy set A can be viewed as a fuzzy value that we assign to a variable. Viewed as a possibility restriction A is the fuzzy set of nonfuzzy values that can possibly be assigned to v.

Example In a nonfuzzy case (Zadeh, Reference from IV.2, 1977b), consider the variable *sister* (Dedre) to which we assign a set or a possibility distribution, setting

sisters(Dedre) = {Sue, Jane, Lorraine}   (set);

sister(Dedre) ∈ {Sue, Jane, Lorraine}   (possibility distribution).

N.B.: 2. The meaning of π entails: "it is impossible that v belongs to the complement of supp A in X." That is, "it is necessary that v belongs to supp A" because we suppose that we are sure that v takes its values on X and only on X. However, v can be any element of A with a given possibility. π does not model "it is possible that v belongs to supp A" but "each element of supp A and only of supp A is a possible value for v."

b. Possibility of a Nonfuzzy Event

Let π be a possibility distribution induced by a fuzzy set F in X. Let A a nonfuzzy set of X; the possibility that x belongs to A is $\Pi(A)$ where Π is the possibility measure induced by π, and we have (see A.a.ζ)

$$\Pi(A) = \sup_{x \in A} \mu_F(x) = \sup_{x \in A} \pi(x). \tag{9}$$

As pointed out by Nguyen (1977c), $\Pi(A)$ is generally a Choquet (NF 1953) strong precapacity.

Similarly, if p is a probability distribution over X, the probability that x belongs to A is

$$P(A) = \int_A dP = \begin{cases} \int_A p(x)\,dx & \text{if } X = \mathbb{R} \\ \sum_{x \in A} p(x) & \text{if } X \text{ is finite.} \end{cases}$$

Note that in (9) sup acts as a Lebesgue integral. Indeed, it is a fuzzy integral in the sense of Sugeno:

$$\Pi(A) = \oint_X \mu_A(x) \circ \Pi(\cdot).$$

Moreover, $\forall A, B \in \mathscr{P}(X), \Pi(A \cup B) = \max(\Pi(A), \Pi(B))$, which corresponds to $P(A \cup B) = P(A) + P(B)$ when $A \cap B = \varnothing$.

"Intuitively, possibility relates to our perception of the degree of feasibility or ease of attainment whereas probability is associated with a degree of likelihood, belief, frequency or proportion" (Zadeh, Reference from IV.2, 1977b).

c. Consistency Principle

As pointed out by Zadeh (1978), it seems quite natural to think that "what is possible may not be probable and what is improbable need not to be impossible." Proceeding further, we may state that what is probable is certainly possible and what is inevitable (necessary) is certainly probable. This informal principle may be translated as: the degree of possibility of an event is greater than or equal to its degree of probability, which must be itself greater than or equal to its degree of necessity. To calculate a degree of necessity we may think of using a consonant belief measure as hinted by eq. (4). A consequence of the above principle would be that $\Pi(X) = 1$ as soon as $P(X) = 1$, which is usually taken for granted. This means that the possibility distribution should be normalized, i.e., $\sup_{x \in X} \pi(x) = 1$. Hägg (Reference from IV.3) suggests the use of a nonnormalized probability distribution, and $P(X) < 1$ is assumed to be equal to the rate of possibility of X, i.e., $P(X) = \Pi(X)$. $1 - P(X)$ is interpreted as the probability of occurrence of an event outside the universe X.

d. Conditional Possibilities

Let X and Y be two universes, and u, v two variables. Let $\pi_{(u, v)}(x, y)$ be a possibility distribution associated with (u, v). $\pi_u(x)$ and $\pi_v(y)$ respectively denote the projection of $\pi_{(u, v)}(x, y)$ on X and Y:

$$\pi_u(x) = \sup_y \pi_{(u, v)}(x, y); \qquad \pi_v(y) = \sup_x \pi_{(u, v)}(x, y).$$

$\pi_u(x)$ and $\pi_v(y)$ are said to be *marginal possibility distributions*. Recall that the separability of $\pi_{(u, v)}(x, y)$ means that

$$\pi_{(u, v)}(x, y) = \min(\pi_u(x), \pi_v(y)).$$

Note that the following formula always holds:

$$\pi_u(x) = \sup_y \min(\pi_{(u, v)}(x, y), \pi_v(y)). \tag{10}$$

When $\pi_{(u, v)}(x, y)$ is separable, it becomes

$$\pi_u(x) = \sup_y \min(\pi_u(x), \pi_v(y)).$$

$\pi_{(u, v)}(x, y)$ can be interpreted as a conditional possibility distribution.

Let us investigate the relationship between (10) and Sugeno's fuzzy integral.

The fuzzy integral of a function h from X to $[0, 1]$ over a nonfuzzy domain D with respect to a possibility measure Π is

$$I = \int_D h(x) \circ \Pi(\cdot) = \sup_{\alpha \in [0, 1]} \min\left[\alpha, \sup_{x \in D \cap H_\alpha} \pi(x)\right]$$

where $H_\alpha = \{x, h(x) \geqslant \alpha\}$. Let us transform I:

$$I = \sup_{\substack{\alpha \in [0, 1]}} \sup_{x \in D \cap H_\alpha} \min(\alpha, \pi(x)) = \sup_{\substack{\alpha \in [0, 1] \\ x \in D}} \min(\alpha, \mu_{H_\alpha}(x), \pi(x))$$

$$I = \sup_{x \in D} \min\left(\pi(x), \sup_{\alpha \in [0, 1]}(\alpha, \mu_{H_\alpha}(x))\right) = \sup_{x \in D} \min(\pi(x), h(x)). \quad (11)$$

Hence, I is the degree of consistency of π and h.

Now let us prove that

$$\Pi_u(A) = \int_Y \Pi_{(u, v)}(A, y) \circ \Pi_v(\cdot)$$

where $\Pi_{(u, v)}(A, y) = \sup_{x \in A} \pi_{(u, v)}(x, y)$; $\Pi_u(A) = \sup_{x \in A} \pi_u(x)$.

Proof: Using (11), the right-hand side is equal to

$$\sup_{y \in Y} \min\left(\sup_{x \in A} \pi_{(u, v)}(x, y), \pi_v(y)\right) = \sup_{x \in A} \sup_{y \in Y} \pi_{(u, v)}(x, y) = \Pi_u(A).$$

For $A = \{x\}$, we recover (10), which thus proves to be a particular case of (7). Q.E.D.

The analogy between conditional possibilities and conditional probabilities was developed by Nguyen (1977b), who introduced the notion of a "normalized" conditional possibility distribution. Denote such a distribution by $\pi(x \mid y)$. $\pi(x \mid y)$ is assumed to be expressed as

$$\pi(x \mid y) = \pi_{(u, v)}(x, y) \cdot \alpha(\pi_u(x), \pi_v(y))$$

where $\alpha(\cdot, \cdot)$ is a normalization function. α is determined from two requirements:

(i) $\pi(x \mid y) \in [0, 1]$;
(ii) $\min(\pi_u(x), \pi_v(y)) \cdot \alpha(\pi_u(x), \pi_v(y)) = \pi_u(x)$.

(ii) means that when $\pi_{(u, v)}(x, y)$ is separable, the normalized conditional possibility distribution equals the projection $\pi_u(x)$. This situation is similar to that of a conditional probability $P(E \mid F)$, which equals $P(E)$ if E and F are independent ($P(E \cap F) = P(E)P(F)$). Hence the notion of noninteractivity for possibilities may play the same role as independence for probabilities.

(i) and (ii) lead to the expression

$$\pi(x \mid y) = \begin{cases} \pi_{(u, v)}(x, y) & \text{if } \pi_u(x) \leqslant \pi_v(y), \\ \pi_{(u, v)}(x, y) \dfrac{\pi_u(x)}{\pi_v(y)} & \text{if } \pi_u(x) > \pi_v(y). \end{cases} \quad (12)$$

Nguyen [7] showed that

$$\pi_u(x) = \sup_y \min(\pi(x \mid y), \pi_v(y)).$$

Moreover, we can state this equality together with (10) for possibility measures:

$$\Pi_u(A) = \sup_y \min(\Pi_{(u,v)}(A, y), \pi_v(y))$$

$$= \sup_y \min(\Pi_{(u,v)}(A \mid y), \pi_v(y))$$

where $\Pi(A \mid y) = \sup_{x \in A} \pi(x \mid y)$.

This equality is similar to $P(E) = \sum_{y \in Y} P(E \mid y) p(y)$ (Y finite) in probability theory.

Lastly, formula (8) can be written for possibilities:

$$\forall A \in \mathcal{P}(X), \quad \forall B \in \mathcal{P}(Y)$$

$$\oint_B \Pi_{(u,v)}(A, y) \circ \Pi_v(\cdot) = \oint_A \Pi_{(u,v)}(x, B) \circ \Pi_u(\cdot). \tag{13}$$

Proof:

$$\oint_B \Pi_{(u,v)}(A, y) \circ \Pi_v(\cdot) = \sup_{y \in B} \min(\pi_v(y), \Pi_{(u,v)}(A, y))$$

$$= \sup_{y \in B} \min\left(\pi_v(y), \sup_{x \in A} \pi_{(u,v)}(x, y)\right)$$

$$= \sup_{\substack{x \in A \\ y \in B}} \min(\pi_v(y), \pi_{(u,v)}(x, y))$$

$$= \sup_{\substack{x \in A \\ y \in B}} \pi_{(u,v)}(x, y) = \Pi_{(u,v)}(A, B)$$

because $\pi_v(y) = \sup_x \pi_{(u,v)}(x, y) \geq \pi_{(u,v)}(x, y), \forall y$. Obviously, the right-hand side of (13) gives the same result.   Q.E.D.

(13) can be viewed as a Bayes theorem for possibilities; $\Pi_{(u,v)}(A, B)$ is similar to $P(A \cap B)$ in probabilities.

Other considerations on conditional possibility distributions have been recently developed by Hisdal (1978).

C. FUZZY EVENTS

Events are often ill defined. The question of the probability, of the possibility of such fuzzy events may arise. For instance, What is the probability/possibility of a warm day tomorrow? The extension of probability theory in order to deal with fuzzy events was introduced by Zadeh (1968), who considered nonfuzzy probabilities of fuzzy events. However, fuzzy probabilities of the same fuzzy events can also be defined. Both points of view can also be applied to possibility calculus.

a. Nonfuzzy Probability/Possibility of a Fuzzy Event

We shall assume for simplicity that X is the Euclidian n-space $\mathbb{R}^n$. Let $\mathcal{B}$ be a Borel field in $\mathbb{R}^n$ and P a probability measure on $\mathcal{B}$. A fuzzy event in $\mathbb{R}^n$ is a fuzzy set A on $\mathbb{R}^n$ whose membership function is measurable. The probability of a fuzzy event A is defined by the Lebesgue–Stieltjes integral

$$P(A) = \int_{\mathbb{R}^n} \mu_A(x)\, dP. \tag{14}$$

Note that when A is nonfuzzy, we obtain the usual probability of A. The probability of a fuzzy event is the expectation of its membership function.

$P(A)$ evaluates the degree with which the sample set $\mathbb{R}^n$ has the fuzzy property A. The corresponding experiment is a random selection of elements x_i more or less belonging to A. At each trial a membership value $\mu_A(x_i)$ is provided. $P(A)$ is

$$\lim_{m \to \infty} \frac{\sum\limits_{i=1}^{m} \mu_A(x_i)}{m}$$

where m is the number of trials. Thus, $P(A)$ can be interpreted as a proportion of elements of $\mathbb{R}^n$ "belonging" to A.

It is easy to see that $\forall A, B \in \hat{\mathcal{P}}(\mathbb{R}^n)$ (μ_A, μ_B measurable), if $A \subseteq B$, then $P(A) \leqslant P(B)$, $P(A \hat{+} B) = P(A) + P(B) - P(A \cdot B)$, and $P(A \cup B) = P(A) + P(B) - P(A \cap B)$.

Two fuzzy events A and B are *independent* iff $P(A \cdot B) = P(A)P(B)$. An immediate consequence of the above definition is the following. Let $X_1 = \mathbb{R}^m$, $X_2 = \mathbb{R}^p$, and P be the product measure $P_1 \times P_2$ where P_1 and P_2 are probability measures on X_1 and X_2, respectively. Let A_1 and A_2 be events in X_1 and X_2 characterized by the membership functions $\mu_{A_1}(x^1, x^2) = \mu_{A_1}(x^1)$ and $\mu_{A_2}(x^1, x^2) = \mu_{A_2}(x^2)$ respectively. Then A_1 and A_2 are independent events. This would not be true if independence were defined in terms of $P(A \cap B)$ rather than $P(A \cdot B)$ (Zadeh, 1968).

The conditional probability of a fuzzy event A given B is then defined by $P(A \mid B) = P(A \cdot B)/P(B)$ provided $P(B) > 0$. Note that if A and B are independent, then $P(A \mid B) = P(A)$, as in the nonfuzzy case.

The notions of mean, variance, and entropy of fuzzy events can be defined in a similar way (for instance, the mean is

$$\frac{1}{P(A)} \int_{\mathbb{R}^n} x\mu_A(x)\,dP$$

(Zadeh, 1968)).

The possibility of a fuzzy event A in a universe X with respect to the possibility measure Π can be defined analogously as

$$\Pi(A) = \oint_A 1 \circ \Pi(\cdot) = \oint_X \mu_A(x) \circ \Pi(\cdot).$$

According to (11), $\Pi(A) = \sup_{x \in X} \min(\mu_A(x), \pi(x))$ (Zadeh, 1978) where $\pi(x)$ is the possibility distribution associated with Π. The possibility of a fuzzy event is thus the degree of consistency of this fuzzy event with a possibility distribution. As in the nonfuzzy case, we have $\Pi(A \cup B) = \max(\Pi(A), \Pi(B))$ and $A \subseteq B$ implies $\Pi(A) \leqslant \Pi(B)$.

Let X and Y be two universes, A, B be two fuzzy events in X, Y, and $\pi(x, y)$ a possibility distribution over $X \times Y$. Two fuzzy sets A and B will be said to be *noninteractive* iff

$$\Pi(A \cap B) = \min(\Pi(A), \Pi(B)).$$

In particular, when $\pi(x, y)$ is separable, i.e.,

$$\pi(x, y) = \min(\pi_X(x), \pi_Y(y));$$

and if A is a fuzzy set on X and B on Y, then considering the cylindrical extensions $c(A)$, $c(B)$ we have

$$\Pi(c(A) \cap c(B)) = \min(\Pi_X(A), \Pi_Y(B)).$$

Proof:

$$\Pi(c(A) \cap c(B)) = \sup_{x, y} \min(\mu_A(x), \mu_B(y), \pi(x, y))$$

$$= \sup_{x, y} \min(\min(\mu_A(x), \pi_X(x)), \min(\mu_B(y), \pi_Y(y)))$$

$$= \min\left(\sup_x (\mu_A(x), \pi_X(x)), \sup_y (\mu_B(y), \pi_Y(y))\right),$$

and A and B are noninteractive.   Q.E.D.

Lastly, there holds Bayes's theorem for possibilities of fuzzy events:

$$\Pi(A,B) = \oint_B \Pi(A,y) \circ \Pi_Y(\cdot) = \oint_A \Pi(x,B) \circ \Pi_X(\cdot)$$

where $A \in \tilde{\mathcal{P}}(X)$, $B \in \tilde{\mathcal{P}}(Y)$.

$$\Pi(A,y) = \sup_{x \in X} \min(\mu_A(x), \pi(x,y)),$$

$$\Pi(x,B) = \sup_y \min(\mu_B(y), \pi(x,y)).$$

Π_y (resp. Π_x) is the possibility measure associated to the projection on Y (resp. on X) of the (separable or not) distribution $\pi(x,y)$:

$$\Pi(A,B) = \sup_{\substack{x \in X \\ y \in Y}} \min(\mu_A(x), \mu_B(y), \pi(x,y)).$$

The proof is similar to that of (13).

b. Fuzzy Probability/Fuzzy Possibility of a Fuzzy Event

We give here only basic definitions and a rationale. Instead of evaluating the proportion of elements of a sample space "belonging" to a fuzzy set C, we may calculate the possibility level that there exists a nonfuzzy event matching C, which occurs with a given probability. In the following we assume $X = \mathbb{R}$. Let $p(x)$ be a probability distribution and (A,B) a fuzzy interval bounded by two nonoverlapping convex normalized fuzzy sets on $\mathbb{R}$. According to 4.C.b., the fuzzy probability of the fuzzy event "x belongs to the fuzzy interval (A,B)" is $FP[(A,B)]$ (Dubois and Prade, 1978)

$$FP[(A,B)] = \int_A^B p(x)\,dx = P(B) \ominus P(A).$$

When A and B are ordinary numbers, the above formula becomes the usual definition of probability that $x \in [A,B]$. Here, the result is a fuzzy set of $[0,1]$ which can be interpreted as a linguistic probability (Zadeh, 1975).

Analogously, fuzzy possibilities of fuzzy events can be defined through the extension principle: $F\Pi[(A,B)]$ has membership function

$$\mu_{F\Pi[(A,B)]}(z) = \sup_{\substack{x,y \\ z = \max \pi(t) \\ t \in [x,y]}} \min(\mu_A(x), \mu_B(y)),$$

which is a particular case of Sugeno's integral extended by the extension principle. π is a possibility distribution over $\mathbb{R}$. This formula can be simplified according to the respective positions of π (assumed normalized and convex) and (A, B).

N.B.: The existence of two points of view on "fuzzification," yielding either nonfuzzy results or fuzzy ones, seems to be very general. Other examples are the power of a fuzzy set and fuzzy cardinality (1.D.), extremum of a function on a fuzzy domain (4.B.), fuzzy or nonfuzzy integration over a fuzzy interval (4.C.b.).

D. FUZZY DISTRIBUTIONS

a. Probabilities

A probability distribution cannot always be precisely identified. Thus, probability and possibility values are often rather subjectively assessed. A linguistic probability will be modeled by a fuzzy set on $[0, 1]$ (Zadeh, 1975).

Let $X = \{x_1, \ldots, x_n\}$. To each x_i is assigned a linguistic probability $\tilde{p}_i \in \tilde{\mathcal{P}}([0, 1])$ and a variable $P(\{x_i\})$ restricted by $\tilde{p}_i$. The linguistic probabilities $\tilde{p}_i$, $i = 1, n$, are β-interactive (3.A.b) because of the normalization constraint on the possible nonfuzzy values of $P(\{x_i\})$ ($\tilde{p}_i$ is viewed as inducing a possibility distribution on the values of $P(\{x_i\})$). The fuzzy restriction associated with $P(\{x_1\}), \ldots, P(\{x_n\})$ is $R(P(\{x_i\})), \ldots, P(\{x_n\})) = R(P(\{x_1\})) \times \cdots \times R(P(\{x_n\})) \cap Q$ where Q is the nonfuzzy relation $Q(u_1, \ldots, u_n) = 1$ iff $\sum_{i=1}^{n} u_i = 1$, and $R(P(\{x_i\})) = \tilde{p}_i$.

Now consider the interactive sum σ of the $\tilde{p}_i$ ($i = 1, n$). Its membership function is given by

$$\mu_\sigma(z) = \sup_{\substack{z = u_1 + \cdots + u_n \\ 1 = u_1 + \cdots + u_n}} \min_i \mu_{\tilde{p}_i}(u_i).$$

Obviously, $\mu_\sigma(z) = 0$ for $z \neq 1$; and $\mu_\sigma(1)$ evaluates the mutual consistency of the $\tilde{p}_i$ with respect to the normalization constraint. We shall admit that an n-tuple $(\tilde{p}_1, \ldots, \tilde{p}_n)$ of linguistic probabilities is totally consistent whenever $\mu_\sigma(1) = 1$, i.e., $\exists(u_1, \ldots, u_n) \in [0, 1]^n$ such that $\mu_{\tilde{p}_i}(u_i) = 1$, $\forall i$ and $\sum_{i=1}^{n} u_i = 1$.

The fuzzy probability of a subset of X, say $X' = \{x_1, \ldots, x_k\}, k \leqslant n$, is a fuzzy interactive sum σ_k of the $\tilde{p}_i$ ($i = 1, k$), such that

$$\mu_{\sigma_k}(z) = \sup_{\substack{z = u_1 + \cdots + u_k \\ u_1 + \cdots + u_k \leqslant 1}} \min_{i = 1, k} \mu_{\tilde{p}_i}(u_i) = \widetilde{\min}(1, \tilde{p}_1 \oplus \cdots \oplus \tilde{p}_k).$$

The fuzzy expectation of a random variable V taking values in a set

$\{a_1, \ldots, a_n\} \subset \mathbb{R}$ with linguistic probabilities $\tilde{p}_i$, $i = 1, n$ is the interactive sum $E(V)$ such that

$$\mu_{E(V)}(z) = \sup_{\substack{a_1u_1 + \cdots + a_nu_n = z \\ u_1 + \cdots + u_n = 1}} \min_{i=1,n} \mu_{\tilde{p}_i}(u_i).$$

The calculation of $\mu_{E(V)}$ may be tricky when $n > 2$. It is equivalent to the mathematical programming problem: maximize θ under the constraints

$$\mu_{\tilde{p}_i}(u_i) \geqslant \theta, \quad i = 1, n; \qquad \sum_{i=1}^{n} a_i u_i = z; \qquad \sum_{i=1}^{n} u_i = 1.$$

The case $n = 2$ was already solved in Section 3.A.b. When the a_i are membership values that characterize a fuzzy event A, $E(V)$ is the linguistic probability of the fuzzy event A in the sense of Zadeh (1975).

When $X = \mathbb{R}$, the probability distribution becomes a fuzzy function $\tilde{p}$ of a nonfuzzy real variable. Two points of view exist because we may choose a fuzzy bunch or a fuzzifying function (see 4.A.c). If $\tilde{p}$ is a fuzzy bunch $\tilde{p} = \int \alpha/p_\alpha$, it will be a fuzzy set of probability distributions, i.e.,

$$\int_{\mathbb{R}} p_\alpha(x)\, dx = 1 \quad \forall \alpha.$$

Note that this point of view would not be, in the discrete case, equivalent to that of linguistic probabilities. The corresponding approach would be a set of n-tuples $\{p_{i\alpha}\}_{i=1,n}$ with $\sum_{i=1}^{n} p_{i\alpha} = 1$ $\forall \alpha \in [0,1]$, where α is a membership value for the n-tuple. A drawback of this approach is the possible existence of ambiguities in the value of the membership function of the probability of some fuzzy events. (It is possible to have $\sum_{i \in I} a_i p_{i\alpha} = \sum_{i \in I} a_i p_{i\alpha'}$ for $\alpha \neq \alpha'$ and some $I \subset (1, \ldots, n)$: then we may use $\sup(\alpha, \alpha')$ to solve the ambiguity; see 4.A.c.β.) The point of view equivalent to linguistic probabilities is the use of a fuzzifying function $x \mapsto \tilde{p}(x)$. To be sure that $\tilde{p}$ is in some sense a "fuzzified" probability distribution, we may impose its 1-level curve p_1 (4.C.a.α) to be such that $\int_{\mathbb{R}} p_1(x)\, dx = 1$. The probability measure of an interval $[a,b]$ will be $\widetilde{\min}(1, \int_a^b \tilde{p}(x)\, dx)$ (see 4.C.A.β, γ).

b. Possibilities

Let X be a finite set. A fuzzy possibility value $\tilde{\pi}(x)$ can be assigned to each x in X. $\tilde{\pi}(x) \in \tilde{\mathcal{P}}([0,1])$. Such a fuzzy possibility distribution will be said to be normalized iff $\widetilde{\max}_{x \in X} \tilde{\pi}(x) = \tilde{\Pi}(X)$ is such that $\mu_{\tilde{\Pi}(X)}(1) = 1$. This happens whenever $\exists x \in X$ such that $\mu_{\tilde{\pi}(x)}(1) = 1$. Hence, the fuzzy possibilities $\tilde{\pi}(x)$ are not β-interactive. The fuzzy possibility of a nonfuzzy event (set) A of X will be

$$\tilde{\Pi}(A) = \widetilde{\max_{x \in A}} \tilde{\pi}(x).$$

If A is fuzzy, $\tilde{\Pi}(A) = \widetilde{\max}_{x \in X} \widetilde{\min}(\tilde{\pi}(x), \mu_A(x))$ in the sense of C.a.

When X is nonfinite, we need an extended sup, $\widetilde{\text{sup}}$, to carry out the same approach.

Obviously fuzzy possibilities may model linguistic possibility values; this together with linguistic probabilities will be studied from a logical or semantic point of view further on.

REFERENCES

Banon, G. (1978). "Distinction entre Plusieurs Sous-Ensembles de Mesures Floues," Note interne, No. 78.I.11. L.A.A.S.-A.S., Toulouse, France. (Also in *Proc. Colloq. Int. Théorie Appl. Sous-Ensembles Flous, Marseille 1978.*)

Dubois, D., and Prade, H. (1978). Toward fuzzy analysis: Integration and derivation of fuzzy functions. In "Fuzzy Algebra, Analysis, Logics," Tech. Rep. TR-EE 78/13, Part C. Purdue Univ., Lafayette, Indiana. (Reference from II.4.)

Hisdal, E. (1978). Conditional possibilities. Independence and non-interactivity. *Int. J. Fuzzy Sets Syst.* **1**, 283–297.

Kandel, A. (1977). "Fuzzy Statistics and Its Applications to Fuzzy Differential Equations." Ph.D. Thesis, Univ. of New Mexico, Albuquerque.

Kandel, A. (1978). Fuzzy statistics and forecast evaluation. *IEEE Trans. Syst., Man Cybern.* **8**, No. 5, 396–401.

Kandel, A., and Byatt W. J. (1978). Fuzzy sets, fuzzy algebra, and fuzzy statistics. *Proc. IEEE* **68**, 1619–1639.

Kwakernaak, H. (1978). Fuzzy random variables. I. Definitions and theorems. *Inf. Sci.* **15**, 1–29.

Nahmias, S. (1978). Fuzzy variables. *Int. J. Fuzzy Sets Syst.* **1**, No. 2, 97–110. (Reference from II.2.)

Nguyen, H. (1977a). On fuzziness and linguistic probabilities. *J. Math. Anal. Appl.* **61**, 658–671.

Nguyen, H. (1977b). On Conditional Possibility Distributions. Memo UCB/ERL M77/52. Univ. of California, Berkeley. [Also in *Int. J. Fuzzy Sets Syst.* **1**, No. 4, 299–310 (1978).]

Nguyen, H. (1977c). Some mathematical tools for linguistic probabilities. *Proc. IEEE Conf. Decision Control, New Orleans* pp. 1345–1350. [Also in *Int. J. Fuzzy Sets. Syst.* **2**, No. 1, 53–66 (1979).]

Nguyen, H. (1978). On random sets and belief functions. *J. Math. Anal. Appl.* **65**, 531–542.

Ponsard, C. (1977). Aléa et flou. In "Mélanges Offerts à H. Guitton," pp. 287–299. Dalloz Sirey, Paris.

Sugeno, M. (1974). "Theory of Fuzzy Integral and Its Applications." Ph.D. Thesis, Tokyo Inst. of Technol., Tokyo.

Sugeno, M. (1977). Fuzzy measures and fuzzy integrals: A survey. In "Fuzzy Automata and Decision Processes" (M. M. Gupta, G. N. Saridis, and B. R. Gaines, eds.), pp. 89–102. North-Holland Publ., Amsterdam.

Terano, T., and Sugeno, M. (1975). Conditional fuzzy measures and their applications. In "Fuzzy Sets and Their Applications to Cognitive and Decision Processes" (L. A. Zadeh, K. S. Fu, K. Tanaka, and M. Shimura, eds.), pp. 151–170. Academic Press, New York.

Tsichritzis, D. (1971). Participation measures. *J. Math. Anal. Appl.* **36**, 60–72.

Zadeh, L. A. (1968). Probability measures of fuzzy events. *J. Math. Anal. Appl.* **23**, 421–427.

Zadeh, L. A. (1975). The concept of a linguistic variable and its application to approximate reasoning. Part 3. *Inf. Sci.* **9**, 43–80. (Reference from II.2.)

Zadeh, L. A. (1978). Fuzzy sets as a basis for a theory of possibility. *Int. J. Fuzzy Sets Syst.* **1**, No. 1, 3–28.

Part **III**

FUZZY MODELS AND
FORMAL STRUCTURES

It seems that a lot of researchers have focused their attention on fuzzy formal structures, i.e., models of static, deductive, algorithmic, and dynamic fuzzy systems. Most of these are extensions of already existing nonfuzzy structures. However, a few depart from classical approaches.

Chapter 1 deals principally with fuzzy logic, i.e., fuzzy switching logic, multivalent logics as underlying fuzzy set theory, and approximate reasoning.

Chapter 2 is devoted to fuzzy dynamical systems. Our constant concern is to keep clear the semantic interpretation of the formal developments.

Chapter 3 first surveys the past and current research on fuzzy formal languages and grammars and their relations to automata. Then, two points of view on fuzzy algorithms are presented.

Chapter 4 reflects the first attempts to apply fuzzy set theory to operations research models.

Chapter 1

MULTIVALENT AND FUZZY LOGICS

This chapter is devoted to semantical aspects of non-Boolean logics in correlation with fuzzy set theory. The first section gives an account of fuzzy switching logic focusing on the fuzzy version of a well-known problem for Boolean functions, that of their canonical and minimal representations in terms of conjunction and disjunction. Section B provides a systematic presentation of multivalent logics as underlying fuzzy set theory. Most of these were developed in the 1920s and 1930s without any set-theoretic interpretation. Applications of fuzzy set theory to modal logic are briefly sketched in Section C. However, the link between possibility theory and modal logic has not been made completely clear yet. We deal then with the extension of multivalent logics to fuzzy truth values. Lastly, Zadeh's recent theory of approximate reasoning is emphasized. It contrasts with multivalent and fuzzy-valued logics in that a proposition is now viewed as associated with a possibility distribution that fuzzily restricts the values of the variables involved in the proposition. Although this approach is very new, it already appears to be a promising methodology for modeling human reasoning.

A. FUZZY SWITCHING LOGIC

One of the major fields of application of Boolean logic is the theory of electronic switching circuits. Such circuits are modeled by Boolean expres-

151

sions which may involve only negation, disjunction, and conjunction connectives. The problem of finding a minimal representation for these expressions has been considered at length in the literature (McCluskey, NF 1965). We are concerned here with the representation and minimization of fuzzy logical expressions. Many works have already been published on this very specific topic as shown by the extensive bibliography at the end of the chapter. For a more detailed presentation, see Lee and Kandel (1978).

a. Fuzzy Expressions

Let x, x_1, x_2 be variables taking their values in $[0, 1]$. The following notations are adopted: $\neg x = 1 - x$ (negation); $x_1 \vee x_2 = \max(x_1, x_2)$ (disjunction); $x_1 \wedge x_2 = \min(x_1, x_2)$ (conjunction).

Recall that $(\{0, 1\}, \vee, \wedge, \neg)$ is a Boolean lattice (see II.1.B.d), whereas $([0, 1], \vee, \wedge, \neg)$ is only a pseudocomplemented distributive lattice (see II.1.B.d). In particular, $\forall x \in]0, 1[, \ x \wedge (\neg x) \neq 0, \ x \vee (\neg x) \neq 1$, which contrasts with the Boolean case.

A *fuzzy expression* is a function from $[0, 1]^n$ to $[0, 1]$ defined by the following rules only:

  (i)   0, 1, and variables $x_i, i = 1, n$, are fuzzy expressions;
  (ii) if f is a fuzzy expression, then $\neg f$ is a fuzzy expression;
  (iii) if f and g are fuzzy expressions, then $f \wedge g$ and $f \vee g$ are too.

Note that all Boolean expressions, once their domain is extended to $[0, 1]^n$, can be fuzzy expressions. This is because all Boolean expressions can be expressed only in terms of $\neg, \wedge, \vee$. However, this is not the only way to extend Boolean expressions; operators different from max and min can be used, as will be seen in B.

The fact that we consider here only fuzzy expressions is what makes this section rather specific from a logical point of view. Its interest lies in its practical attractiveness for switching-circuit specialists.

A *literal* is a variable x_i or its negation $\neg x_i$. A *phrase* is a conjunction of literals. A disjunction of literals is called a *clause*. Owing to the mutual distributivity of $\vee$ and $\wedge$, any fuzzy expression can be transformed into a disjunction of phrases or a conjunction of clauses.

b. Some Properties of Fuzzy Expressions

α. *Monotonicity with Respect to Ambiguity*   (Mukaidono, 1975a)

The partial ordering relation that describes ambiguity is A such that: $\forall a_i, a_j \in [0, 1], a_i A a_j$ iff either $\frac{1}{2} \leqslant a_i \leqslant a_j$ or $\frac{1}{2} \geqslant a_i \geqslant a_j$. We have $\frac{1}{2} A a_i$, $\forall a_i \in [0, 1]$. Moreover, $a_i \in]\frac{1}{2}, 1]$ and $a_j \in [0, \frac{1}{2}[$ cannot be compared. $a_i A a_j$

means: "a_i is more ambiguous than a_j." A is extended to $[0,1]^n$ as follows:

$$\forall a = (a_1, \ldots, a_n) \in [0,1]^n, \quad \forall b = (b_1, \ldots, b_n) \in [0,1]^n,$$

$$a A b \quad \text{iff} \quad \forall i = 1, n, \quad a_i A b_i.$$

Theorem (Mukaidono, 1975a) Let f be a fuzzy expression mapping $[0,1]^n \to [0,1]$; if $a A b$, then $f(a) A f(b)$.

Proof: The results trivially holds for $0, 1$, and any variable. Now it is easy to show that if it holds for f and g, fuzzy expressions, then it also holds for $\neg f, f \wedge g, f \vee g$. Q.E.D.

As a consequence $\forall b$ such that $b A a$, if $f(a) = \frac{1}{2}$, then $f(b) = \frac{1}{2}$. Replacing terms in a by others that are closer to $\frac{1}{2}$ does not change $f(a)$ in that case. Moreover, if $f(a) = \omega$, $\omega \in \{0,1\}$, then $f(b) = \omega$, $\forall b$ such that $a A b$.

β. *Canonical Disjunctive Form of a Fuzzy Expression* (Davio and Thayse, 1973)

Since excluded-middle laws no longer hold on $([0,1], \wedge, \vee, \neg)$, there is no unique way to represent a fuzzy expression as a disjunction of phrases. There are two kinds of phrases:

simple phrases in which a variable appears at most once, as a literal;
contradictory phrases in which conjunctions such as $x_i \wedge (\neg x_i)$ appear.

If P is a contradictory phrase, then $P(a) \leqslant \frac{1}{2} \forall a \in [0,1]^n$. Hence, since $x_i \vee (\neg x_i) \geqslant \frac{1}{2}$,

$$P \wedge (x_i \vee (\neg x_i)) = P = [P \wedge x_i] \vee [P \wedge (\neg x_i)].$$

The latter expression and P are two forms of the same fuzzy expression. The above manipulation clearly indicates that any contradictory phrase can be expanded into a disjunction of contradictory phrases which contain each variable at least once. Such phrases are called *completed*.

The canonical disjunctive form of a fuzzy expression is a disjunction of simple or completed contradictory phrases. The proof of the uniqueness of this form can be found in Mukaidono (1975a). An algorithm for obtaining the canonical disjunctive form is (Davio and Thayse, 1973):

expand the expression into a fuzzy disjunctive form;
expand the contradictory phrases into a disjunction of completed contradictory phrases;
suppress redundant phrases using absorption laws (see II.1.B.d).

N.B.: Dual results on a canonical conjunctive form involving only simple and completed "tautological" clauses obviously hold.

γ. *Fuzzy Expressions and Ternary Logic*

The number of fuzzy expressions involving n variables is finite. This was proved by Preparata and Yeh (1972) who gave the following theorem, here stated in the terminology of Mukaidono (1975a): If fuzzy expressions f and g satisfy $f(a) = g(a)$ $\forall a \in \{0, \frac{1}{2}, 1\}^n$, then $f(a) = g(a)$ $\forall a \in [0, 1]^n$.

Proof: Denote by x_i^* either x_i or $\neg x_i$. The domain of f can be partitioned into subdomains characterized by the constraint $0 \leqslant x_{i_1}^* \leqslant \cdots \leqslant x_{i_n}^* \leqslant \frac{1}{2}$ where $(i_1, \ldots, i_n)$ is a permutation of $(1, \ldots, n)$. Clearly, for a given permutation there are 2^n ways of choosing an n-tuple of literals, so that the number of subdomains is $n! \cdot 2^n$. Now consider a given subdomain. The value of a phrase is that of a unique literal over the whole subdomain. Viewing f as a disjunction of phrases, the value of f is also that of a unique literal over this subdomain, say either x_l or $\neg x_l$. Each subdomain is a convex polyhedron whose vertices are $\{v^k\}_{k=0,n}$ such that

$$v^0 \quad \text{is defined by} \quad x_i^* = 0, \quad\quad i = 1, n;$$

$$v^k \quad \text{is defined by} \quad x_{i_j}^* = 0, \quad\quad j = 1, n - k;$$

$$x_{i_j}^* = \tfrac{1}{2}, \quad \forall j = n - k + 1, n \ (k > 0);$$

$$v^n \quad \text{is defined by} \quad x_i = \tfrac{1}{2}, \quad\quad i = 1, n.$$

Any element a of the subdomain is a convex combination of the v_i, say $a = \sum_{i=0}^n \alpha_i v^i$. Now on the subdomain assume $f(a) = a_l \in [0, 1]$, then

$$f(a) = a_l = \sum_{i=0}^n \alpha_i v_l^i = \sum_{i=0}^n \alpha_i f(v^i)$$

(with $v^i = (v_1^i, \ldots v_l^i, \ldots v_n^i)$). If $f(a) = \neg a_l$, the same result holds because $\sum_{i=0}^n \alpha_i = 1$.   Q.E.D.

This proof was given by Preparata and Yeh (1972). The value of f for any element of $[0, 1]^n$ is thus determined by its values for $n + 1$ elements of $\{0, \frac{1}{2}, 1\}^n$.

Kaufmann (1975) suggested the use of the subdomain defined above to check the equality of two fuzzy expressions. However, using the above result, this checking is easier by enumeration of elements in $\{0, \frac{1}{2}, 1\}^n$.

N.B.: Some authors have tried to evaluate the number of fuzzy expressions involving n variables. Until now, only upper and lower bounds—and not very good ones—have been found (see Kandel, 1974b; Kaufmann, 1975; Kameda and Sadeh, 1977).

c. Minimization of Fuzzy Expressions

The minimization of Boolean expressions has already been completely discussed in the literature. It usually proceeds in two steps. First, obtain a set of prime implicants; second, select the minimal set of implicants whose disjunction is equal to the Boolean expression. Because of the lack of excluded-middle laws, implicants of fuzzy expressions may be contradictory phrases. Thus, Boolean methods are no longer valid for determining prime implicants. However, the minimal form is still the disjunction of a minimal subset of prime implicants, which are said to be essential.

Definitions An *implicant* P of a fuzzy expression is a phrase such that $\forall a \in [0, 1]^n$, $P(a) \leqslant f(a)$, which is denoted $P \Rightarrow f$.

N.B.: Actually, $\Rightarrow$ is not a natural implication connective for fuzzy switching logic, which is nothing but K-SEQ (see B.b.α)).

A *prime implicant* P of a fuzzy expression f is an implicant such that for any phrase P', if $P \Rightarrow P'$ and $P' \Rightarrow f$, then $P = P'$ or $P' = f$. Hence a prime implicant is a "greatest" implicant of f.

First algorithms for generating all prime implicants were proposed by Lee and Chang (1971) and by Siy and Chen (1972). These methods were criticized (Kandel, 1973a; Negoita and Ralescu, 1976). Then, Kandel (1973b) presented a method based on the notion of fuzzy consensus, which was extended (Kandel, 1974c) to incompletely specified fuzzy expressions. Further critiques and refinements of this method can be found in Kandel (1973c, 1976a, 1977), Mukaidono (1975a), and Lee (1977). We follow here Mukaidono (1975a).

Let P and P' be phrases over the set of variables $\{x_1, x_2, \ldots, x_n\}$. A *fuzzy consensus* of P and P' is a contradictory phrase Φ built as follows:

find x_i with $P = x_i^* \wedge Q$, $P' = (\neg x_i^*) \wedge Q'$ where Q, Q' do not contain the variable x_i;

$\Phi = Q \wedge Q'$ iff $Q \wedge Q'$ is a contradictory phrase;

$\Phi = Q \wedge Q' \wedge x_i \wedge (\neg x_i)$ iff $Q \wedge Q'$ is a simple phrase;

(x_i^* means either x_i or $\neg x_i$.)

Theorem (Mukaidono, 1975a) A disjunctive form of a fuzzy expression f contains all its prime implicants iff:

(i)   there is no phrase that is an implicant of another phrase of f;

(ii) the fuzzy consensuses of any two phrases either do not exist or are implicants of at least one phrase of f.

Hence, there can be developed an algorithm that works on a disjunctive form, calculates and adds all the fuzzy consensuses of pairs of phrases until the conditions of the above theorem are satisfied; the set of all prime implicants of f is thus generated.

Another kind of method serving the same purpose was initiated by Preparata and Yeh (1972); it is based on the distinction between simple phrases and contradictory phrases of the canonical disjunctive form of a fuzzy expression. In particular, simple phrases are prime implicants. More about this approach can be found in Davio and Thayse (1973), Mukaidono (1975b), and Negoita and Ralescu (1975). Another approach is that of Benlahcen *et al.* (1977); this uses decomposition into subdomains as in b.γ). Lastly, Neff and Kandel (1977) have proposed a very fast algorithm that generates at once *essential* prime implicants, i.e., those prime implicants whose disjunction realizes a minimal disjunctive form for the fuzzy expression under consideration.

N.B.: 1. Boolean Karnaugh maps have also been extended to deal with fuzzy expressions (see Malvache and Willaeys, 1974; Kandel, 1976b; Schwede and Kandel, 1977).

N.B.: 2. Dually, fuzzy implicates could be sought to build a minimal conjunctive form for a fuzzy expression (see Davio and Thayse, 1973; Negoita and Ralescu, 1975).

d. Analysis and Synthesis of Fuzzy Expressions

These problems were investigated by Marinos (1969). To analyze a fuzzy expression f is to find a range for each of its variables such that $f(x) \in [\alpha, \beta[\subset [0, 1]$, where $x = (x_1, \ldots, x_n)$. f is assumed to be in conjunctive or disjunctive form. For both forms, Marinos proposed automatic rules for stating the conditions that the variables must satisfy. These conditions can be separated into two dual groups, one of which corresponds to $f(x) \geq \alpha$, the other to $f(x) < \beta$.

The converse problem, i.e., find f from knowledge of the ranges of the variables and of $[\alpha, \beta[$, is called synthesis. The structure of the fuzzy expression crucially depends on that of the groups of conditions. Marinos gives a method for simulating a fuzzy expression using analogue devices. A detailed presentation of this method is provided in Negoita and Ralescu (1975) and Kaufmann's (1975) book, in which numerous examples are discussed.

A similar attempt is that of Srini (1975) who realized fuzzy expressions by means of networks of electronic binary switching devices. He used values between 0 and 1 in the form

$$x = \sum_{i=1}^{\infty} p_i 2^{-i}, \qquad p_i \in \{0, 1\},$$

and then approximated as $\sum_{i=1}^{n} p_i 2^{-i}$. The p_i are inputs and outputs of the binary devices.

e. Detection of Hazards

Let f_B be a Boolean expression. Consider the set $F(f_B)$ of all fuzzy expressions f compatible with f_B, i.e., such that $\forall a \in \{0, 1\}^n$, $f_B(a) = f(a)$. Davio and Thayse pointed out that $F(f_B)$ was a lattice for $\vee$, $\wedge$ (see Kameda and Sadeh, 1977), more specifically a sublattice of the lattice of fuzzy expressions. This sublattice is, of course, distributive. The following result holds (Kameda and Sadeh, 1977).

Let f_m and f_M be the canonical conjunctive and disjunctive forms of f_B, then f_m and f_M are the minimal and maximal elements of $F(f_B)$, respectively.

Mukaidono (1975b) showed that $F(f_B)$ was also a complete distributive lattice in the sense of the ambiguity relation A (see b.α). The minimal element is the disjunction of all prime implicants of f_B. The maximal element is f such that $f \in F(f_B)$ and $f(a) = \frac{1}{2}$ $\forall a \in \{0, \frac{1}{2}, 1\}^n - \{0, 1\}^n$. Davio and Thayse (1973) gave a binary parametric representation of $F(f_B)$ which led to the design of a logic module capable of realizing any fuzzy expression compatible with a given Boolean expression. They also hinted that $F(f_B)$ could model the possible transient behavior of a switching circuit realizing f_B.

Kandel (1974a), then Hughes and Kandel (1977) indeed used fuzzy switching logic to detect hazard in combinatorial switching circuits. Formerly, the mathematical tool for hazard detection was ternary logic (see, for instance, Mukaidono (1972) and Kandel (1974a) for a bibliography).

Two binary vectors $a = (a_1, \ldots, a_n)$ and $b = (b_1, \ldots, b_n)$ in $\{0, 1\}^n$ are said to be adjacent iff $\exists! j \in \{1, \ldots, n\}$ such that $b = (a, \ldots, a_{j-1}, \neg a_j, a_{j+1}, \ldots, a_n)$. The device under study is assumed to have n inputs and one output whose value is $f(a)$ when the inputs are $a_1, \ldots, a_n$. f is a fuzzy expression. We consider the case when an input j switches from a_j to $\neg a_j \in \{0, 1\}$. A transient value of j is $t_j \in]0, 1[$ and $t = (a_1, \ldots, a_{j-1}, t_j, a_{j+1}, \ldots, a_n)$. Assume the switching of input j does not modify the steady state of the output, i.e., $f(a) = f(b)$. t is then called a hazard iff $f(t) \neq f(a) = f(b)$, i.e., the output is not steady during the transient phase. If $f(a) = f(b) = 1$, t is said a 1-hazard; if $f(a) = f(b) = 0$, it is said a 0-hazard.

Now $f(x)$ can be written

$$f(x) = \left(x_j \wedge f_1(x^j) \right) \vee \left(\left(\neg x_j \right) \wedge f_2(x^j) \right) \vee \left(\left[\neg x_j \right] \wedge x_j \wedge f_3(x^j) \right) \vee f_4(x^j)$$

$$(1)$$

with $x^j = (x_1, \ldots, x_{j-1}, x_{j+1}, \ldots, x_n)$ and f_1, f_2, f_3, f_4 fuzzy expressions of $n - 1$ variables.

The following theorems, due to Kandel (1973c), give conditions on the steady input states for the output to be disturbed during the switching of input j:

t is a 1-hazard   iff $f_1(a^j) = f_2(a^j) = 1$ and $f_4(a^j) = 0$.

t is a 0-hazard   iff $f_1(a^j) = f_2(a^j) = f_4(a^j) = 0$ and $f_3(a^j) = 1$.

Proof: Assume t is a 1-hazard; $a_j = 1$.
Initial state: $a_j = 1$, hence $f_1(a^j) \vee f_4(a^j) = 1$.
Final state: $a_j = 0$, hence $f_2(a^j) \vee f_4(a^j) = 1$.
Transition state: by hypothesis $f(t) \in \,]0, 1[$. Hence, from (1), $f_4(a^j) \neq 1$, i.e., $f_4(a^j) = 0$, which yields the result. For 0-hazards the proof is very similar and omitted. The "if" parts of the theorems are obvious.   Q.E.D.

Lastly, Hughes and Kandel (1977) generalized this approach to detect hazards when several inputs switch simultaneously.

B.   MULTIVALENT LOGICS

Three fuzzy set theories were presented in Chapter 1 of Part II: $(\tilde{\mathcal{P}}(X), \cup, \cap, \bar{\ })$, $(\tilde{\mathcal{P}}(X), \hat{+}, \cdot, \bar{\ })$, and $(\tilde{\mathcal{P}}(X), \cup, \cap, \bar{\ })$. Multivalent logics, which are bases for these set theories and some others, are the topic of this section. We consider here, from a semantic point of view, only indenumerably valued logics whose truth space is the real interval $[0, 1]$. We are not concerned here with the "fuzzification" of binary and finite multivalued logical calculi in the sense of Pinkava (1976) (i.e., to get a functionally complete logical calculus with "generalized" connectives). The exposition uses Piaget's group of transformations, which is first reviewed. A general survey of multivalued logics can be found in Rescher (NF 1969).

a.   Piaget's Group

Let Φ be a propositional variable containing elementary propositions $P, Q, R \ldots$ joined with logical connectives. Φ is a wff symbolically written $\Phi = f(P, Q, R, \ldots)$.

Four transformations can be defined on Φ:

(1) *identity*: $I(\Phi) = \Phi$;
(2) *negation*: $N(\Phi) = \neg\, \Phi$;
(3) *reciprocity*: $R(\Phi) = f(\neg\, P, \neg\, Q, \neg\, R, \ldots)$;
(4) *correlativity*: $C(\Phi) = \neg\, R(\Phi)$;

where $\neg$ denotes the unary connective for negation of a proposition.

These transformations, for a function compositional law, have a Klein group structure whose table is given in Fig. 1

	I	N	R	C
I	I	N	R	C
N	N	I	C	R
R	R	C	I	N
C	C	R	N	I

Figure 1

Piaget showed that, for children, learning of human reasoning demands a perception of these transformations, that is, understanding the difference between sentences such as

$$\Phi = \text{“Good poets are bad husbands.”}$$

$$N(\Phi) = \text{“Good poets are not bad husbands.”}$$

$$R(\Phi) = \text{“Bad poets are good husbands.”}$$

$$C(\Phi) = \text{“Bad poets are not good husbands.”}$$

The mathematical formalization of this group of transformations can be found for instance in "Piaget's theory of development: The main stages" by Hermine Sinclair (in Murray, NF 1972, pp. 68–78).

Let us make explicit the link between these transformations and binary

P	Q	$\overset{\bullet}{PQ}$	$P \vee Q$	$Q \to P$	P	$P \to Q$	Q	$P \leftrightarrow Q$	$P \wedge Q$
1	1	1	1	1	1	1	1	1	1
1	0	1	1	1	1	0	0	0	0
0	1	1	1	0	0	1	1	0	0
0	0	1	0	1	0	1	0	1	0

P	Q	$P\vert Q$	$P \text{ ex } Q$	$\neg Q$	$Q \rightsquigarrow P$	$\neg P$	$P \rightsquigarrow Q$	$P \downarrow Q$	$\overset{\circ}{PQ}$
1	1	0	0	0	0	0	0	0	0
1	0	1	1	1	1	0	0	0	0
0	1	1	1	0	0	1	1	0	0
0	0	1	0	1	0	1	0	1	0

Figure 2

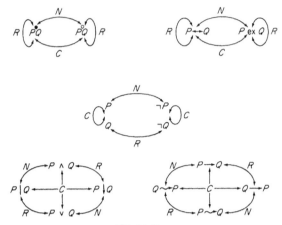

Figure 3

connectives in the case of the binary propositional calculus. The truth tables of the 16 standard binary connectives are given in Fig. 2, where $\cdot$ denotes tautology, $\vee$ disjunction, $\rightarrow$ implication, $\leftrightarrow$ equivalence, $\wedge$ conjunction, $|$ is Sheffer's connective, ex denotes exclusive disjunction, $\downarrow$ is Peirce's connective, and $\circ$ denotes contradiction. $\rightsquigarrow$ has no common name.

These 16 connectives are exchanged through $I, R, N,$ and C as shown in Fig. 3.

b. Multivalent Logics Associated with Fuzzy Set Theories

The semantic truth functions of the three multivalent logics underlying the three fuzzy set theories $(\tilde{\mathcal{P}}(X), \cup, \cap, \bar{\ }), (\tilde{\mathcal{P}}(X), \cup, \cap, \bar{\ }), (\tilde{\mathcal{P}}(X), \hat{+}, \cdot, \bar{\ })$ are now given. Let us denote by $v(P)$ the truth value of a proposition P, $v(P) \in [0, 1]$.

In the three cases, the valuation of the negation is $v(\neg P) = 1 - v(P)$. Hence, $v(\neg \neg P) = v(P)$.

In the three cases, the implication connective $\rightarrow$ is always defined as $v(P \rightarrow Q) = v(\neg P \vee Q)$ and the equivalence as $v(P \leftrightarrow Q) = v[(P \rightarrow Q) \wedge (Q \rightarrow P)]$; ex, $|, \downarrow,$ and $\rightsquigarrow$ are expressed as the negation of $\leftrightarrow, \wedge, \vee,$ and $\leftarrow$, respectively; the tautology and the contradiction are defined respectively as:

$$v(\overset{\bullet}{P}) = v(P \vee \neg P); \qquad v(\overset{\circ}{P}) = v(P \wedge \neg P).$$

More generally,

$$v(\overset{\bullet}{PQ}) = v((P \vee \neg P) \vee (Q \vee \neg Q));$$
$$v(\overset{\circ}{PQ}) = v((P \wedge \neg P) \wedge (Q \wedge \neg Q)).$$

These three multivalent logics are extensions of the classical two-valued logic.

α. *Logic Associated with* $(\tilde{\mathscr{P}}(X), \cup, \cap, \bar{\ })$

The disjunction and the conjunction underlying $\cup$ and $\cap$ (see II.1.A) are respectively

$$v(P \vee Q) = \max(v(P), v(Q)), \qquad v(P \wedge Q) = \min(v(P), v(Q)).$$

It is clear that $\vee$ and $\wedge$ are commutative, associative, idempotent, distributive over one another, and do not satisfy the excluded-middle laws in the sense that $v(P \vee \neg P) \neq 1$ and $v(P \wedge \neg P) \neq 0$; moreover, we have

$$v(P \vee (P \wedge Q)) = v(P); \qquad v(P \wedge (P \vee Q)) = v(P)$$

(absorption);

$$v(\neg(P \wedge Q)) = v(\neg P \vee \neg Q); \qquad v(\neg(P \vee Q)) = v(\neg P \wedge \neg Q)$$

(DeMorgan);

$$v\big[(\neg P \vee Q) \wedge (P \vee \neg Q)\big] = v\big[(P \wedge Q) \vee (\neg P \wedge \neg Q)\big]$$

(equivalence);

$$v\big[(\neg P \wedge Q) \vee (P \wedge \neg Q)\big] = v\big[(P \vee Q) \wedge (\neg P \vee \neg Q)\big]$$

(exclusive disjunction).

Figure 4 gives the valuation of the 16 connectives that have been introduced with $v(P) = p$ and $v(Q) = q$.

P Q	$\overset{\bullet}{P}Q$	$P \vee Q$	$Q \to P$	P
p q	$\max(p, 1-p, q, 1-q)$	$\max(p, q)$	$\max(p, 1-q)$	p

P Q	$P \to Q$	Q	$P \leftrightarrow Q$	$P \wedge Q$
p q	$\max(1-p, q)$	q	$\min(\max(1-p, q), \max(p, 1-q))$	$\min(p, q)$

P Q	$P \mid Q$	P ex Q	$\neg Q$	$Q \nrightarrow P$
p q	$\max(1-p, 1-q)$	$\max(\min(1-p, q), \min(p, 1-q))$	$1-q$	$\min(p, 1-q)$

P Q	$\neg P$	$P \nrightarrow Q$	$P \downarrow Q$	$\overset{\circ}{P}Q$
p q	$1-p$	$\min(1-p, q)$	$\min(1-p, 1-q)$	$\min(p, 1-p, q, 1-q)$

Figure 4

Valuations for quantifiers are straightforwardly defined (coherently with $\wedge$ and $\vee$) as

$$v(\forall x P(x)) = \inf_x (v(P(x))), \qquad v(\exists x P(x)) = \sup_x (v(P(x)))$$

where x denotes an element of the universe of discourse.

This multivalent logic is usually called K-standard sequence logic (K-SEQ), first developed by Dienes. This logic is compatible with Piaget's group of transformations in the sense of Fig. 3.

Moreover, we have the following properties:

for implication, $v[P \rightarrow (Q \rightarrow R)] = v[(P \wedge Q) \rightarrow R]$;
for tautology and contradiction,

$$v(P \rightarrow P) = v(\overset{\bullet}{P}), \qquad v(\overset{\bullet}{P} \rightarrow P) = v(P);$$

$$v(P \rightarrow \overset{\bullet}{P}) = v(\overset{\bullet}{P}); \qquad v(P \leftrightarrow P) = v(\overset{\bullet}{P});$$

$$v(\overset{\circ}{P} \rightarrow P) = v(\overset{\bullet}{P}); \qquad v(P \rightarrow \overset{\circ}{P}) = v(\neg P);$$

$$v(P \leftrightarrow \neg P) = v(\overset{\circ}{P});$$

for Sheffer's and Peirce's connectives,

$$v(\neg P) = v(P|P); \qquad v(P \rightarrow Q) = v(P|(Q|Q));$$

$$v(\overset{\bullet}{P}) = v(P|(P|P)).$$

Sheffer's connective alone (or Peirce's) is sufficient to build every binary and unary connective in standard binary logic. This result remains valid for the "extended" connectives of K-SEQ.

The implication $\rightarrow$ is clearly related to the weak set inclusion introduced in II.1.E.c.α, and ex to the symmetrical difference $\triangle$ (II.1.B.f).

Gaines (1976b) has shown that this multivalued logic was nothing but the fuzzification (in the sense of the extension principle) of standard propositional calculus. Each proposition P is associated with a normalized fuzzy set on $\{0, 1\}$, i.e. a pair ($\mu_P(0)$, $\mu_P(1)$) where $\mu_P(0)$ may be interpreted as a degree of falsity and $\mu_P(1)$ as a degree of truth. Since the logical connectives of the standard propositional calculus are truth functional, i.e., may be represented as functions, they can be fuzzified. Defining $v(P) = (1 - \mu_P(0) + \mu_P(1))/2 \in [0, 1]$, Gaines gets the multivalued logic described above. For proofs, the reader is referred to Gaines (1976a, b). Lastly, this multivalued logic is basically trivalent in the sense that when two wffs, built with the above connectives, coincide on $\{0, \frac{1}{2}, 1\}$, then they coincide on $[0, 1]$ (see A.b.γ).

β. Logic Associated with $(\tilde{\mathscr{P}}(X), \cup, \cap, \bar{\ })$

The disjunction and the conjunction underlying $\cup$ and $\cap$ (see II.1.B.e) are respectively

$$v(P \vee Q) = \min(1, v(P) + v(Q)),$$
$$v(P \wedge Q) = \max(0, v(P) + v(Q) - 1).$$

It is clear that $\vee$ and $\wedge$ are commutative, associative, but are not idempotent and not distributive over one another; they satisfy

$$v(\neg(P \wedge Q)) = v(\neg P \vee \neg Q);$$
$$v(\neg(P \vee Q)) = v(\neg P \wedge \neg Q) \qquad \text{(De Morgan)}$$
$$v(P \vee \neg P) = 1; \qquad v(P \wedge \neg P) = 0 \qquad \text{(excluded-middle laws)}.$$

Figure 5 gives the valuation of the 16 connectives that have been introduced ($v(P) = p$; $v(Q) = q$). (To avoid confusion, $\vee$, $\rightarrow$, $\leftrightarrow$, $\wedge$, $|$, ex, $\rightsquigarrow$, $\downarrow$ are denoted in this logic $\vee$, $\Rightarrow$, $\Leftrightarrow$ $\wedge$, $\|$, (ex), $\Rrightarrow$, $\Downarrow$.)

P Q	$\overset{\bullet}{PQ}$	$P \vee Q$	$Q \Rightarrow P$	P
p q	1	$\min(1, p + q)$	$\min(1, p + 1 - q)$	p

P Q	$P \Rightarrow Q$	Q	$P \Leftrightarrow Q$	$P \wedge Q$
p q	$\min(1, 1 - p + q)$	q	$1 - \lvert p - q \rvert$	$\max(0, p + q - 1)$

P Q	$P \| Q$	P (ex) Q	$\neg Q$	$Q \Rrightarrow P$
p q	$\min(1, 1 - p + 1 - q)$	$\lvert p - q \rvert$	$1 - q$	$\max(0, p - q)$

P Q	$\neg P$	$P \Rrightarrow Q$	$P \Downarrow Q$	$\overset{\circ}{PQ}$
p q	$1 - p$	$\max(0, q - p)$	$\max(0, 1 - p - q)$	0

Figure 5

This logic is compatible with Piaget's group of transformations in the sense of Fig. 3.

Moreover, we have the following properties for tautology and contradiction:

$$v(P \Rightarrow P) = v(\overset{\bullet}{P}); \qquad v(\overset{\bullet}{P} \Rightarrow P) = v(P);$$
$$v(P \Rightarrow \overset{\bullet}{P}) = v(\overset{\bullet}{P}); \qquad v(P \Leftrightarrow P) = v(\overset{\bullet}{P});$$
$$v(\overset{\circ}{P} \Rightarrow P) = v(\overset{\bullet}{P}); \qquad v(P \Rightarrow \overset{\circ}{P}) = v(\neg P).$$

The implication $\Rightarrow$ is clearly related to the usual inclusion (in the sense of Zadeh) for fuzzy sets (II.1.E.a). $\widehat{\text{ex}}$ and $\twoheadrightarrow$ correspond respectively to the set operators $\triangledown$ (symmetrical difference) and $|-|$ (bounded difference) introduced in II.1.B.f. Lastly, we have

$$v(P \to Q) = v(\neg P \vee (P \wedge Q))$$

and

$$v(P \Rightarrow Q) = v(\neg P \vee (P \wedge Q)).$$

γ.   *Logic Associated with* $(\widetilde{\mathscr{P}}(X), \dot{+}, \cdot, \bar{\ })$

The disjunction and the conjunction underlying $\dot{+}$ and $\cdot$ (see II.1.B.e) are respectively

$$v(P \gamma Q) = v(P) + v(Q) - v(P) \cdot v(Q);$$
$$v(P \,\&\, Q) = v(P) \cdot v(Q).$$

It is clear that γ and $\&$ are commutative, associative, but are not idempotent and not distributive over one another; they satisfy

$$v(\neg(P \,\&\, Q)) = v(\neg P \gamma \neg Q);$$
$$v(\neg(P \gamma Q)) = v(\neg P \,\&\, \neg Q) \qquad \text{(De Morgan)}.$$

Although it is easy to build the valuation of the 16 connectives as in both preceding logics, we give only some of them for the sake of briefness, with $v(P) = p$, $v(Q) = q$:

implication: $v(P \leftrightarrow Q) = 1 - p + pq$;
tautology: $v(\dot{P}) = 1 - p(1 - p)$;
contradiction: $v(\overset{\circ}{P}) = p(1 - p)$.

Note also we have the hybrid formulas

$$v(P \leftrightarrow Q) = v(\neg P \vee (P \,\&\, Q)) \qquad \text{and}$$
$$v[(P \leftrightarrow Q) \wedge (Q \leftrightarrow P)] = v[(P \,\&\, Q) \vee (\neg P \,\&\, \neg Q)]$$
$$= v[(P \,\&\, Q) \Leftrightarrow (P \gamma Q)].$$

This logic is compatible with Piaget's group of transformations in the sense of Fig. 3. This logic is often called stochastic logic.

Let us examine in what situations $\&$ coincides with $\wedge$ or $\wedge$ (or γ with $\vee$ or $\vee$). First, note that

$$0 \leqslant \max(0, p + q - 1) \leqslant pq \leqslant \min(p, q),$$
$$1 \geqslant \min(1, p + q) \geqslant p + q - pq \geqslant \max(p, q).$$

Then it is easy to check that

$$v(P \& Q) = v(P \wedge Q),$$

i.e., $pq = \min(p, q)$ iff the truth value of P or of Q is equal to 0 or to 1. $v(P \gamma Q) = v(P \vee Q)$ holds under the same conditions. And it is the same for $v(P \& Q) = v(P \wedge Q)$ and $v(P \gamma Q) = v(P \vee Q)$.

Thus, two of three connectives for conjunction (resp. disjunction) coincide iff the truth value of P or of Q is equal to 0 or to 1. In that case the three connectives for conjunction (resp. disjunction) coincide.

Remark We have the following inequalities:

$$v(P) + v(Q) = v(P \wedge Q) + v(P \vee Q) = v(P \wedge Q) + v(P \vee Q)$$
$$= v(P \& Q) + v(P \gamma Q),$$

i.e., v is a valuation (Birkhoff, NF 1948) in the lattice sense for the three logics.

c. Other Multivalent Logics; Other Implications

Assembling the already introduced semantic truth functions differently, other multivalent logics may be defined, for instance Lukasiewicz logic. In Lukasiewicz logic the semantic truth functions for conjunction, disjunction, and quantifiers are those of K-SEQ (i.e., $v(P \wedge Q) = \min(v(P), v(Q))$; $v(P \vee Q) = \max(v(P), v(Q))$; $v(\forall x P(x)) = \inf_x v(P(x))$; $v(\exists x P(x)) = \sup_x v(P(x))$. The implication and the equivalence are those of the logic associated with $(\tilde{\mathscr{P}}(X), \cup, \cap, ^-)$ (i.e., $v(P \Rightarrow Q) = \min(1, 1 - v(P) + v(Q))$, $v(P \Leftrightarrow Q) = 1 - |v(P) - v(Q)|$). The negation is classically $v(\neg P) = 1 - v(P)$. This logic is called $L_{\aleph_1}$. $L_{\aleph_1}$ is the multivalent logic underlying Zadeh's ordinary fuzzy set theory, i.e., $\cup$ for union, $\cap$ for intersection, and $\subseteq$ for inclusion (see II.1.B.a and II.1.E.a). This logic is obviously compatible with Piaget's group of transformations in the sense of Fig. 3. However, the link between disjunction and implication is now $v(P \vee Q) = v((P \Rightarrow Q) \Rightarrow Q)$. Similarly to the stochastic implication $v(P \leftrightarrow Q) = 1 - v(P) + v(P \& Q)$, we have here $v(P \Rightarrow Q) = 1 - v(P) + v(P \wedge Q)$. Note that we have also $v(P \to Q) = 1 - v(P) + v(P \wedge Q)$.

Giles (1976a) has proposed an interpretation of $L_{\aleph_1}$ in terms of risk. Every chain of reasoning is seen as a dialogue between speakers whose assertions entail a commitment about their truth.

Other semantic truth functions for implication may be found in the literature:

(1)

$$v\left(P \overset{1}{\rightarrow} Q\right) = \begin{cases} 1 & \text{if } v(P) \leqslant v(Q), \\ 0 & \text{otherwise.} \end{cases}$$

The associated equivalence is

$$v\left(P \overset{1}{\leftrightarrow} Q\right) = \begin{cases} 1 & \text{if } v(P) = v(Q), \\ 0 & \text{otherwise.} \end{cases}$$

With the semantic truth functions of K-SEQ for conjunction, disjunction, negation, and the quantifiers, we get another standard sequence logic, called R-SEQ (see Maydole, 1975).

(2)

$$v\left(P \overset{2}{\rightarrow} Q\right) = \begin{cases} 1 & \text{if } v(P) \leqslant v(Q), \\ v(Q) & \text{otherwise.} \end{cases}$$

The associated equivalence is

$$v\left(P \overset{2}{\leftrightarrow} Q\right) = \begin{cases} 1 & \text{if } v(P) = v(Q), \\ \min(v(P), v(Q)) & \text{otherwise.} \end{cases}$$

The implication $\overset{2}{\rightarrow}$, sometimes called Brouwerian implication, is nothing but the operator α introduced in II.1.G.a and used by Sanchez, (Reference from II.3, 1976) (see II.3.E). With the semantic truth functions of K-SEQ for conjunction, disjunction and the quantifiers, and the negation

$$v(\neg P) = \begin{cases} 1 & \text{if } v(P) = 0, \\ 0 & \text{otherwise,} \end{cases}$$

we get the indenumerably valued Gödelian logic (see Maydole, 1975).

(3) $v(P \overset{3}{\rightarrow} Q) = \max(1 - v(P), \min(v(P), v(Q)))$. Note that

$$v(P \rightarrow Q) = \max(1 - v(P), v(Q)) \neq v(\neg P \vee (P \wedge Q)) = v\left(P \overset{3}{\rightarrow} Q\right).$$

This implication was considered by Zadeh (Reference from III.3, 1973).

(4) $v(P \overset{4}{\rightarrow} Q) = \min(1, v(Q)/v(P))$. This implication was introduced by Goguen (1969). Gaines (1976b) noticed that this implication was closely related to conditional probability since $v(P \overset{4}{\rightarrow} Q) = v(P \wedge Q)/v(P)$.

In order to compare all the introduced implications it should be noticed that the following inequalities hold:

$\forall P, Q,$

$$v(P \Rightarrow Q) \geqslant v(P \leftrightarrow Q) \geqslant v(P \rightarrow Q) \geqslant v\left(P \overset{3}{\rightarrow} Q\right) \geqslant \min(v(P), v(Q));$$

$$\forall P, Q, \quad v(P \Rightarrow Q) \geqslant v\left(P \overset{4}{\rightarrow} Q\right) \geqslant v\left(P \overset{2}{\rightarrow} Q\right) \geqslant \min(v(P), v(Q));$$

$$\forall P, Q, \quad v\left(P \overset{2}{\rightarrow} Q\right) \geqslant v\left(P \overset{1}{\rightarrow} Q\right).$$

Thus, the implication corresponding to Zadeh fuzzy set inclusion has the greatest valuation of the implications introduced.

The implications $\overset{2}{\rightarrow}$, $\overset{3}{\rightarrow}$, $\overset{4}{\rightarrow}$ are not compatible with Piaget's group of transformations in the sense of Fig. 3.

Moreover, Maydole (1975) generates paradoxes for R-SEQ and Godelian logic—which use $\overset{1}{\rightarrow}$ and $\overset{2}{\rightarrow}$, respectively.

d. Detachment Operations; Modus Ponens

The modus ponens rule allows Q to be inferred from P and $P \mapsto Q$ in propositional calculus. In multivalent logics the problem is to compute $v(Q)$ given $v(P)$ and $v(P \mapsto Q)$ where $\mapsto$ is any given multivalent implication. Several authors, especially Goguen (1969), Kling (Reference from IV.2), LeFaivre (Reference from IV.2, 1974a), have looked for a detachment operation * such that

$$v(P)^* v(P \mapsto Q) \leqslant v(Q),$$

to have $v(Q)$ as large as possible. Note that the situation is similar to probabilistic inference where if $P(A) \geqslant \alpha$ and $P(B|A) \geqslant \beta$, then $P(B) \geqslant \alpha\beta$ since $P(B) = P(B|A)P(A) + P(B|\neg A)P(\neg A)$.

For $v(P \mapsto Q) = v(P \rightarrow Q) = \max(1 - v(P), v(Q))$, * can be the min operation since we have, if $\min(v(P), v(P \rightarrow Q))) > 0.5$,

$$\min(v(P), v(P \rightarrow Q)) \leqslant v(Q) \leqslant \max(v(P), v(P \rightarrow Q)).$$

More precisely, if $v(P) \geqslant \alpha$ and $v(P \rightarrow Q) \geqslant \beta$ with $\alpha + \beta > 1$, then $v(Q) \geqslant \beta$. In particular, if $v(P) > \frac{1}{2}$ and $v(P \rightarrow Q) \geqslant \frac{1}{2}$, then $v(Q) \geqslant \frac{1}{2}$; but if $v(P) \geqslant \frac{1}{2}$ instead of $v(P) > \frac{1}{2}$, then $v(Q)$ is indeterminate. The validity of a chain of implications $\rightarrow$ when * = min is not less than the validity of the least valid element in the chain. $]\frac{1}{2}, 1]$ is called the designated set of K-SEQ.

Moreover, every axiom or theorem in standard propositional calculus has a truth value greater than or equal to $\frac{1}{2}$ when we use the semantic truth functions of K-SEQ. Reciprocally, if a wff has always a truth value greater or equal to $\frac{1}{2}$ in K-SEQ, then it is a theorem in standard propositional calculus.

Proof: Let Φ be a theorem. If $P, Q, R, \ldots$ are elementary propositions involved in Φ, then $\forall v(P), v(Q), v(R), \ldots \in \{0, 1\}, v(\Phi) = 1$. Let us assume $\exists p^*, q^*, r^*, \ldots$ such that $v(\Phi) = f(p^*, q^*, t^*, \ldots) < \frac{1}{2}$. f can be stated as a conjunction of disjunctions (i.e., of clauses). From the assumption, one of the clauses is strictly smaller than $\frac{1}{2}$. Thus if p is involved in this clause, $1 - p$ is not and conversely. We set to 0 every elementary truth value in the clause—which becomes null—and we give arbitrary truth values 0 or 1 to the other elementary propositions. Thus, we get $v(\Phi) = 0$, which contradicts the assumption. The converse is obvious.   Q.E.D.

Using K-SEQ, R. C. T. Lee (1972) proved that if the most reliable clause of a given set of clauses has truth value a and the most unreliable clause has truth value b, then all the logical consequences obtained by repeatedly applying the resolution principle (see, e.g., Robinson, NF 1965) will have a truth value between a and b.

If we use Goguen's implication $\overset{4}{\rightarrow}$, a detachment operation is now the product since

$$v\left(P \overset{4}{\rightarrow} Q\right) \cdot v(P) = \min(1, v(Q)/v(P)) \cdot v(P) \leqslant v(Q).$$

The validity of a chain of implications $\mapsto$, when $*$ is the product, may decrease with the length of the chain.

Lastly, with $v(P \Rightarrow Q) = \min(1, 1 - v(P) + v(Q))$, we may observe that:

if $v(P) = \alpha$ and $v(P \Rightarrow Q) = 1$, then $v(Q) \geqslant \alpha$;
if $v(P) = \alpha$ and $v(P \Rightarrow Q) = 1 - \epsilon < 1$, then $v(Q) = \alpha - \epsilon$.

At the end of n inferences whose truth values are equal to $1 - \epsilon$, the truth value of the premise being α, the conclusion has a truth value equal to $\alpha - n\epsilon$. A detachment operation for $\Rightarrow$ is $\wedge$. Gaines (1976b) uses $\overset{4}{\rightarrow}$ or $\Rightarrow$ to explain the paradox of the bald man (if a man who has n hairs is bald, then a man who has $n + 1$ hairs is still bald).

N.B.:   Conversely, given an operation $*$, the appropriate formal definition of $\mapsto$ is $v(P \mapsto Q) = \sup_x \{x, v(P) * x \leqslant v(Q)\}$. Note that when $* = \min$, $\mapsto$ is the Brouwerian implication $\overset{2}{\rightarrow}$; when $* = \text{product}$, $\mapsto = \overset{4}{\rightarrow}$.

e. Comments

With so many multivalent logics and connectives, we may need some points of view for comparison; perhaps the most interesting ones are the compatibility with Piaget's group, the existence of paradoxes, the presence or lack of important structural properties, the validity of a chain of inference, and the associated set theories.

Many authors have used the expression "fuzzy logic" to denominate some multivalent logics, especially $L_{\aleph_1}$ which underlies Zadeh fuzzy set theory. Zadeh employs "fuzzy logic" to designate a logic on which a theory of approximate reasoning is based (see Section E). However, multivalent logics may be viewed as fuzzy logics in the sense that there are no longer only crisp truth values like 0 or 1, but also intermediate ones. Lakoff (1973) generalizes this point of view when he proposes assigning to each proposition a 3-tuple (α, β, γ) such that $\alpha + \beta + \gamma = 1$ and where α, β, γ are interpreted as degrees of truth, falsity, and nonsense, respectively. (If $v(P) = (\alpha, \beta, \gamma)$, then $v(\neg P) = (\beta, \alpha, \gamma)$.)

C. FUZZY MODAL LOGIC

Until now there have been very few works in the domain of fuzzy modal logics. Perhaps this is because "possibility" has been investigated in another way by Zadeh (see Part II, Chapter 5). Thus, this section will be very short, just providing what has been done. For an introduction to modal logic, the reader may consult Hughes and Cresswell (NF 1972).

Lakoff (1973) obtained a fuzzy modal logic by adding to a set of semantic truth functions for connectives and quantifiers the following valuations for the modal operators $\Box$ and $\Diamond$:

$$v(\Box P, w) = \inf_{wRw'} v(P, w'), \qquad v(\Diamond P, w) = \sup_{wRw'} v(P, w'),$$

$w, w' \in W$, and where $v(P, w)$ is the truth value of P in the world w, and R is an alternativeness (or accessibility) reflexive relation between the "possible worlds." W is the set of "possible worlds." Note that the valuations are coherent with the identity $\Box P = \neg \Diamond \neg P, (v(\neg P, w) = 1 - v(P, w))$. $v(\Box P, w)$ is interpreted as the degree of necessary truth of P in w; $v(\Box P, w) = \alpha$ means that the truth value of P never falls below α in any world alternative to w. Lakoff gives the following example of a statement that is necessarily true to a degree: "Approximately half of the prime numbers are of the form $4N + 1$."

Schotch (1975) has applied fuzzy set theory to modal logic in the following way. Let us consider the relational model consisting of a binary relation R on W (intuitively the set of possible worlds) and a valuation V that assigns to each elementary proposition P the set of worlds in which P is true. By definition, we have

$$V(\neg P) = W - V(P), \qquad V(P \wedge Q) = V(P) \cap V(Q)$$

$$V(\Diamond P) = \{w \in W, wRw' \text{ and } w' \in V(P)\},$$

$$V(\Box P) = \{w \in W, wRw' \text{ implies } w' \in V(P)\},$$

which is coherent with $\Box P = \neg \Diamond \neg P$.

This model can be fuzzified in two ways: using fuzzy valuations and/or fuzzy relations.

First, let us consider the case of a fuzzy valuation V $\tilde{V}$ assigns to each elementary proposition P the fuzzy set $\tilde{V}(P)$ of worlds in which P is more or less true. $\mu_{\tilde{V}(P)}(w)$ is the degree to which P is true in the world $w \in W$. $\tilde{V}$ is extended to any wff by

$$\tilde{V}(\neg P) = \overline{\tilde{V}(P)} \qquad (\mu_{\tilde{V}(\neg P)}(w) = 1 - \mu_{\tilde{V}(P)}(w)),$$

$$\tilde{V}(P \wedge Q) = \tilde{V}(P) \cap \tilde{V}(Q) \qquad (\mu_{\tilde{V}(P \wedge Q)}(w)) = \min(\mu_{\tilde{V}(P)}(w), \mu_{\tilde{V}(Q)}(w)).$$

$\mu_{\tilde{V}(\Diamond P)}$ is the (two-valued) characteristic function of the set $\{w \in W, wRw' \text{ and } \mu_{\tilde{V}(P)}(w') \neq 0\}$ and $\mu_{\tilde{V}(\Box P)}$ is the two-valued characteristic function of the set $\{w \in W, wRw' \text{ implies } \mu_{\tilde{V}(P)}(w') = 1\}$.

Let us suppose now that the valuation is no longer fuzzy but that R is a fuzzy relation. $V(\Diamond P)$ is now defined as

$$V(\Diamond P) = \{w \in W, \mu_R(w, w') = 1 \quad \text{and} \quad w' \in V(P)\}.$$

Several kinds of stipulations may be imposed on R according to the classical modal system we want. Moreover, another modal operator, denoted M, may be defined:

$$V(MP) = \{w \in W, \mu_R(w, w') \neq 0 \quad \text{and} \quad w' \in V(P)\}.$$

MP means "it might be possible that P." Then we have $\Diamond P \mapsto MP$; $\neg M \neg P \mapsto \Box P$. Note that $MP \neq \Diamond \Diamond P$.

More generally, we may consider more baroque models where the valuation and the relation are fuzzy (see Schotch, 1975).

Remark Dana Scott has suggested (see Lakoff, 1973) a method for relating modal and many-valued logics. Let $V(P, i) = 1$ stand for "P is true in the valuation i," i.e., $v(P) \geqslant i$, $i \in [0, 1]$. The alternativeness relation R is here $\geqslant$. The set of valuations is constrained by: if $V(P, i) = 1$, then $\forall j$,

$i \geqslant j$, $V(P, j) = 1$. Valuations for $\neg$, $\wedge$, $\vee$ are now defined as

$$V(\neg P, i) = 1 \qquad \text{iff} \qquad \text{not}(V(P, 1 - i) = 1)$$

$$V(P \wedge Q, i) = 1 \qquad \text{iff} \qquad V(P, i) = 1 \quad \text{and} \quad V(Q, i) = 1,$$

$$V(P \vee Q, i) = 1 \qquad \text{iff} \qquad V(P, i) = 1 \quad \text{or} \quad V(Q, i) = 1.$$

D. FUZZY-VALUED LOGICS

A fuzzy-valued logic is a many-valued logic where the truth space is the set of the fuzzy numbers (i.e., convex normalized piecewise continuous fuzzy sets) on the real interval $[0, 1]$; i.e., the truth value of a proposition is a fuzzy number whose support is included in $[0, 1]$. Such fuzzy numbers may model linguistic truth values whose names are "true," "very true," "borderline," "false," etc. Figure 6 sketches the shape of their membership functions.

Fuzzy-valued logics clearly underlie type 2 fuzzy set theories where grades of membership are fuzzy numbers (see II.1.G.d and II.2.C.b). Thus, the semantic truth functions for the connectives of negation, conjunction, and disjunction (underlying $\overset{\mathsf{H}}{}$, $\sqcap$, and $\sqcup$ respectively) are

$$\tilde{v}(\neg P) = 1 \ominus \tilde{v}(P), \qquad \tilde{v}(P \wedge Q) = \widetilde{\min}(\tilde{v}(P), \tilde{v}(Q)),$$

$$\tilde{v}(P \vee Q) = \widetilde{\max}(\tilde{v}(P), \tilde{v}(Q)),$$

where $\tilde{v}(P)$ is a fuzzy number on $[0, 1]$. For the definition and methods for rapid computation of $\ominus$, $\widetilde{\min}$, $\widetilde{\max}$, the reader is referred to II.2.B.d and II.2.B.e.

$\tilde{v}(\neg P)$ is generally called the *antonym* of $\tilde{v}(P)$. Thus, "false" will be defined as the antonym of "true."

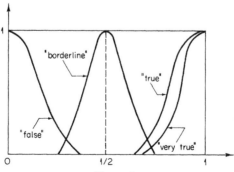

Figure 6

The other semantic truth functions for the connectives introduced in B.b can be extended (by means of the extension principle) in the same way; for instance,

$$\tilde{v}(P \to Q) = \widetilde{\max}(1 \ominus \tilde{v}(P), \tilde{v}(Q)),$$
$$\tilde{v}(P \Rightarrow Q) = \widetilde{\min}(1, 1 \ominus \tilde{v}(P) \oplus \tilde{v}(Q)),$$
$$\tilde{v}(P \vee Q) = \widetilde{\min}(1, \tilde{v}(P) \oplus \tilde{v}(Q)),$$
$$\tilde{v}(P \Leftrightarrow Q) = \text{abs}(\tilde{v}(P) \ominus \tilde{v}(Q))$$

(for $\oplus$ and abs, see II.2.B.d).

It is clear that with these extended valuations, $\vee$ and $\wedge$ are still commutative, associative, idempotent, mutually distributive, and satisfy absorption and De Morgan laws; negation is still involutive; and

$$\tilde{v}(P \wedge Q) \oplus \tilde{v}(P \vee Q) = \tilde{v}(P) \oplus \tilde{v}(Q),$$

provided that $\forall P, Q, \tilde{v}(P), \tilde{v}(Q)$ are fuzzy numbers, i.e., convex and normalized. However,

$$\tilde{v}[(P \wedge Q) \vee (\neg P \wedge \neg Q)] \neq \tilde{v}[(\neg P \vee Q) \wedge (P \vee \neg Q)],$$
$$\tilde{v}[(P \vee Q) \wedge (\neg P \vee \neg Q)] \neq \tilde{v}[(\neg P \wedge Q) \vee (P \wedge \neg Q)].$$

Remark 1   The meaning of "not true and not false" is approximately that of "borderline." However, the membership function of "not true and not false" has a maximum whose value is different from 1. ($\mu_{\text{not true and not false}} = \min(1 - \mu_{\text{true}}, 1 - \mu_{\text{false}})$.) After a renormalization we get a fuzzy number that looks like "borderline." Thus, the classical fuzzy set operations $\cup, \cap, ^-$ can be used in the same way to build new linguistic truth values; on the contrary, $\widetilde{\max}, \widetilde{\min}$, and $1 \ominus (\cdot)$ must be used to valuate composite propositions whose elementary propositions are only fuzzily valued.

2   Other connectives may be worth considering, particularly in a fuzzy-valued logic. For instance, let $\underline{m}$ be the unary connective defined in a multivalent logic by

$$v(\underline{m}P) = [v(P)]^m \qquad m \in \mathbb{R}^+.$$

$\underline{m}$ will be called a modulator because it modulates the affirmation of the proposition P. If $m > 1$, $\underline{m}P$ is a more demanding (stronger) version of P, so its truth value is less than $v(P)$; conversely if $m < 1$, $\underline{m}P$ is a relaxed version of P and has a greater truth value. The extension of $\underline{m}$ to fuzzy-valued logics is rather straightforward because it underlies the mth power operation for a type 2 fuzzy set (see II.2.C.c). For example, for

$m = 2$, $\tilde{v}(2P) = \tilde{v}(P) \odot \tilde{v}(P)$. Notice that if $\tilde{v}(P) = $ true, $\tilde{v}(2P) \neq$ very true where "very true" is modeled by the second power of the type 1 fuzzy set "true" on $[0, 1]$, i.e., $\mu_{\text{very true}}(x) = [\mu_{\text{true}}(x)]^2$, $\forall x \in [0, 1]$ (see II.1.B.f). "Very true" has a mean value equal to 1 but not $\tilde{v}(2P)$. (For a discussion of the modeling of hedges such as "very," see IV.2.B.b.) "Very true" is more precise (less fuzzy) than "true"; $\tilde{v}(2P)$ is less true.

The problem of inference is less straightforward in fuzzy-valued logic than in multivalent logic, i.e., find $\tilde{v}(Q)$ when you know $\tilde{v}(P)$ and $\tilde{v}(P \mapsto Q)$ where $\mapsto$ is some implication connective. For instance, if $\tilde{v}(P) = \tilde{\alpha}$ and $\tilde{v}(P \Rightarrow Q) = v(\neg P \wedge Q) = 1 \ominus \tilde{\epsilon}$, then $\tilde{v}(Q)$ is not a solution of the equation

$$1 \ominus \tilde{\epsilon} = \widetilde{\min}(1, 1 \ominus \tilde{\alpha} \oplus \tilde{x}),$$

which is equivalent to $\tilde{\epsilon} = \tilde{\alpha} \ominus \tilde{x}$ if $\mu_{\tilde{\epsilon}}(0) \neq 0$ because in the equation, when $\tilde{\epsilon}$ is given, the fuzzier $\tilde{\alpha}$ is, the crisper (the less fuzzy) is $\tilde{x}$, and it is counterintuitive that the less precisely defined $\tilde{v}(P)$ is, the more precisely defined is $\tilde{v}(Q)$. The reason is that in fuzzy equations implicit definitions of variables are not equivalent to the corresponding explicit ones (see II.2.B.h)—which are usually the only valid ones. Thus, we must directly fuzzify the nonfuzzy result of the above equation and state $\tilde{v}(Q) = \tilde{\alpha} \ominus \tilde{\epsilon}$.

In conclusion, we notice that in a chain of approximate inferences, truth and precision progress in the same sense, conclusions are always less precise and less true than premises: $\tilde{\alpha} \ominus \tilde{\epsilon}$ is smaller than $\tilde{\alpha}$ (see II.2.B.g) and also more fuzzy.

E. APPROXIMATE REASONING (Zadeh, 1977a)

a. Introduction

"Informally, by *approximate* or, equivalently, *fuzzy reasoning* we mean the process or processes by which a possibly imprecise conclusion is deduced from a collection of imprecise premises. Such reasoning is, for the most part, qualitative rather than quantitative in nature and almost all of it falls outside of the domain of applicability of classical logic" (Zadeh, 1977a).

In Section B we were interested in manipulating statements such as $P \equiv$ "$X \in \hat{A}$" (X is a prescribed element of U) where $\hat{A}$ is a fuzzy subset on a universe U and $v(P) = \mu_{\hat{A}}(X)$ (for example, $P \equiv$ "John is a tall man," i.e., John belongs to the fuzzy set $\hat{A}$ of tall men). In section D $v(P)$ was allowed to be a fuzzy number and was denoted $\tilde{v}(P)$. In this section we consider statements like $P \equiv$ "X is A" where A is a fuzzy set on T inducing a possibility distribution (see II.5.B.a) $\pi_{h(X)} = \mu_A$. h is an attribute of X

and T is the measurement scale of h. For example, in "X is tall," $A = $ "tall" is modeled by a fuzzy set on the universe T of heights. In fact, the statement can be viewed as equivalent to an infinity of statements $P_t \equiv$ "t is the height of X" with $v(P_t) = \mu_A(t)$, $t \in T$, since t belongs to the fuzzy set A of large heights. If t is a fuzzy height $\tilde{t}$ (for instance, "approximately 5'"), $v(P_{\tilde{t}}) = \Pi(\tilde{t}) = \text{hgt}(A \cap \tilde{t})$ where Π is the possibility measure associated with μ_A. In the following the attribute symbol is omitted, and we write π_X instead of $\pi_{h(X)}$, for short.

In order to perform approximate reasoning with statements similar to "X is A," but more complex, we need translation rules so as to model them as possibility distributions, modifier rules in order to perhaps transform them in semantically equivalent possibility distributions, and rules of inference to deduce new possibility distributions. We are not interested here in the question of retranslating these possibility distributions in natural language; for this problem, called "linguistic approximation," see IV.2.B.e. This approach was initiated by Bellman and Zadeh (1977) and developed by Zadeh (1977a).

b. Translation Rules

By *translation rules* is meant a set of rules that yield the translation of a modified composite proposition from the translations of its constituents, e.g., from $P \to \pi_X = \mu_A$ and $Q \to \pi_Y = \mu_B$ deduce $P \wedge Q \to \pi_{(X, Y)}$. The translation of a proposition must be understood as its associated possibility distributions. There are four types of translation rules.

Type 1: modifier rules for simple propositions: Given the proposition $P \equiv$ "X is A" such that $\pi_X = \mu_A$, find $\pi'_X = \mu_{A^*}$ related to "X is mA" where m is a modifier such as "not," "very," "more or less," . . . ; A^* is given by $A^* = mA$. Each modifier is related to a function f such that $\mu_{A^*} = f \circ \mu_A$ ("not": $f(x) = 1 - x$, "very": $f(x) = x^2$, "more or less": $f(x) = \sqrt{x}$ —see IV.2.B.b).

Type 2: composition rules: Composition rules pertain to the translation of a proposition P that is a composition of propositions Q and R, such as conjunction, disjunction, implication. For instance: "If X is A, then Y is B" $\to$ $\pi_{(X, Y)} = \mu_{c(\overline{A}) \cup c(B)}$ if we employ for the implication $v(Q \Rightarrow R) = v(\neg Q \vee R)$. ($\mu_{c(\overline{A}) \cup c(B)}(t, t') = \min(1, 1 - \mu_A(t) + \mu_B(t'))$ where c denotes the cylindrical extension and $t \in T$, $t' \in T'$; T and T' are the respective universes of A and B.)

Type 3: quantification rules: These rules work on propositions of the form $P \equiv$ "FX are A" where F is a fuzzy quantifier (e.g., "most," "many," "few," "some," . . .). F is a fuzzy set on $[0, 1]$ usually. F indicates a fuzzy proportion. There arise two problems which will be dealt with separately.

First, knowing A and F, fuzzy sets respectively on T and on $[0, 1]$, find the possible density functions that are compatible with the statement "FX are A." Let ρ be a density function over the universe T of A. The proportion Prop(A) of Xs that are A is given by the cardinality of A using $\int_{-\infty}^{t} \rho(s)\,ds$ as a measure on T (see II.1.D.a):

$$\text{Prop}(A) = \int_{T} \mu_A(s)\rho(s)\,ds.$$

In fact, the proportion of Xs that are A is fuzzily restricted by F; hence we can induce a fuzzy restriction on the density functions by stating (Zadeh, 1977a)

$$\pi(\rho) = \mu_F\left(\int_{T} \mu_A(s)\rho(s)\,ds\right).$$

$\pi(\cdot)$, the translation of "FX and A," is a possibility distribution on the density functions.

When the universe U to which X belongs is finite—and sufficiently small—we may not use a density function, but directly induce a possibility distribution on the membership values of the elements of U in the fuzzy set $\hat{A}$ of Xs that are A:

$$\mu_F\left(\frac{\sum_U \mu_{\hat{A}}(X)}{|U|}\right) = \pi(\hat{A})$$

(Bellman and Zadeh, 1977).

The second problem is to find F from knowledge of a density function ρ on $T = \mathbb{R}$ made out of a set of measurements $\{h(X), X \in U\}$ and of a fuzzy bound $B \in \tilde{\mathcal{P}}(\mathbb{R})$. The question is, What is the fuzzy proportion F of Xs such that $h(X)$ is greater than or equal to B? For instance, U is a set of men and $h(X)$ the height of X, B is a fuzzy height. F is given by the integral of ρ over the fuzzy interval $(B, +\infty)$ (see II.4.C.b):

$$\mu_F(z) = \sup_{t,\, \int_t^{+\infty} \rho(s)\,ds = z} \mu_B(t).$$

Note that the fuzzy interval $(B, +\infty)$ corresponds to the fuzzy set A in the statement "FX are A."

More generally, we can translate propositions like "FX in C are A" where C is a fuzzy set on U acting as a fuzzy restriction on the values of X. For instance, $F = $"many," C is the fuzzy set of the tall men, A means "fat": "many tall men are fat." We are interested in the proportion of X that are A in C, i.e. (see II.1.E.c.β)

$$|\hat{A} \cap C|/|C| = \text{Prop}(\hat{A} \text{ in } C).$$

The associated possibility distribution is

$$\pi(\hat{A}) = \mu_F(|\hat{A} \cap C|/|C|).$$

Type 4: qualification rules: Among pertinent qualifications for propositions Zadeh (1977a) considered three of them in particular:

linguistic truth qualification,
linguistic probability qualification,
linguistic possibility qualification.

(i) *Truth qualification.* A truth-qualified version of a proposition such as "X is A" is a proposition expressed as "X is A is τ" where τ is a linguistic truth value. We must not confuse τ with the linguistic truth value of a proposition in a fuzzy-valued logic. Here τ is a *local* linguistic truth-value (see Bellman and Zadeh, 1977) rather than an absolute one: τ is defined as the degree of compatibility of the proposition "X is A" with a reference proposition "X is R" (see II.2.A.e.β):

$$\mu_\tau(z) = \sup_{z = \mu_A(t)} \mu_R(t).$$

Here we want to find R from knowledge of A and τ; the greatest R is

$$\mu_R = \mu_\tau \circ \mu_A.$$

The translation of the proposition "X is A is τ" is thus the possibility distribution induced by R. Note that when τ is defined as $\mu_\tau(z) = z \ \forall z \in [0, 1]$ (Zadeh calls such a truth value "u-true"), we have $R = A$.

(ii) *Probability qualification.* A probability-qualified version of a proposition such as "X is A" is a proposition expressed as "X is A is λ" where λ is a linguistic probability value such as "likely," "very likely," ... . This may be interpreted as "$P(A)$ is λ" where A is viewed as a fuzzy event whose probability is $P(A)$. Using the definition of II.5.C.a, we get

$$P(A) = \int_T \mu_A(t)p(t)\,dt$$

where p is a probability distribution. Since $P(A)$ is fuzzily restricted by λ, the probability distribution is fuzzily restricted by the possibility distribution

$$\pi(p) = \mu_\lambda\left(\int_T \mu_A(t)p(t)\,dt\right)$$

This result is formally equivalent to that of the quantification rule (type 3).

The problem of finding λ from knowledge of the fuzzy event A and the probability distribution p was already solved in II.5.C.b, provided that A is modeled as a fuzzy real interval (B, C).

(iii) *Possibility qualifications.* A possibility-qualified version of a proposition such as "X is A" is a proposition expressed "X is A is ω" where ω is a linguistic possibility value such as "possible," "very possible," "almost impossible," .... ω is viewed as a fuzzy restriction on the nonfuzzy possibility values $\Pi(A)$ of the fuzzy event A. Recalling that

$$\Pi(A) = \sup_{t \in T} \min(\mu_A(t), \mathrm{pi}(t))$$

where $\mathrm{pi}(\cdot)$ is the possibility distribution associated with the possibility measure $\Pi(\cdot)$, we get

$$\pi(\mathrm{pi}) = \mu_\omega\left(\sup_{t \in T} \min(\mu_A(t), \mathrm{pi}(t))\right).$$

The translation of "X is A is ω" is a possibility distribution on possibility distributions—which is analogous to (ii).

An alternative interpretation of the proposition "X is A is ω" where $\omega = $ "α-possible" is "It is α-possible that X is A," i.e., "X is A" is contingent to a certain degree α. When $\alpha = 1$, the qualification rule changes A into A^+ such that $\mu_{A^+}(t) = [\mu_A(t), 1]$ with the understanding that the possibility that t qualifies X may be any number in the interval $[\mu_A(t), 1]$. Note that A^+ is an interval-valued fuzzy set (Φ-fuzzy set). $\mu_{A^+}(t)$ is a "degree" of possibility of membership of t in A. More generally, if $\alpha \neq 1$, Zadeh (Reference from IV.2, 1977b) proposes the formula

$$\mu_{A^+}(t) = [\min(\alpha, \mu_A(t)), \min(1, 1 - \mu_A(t) + \alpha)].$$

Sanchez (1978) prefers

$$\mu_{A^+}(t) = [\min(\alpha, \mu_A(t)), \max(1 - \mu_A(t), \alpha)].$$

Both formulas coincide for $\alpha = 1$ and $\alpha = 0$ ("impossible"). Anyway, according to Zadeh (Reference from IV.2, 1977b) these rules should be regarded as provisional in nature. Their relationships to the theory of possibilities and (fuzzy) modal logic have not yet been made clear.

c. **Modifier Rules**   (Zadeh, 1977a)

α.   *Semantic Equivalence and Entailment*

Let P and Q be two propositions and let π_P and π_Q be the possibility distributions induced by P and Q owing to the above translation rules. P and Q are said to be *semantically equivalent* iff $\pi_P = \pi_Q$, which is denoted by $P \Leftrightarrow Q$. This definition could be weakened by means of approximate equalities (II.1.E.c).

While the concept of semantic equivalence relates to the equality of possibility distributions, that of *semantic entailment* relates to inclusion. More specifically, denoting P semantically entails Q by $P \Rightarrow Q$, we have

$$P \Rightarrow Q \qquad \text{iff} \qquad \pi_P \subseteq \pi_Q.$$

β.   *Modifier Rules for Propositions*

The modifier rule that was stated earlier for simple propositions provides the basis for the formulation of a more general modifier rule that applies to propositions translated by rules of type 1, 2, 3, and 4.

This general rule is: if m is a modifier and P is a proposition, then mP is semantically equivalent to the proposition that results from applying m to the possibility distribution induced by P.

(i)   *Simple propositions.* $m(\text{“}X \text{ is } A\text{”}) \Leftrightarrow \text{“}X \text{ is } mA\text{”}$ which is exactly a type 1 translation rule. Examples:

$$m = \text{“not,”} \qquad \mu_{mA} = 1 - \mu_A$$

$$m = \text{“very,”} \qquad \mu_{mA} = \mu_A^2.$$

N.B.: $m(\text{“}X \text{ is } m'A\text{”}) \Leftrightarrow \text{“}X \text{ is } m(m'A)\text{”} \Leftrightarrow mm'(\text{“}X \text{ is } A\text{”}).$

(ii)   *Composed propositions.* $m(\text{“}X \text{ is } A \text{ and } Y \text{ is } B\text{”}) \Leftrightarrow (X, Y) \text{ is } m(A \times B)$. Examples:

not(“X is A and Y is B”) $\Leftrightarrow$ “X is not A or Y is not B”.

very(“X is A and Y is B”) $\Leftrightarrow$ “X is very A and Y is very B”.

(iii)   *Quantified propositions.* $m(\text{“}FX \text{ are } A\text{”}) \Leftrightarrow \text{“}(mF)X \text{ are } A.\text{”}$ Example: $m = \text{“not.”}$ This formula can be employed here to generalize the standard negation rule in predicate calculus:

$$\neg(\forall x)P(x) \qquad \Leftrightarrow \qquad (\exists x)\neg P(x).$$

To see this connection, we first assert the semantic equivalence

“FX are A” $\Leftrightarrow$ “$(\text{ant}\,F)X$ are not A”

where $\text{ant}\,F$ is the antonym $1 \ominus F$ of F.

Proof:

$$\pi(\rho) = \mu_F\left(\int_T \mu_A(s)\rho(s)\,ds\right) = \mu_F\left(1 - \int_T (1 - \mu_A(s))\rho(s)\,ds\right)$$

because $\int_T \rho(s)\,ds = 1$.   Q.E.D.

Thus, we have

"$(\text{not }F)X$ are A" $\Leftrightarrow$ "ant(not F)X are not A"

which for $F = $ "all" gives

"not all X are A" $\Leftrightarrow$ "some X are not A"

with "some" defined as ant(not "all"), meaning "at least some."

(iv) *Qualified propositions.* We consider here only truth-qualified propositions:

$$m(\text{"}X \text{ is } A \text{ is } \tau\text{"})   \Leftrightarrow   \text{"}X \text{ is } A \text{ is } m\tau.\text{"}$$

Example: $m = $ "not," $\tau = $ "true,"

not("X is A is true") $\Leftrightarrow$ "X is A is not true."

On the other hand, we have

"X is not A is τ" $\Leftrightarrow$ "X is A is ant τ."

where if $\tau = $ "true," ant $\tau = $ "false."

N.B.:   For possibility-qualified propositions, we have:

not(X is A is 1-possible)$\Leftrightarrow$"X is A is impossible";
very(X is A is 1-possible)$\Leftrightarrow$"X is very A is 1-possible";

because

$$\forall t \in T, \quad \mu_{\overline{(A^+)}}(t) = [0, 1 - \mu_A(t)];$$

$$\mu_{(\text{very } A)^+}(t) = [\mu_A^2(t), 1] = [\mu_A(t), 1]\odot[\mu_A(t), 1] = \mu_{\text{very}(A^+)}(t).$$

γ. *Example of Inference with Modifier Rules*

Consider as a premise the proposition "FX are A." We want to answer the question, How many X are mA? where $\mu_{mA} = \mu_A^m$ (e.g., $m = 2$, mA means "very A"). The translation of the premise is a possibility distribution

$$\pi(\rho) = \mu_F\left(\int_T \mu_A(s)\rho(s)\,ds\right).$$

The proportion of X that are mA is

$$\text{Prop}(mA) = \int_T \mu_A^m(s)\rho(s)\,ds.$$

What we want to find is a quantifier F' such that "$F'X$ are mA" is the answer to the question. We know only $\pi(\rho)$ and not precisely ρ, so $\pi(\rho)$ induces by the extension principle a possibility distribution on the values of Prop(mA), which is $\mu_{F'}$ such that

$$\mu_{F'}(z) = \sup_{\rho} \pi(\rho) \qquad \text{subject to} \qquad \int_T \mu_A^m(s)\rho(s)\,ds = z.$$

In the finite case this formula becomes (see b, type 3)

$$\mu_{F'}(z) = \sup_{\Sigma_U \mu_A^m(X) = z \odot |U|} \pi(\hat{A}) = \sup_{\Sigma_U \mu_A^m(X) = z \odot |U|} \mu_F\left(\frac{\Sigma_U \mu_A(X)}{|U|}\right).$$

Assume μ_F is increasing on $[0, 1]$, for instance $F =$ "most" and $m = 2$. Then the maximizing values of $\mu_{\hat{A}}(X)$ are $\mu_{\hat{A}}(X) = \sqrt{z}$ $\forall X$. Hence $\mu_{F'}(z) = \mu_F(\sqrt{z})$ or $F' = F \odot F$. This example was given in Bellman and Zadeh (1977).

It can be checked that "most X are A" semantically entails "(most $\odot$ most) X are very A" (Zadeh, 1977a): Let π' be the possibility distribution associated with the last proposition; we have

$$\pi'(\rho) = \mu_{F \odot F}\left(\int_T \mu_A^2(s)\rho(s)\,ds\right)$$

$$= \mu_F\left(\sqrt{\int_T \mu_A^2(s)\rho(s)\,ds}\right)$$

$$\geqslant \mu_F\left(\int_T \mu_A(s)\rho(s)\,ds\right) = \pi(\rho)$$

using Schwarz's inequality and the monotonicity of μ_F.

N.B.: Semantic equivalences or entailments are said to be *strong* (Zadeh, Reference from IV.2, 1977b) as soon as they hold, whatever the fuzzy sets involved in the concerned propositions may be.[†]

δ. *Remark*

Modifiers can be applied to questions such as, What is the fuzzy proportion F of X such that $h(X)$ is greater than or equal to B?, where h measures the Xs and B is a fuzzy number on $\mathbb{R}$ considered as a fuzzy bound, i.e.,

$$\mu_F(z) = \sup_{t, \int_t^{+\infty} \rho(s)\,ds = z} \mu_B(t).$$

[†] For instance, "John is very tall"⇒"John is tall" is a strong semantic entailment, but "John is very tall"⇒"John is not short" depends on the definitions of *tall* and *short*, and hence does not represent a strong semantic entailment.

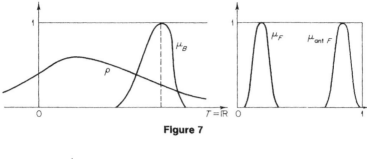

Figure 7

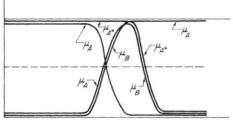

Figure 8

Using, for instance, the modifier "not," we can change this question into the semantically equivalent one, "What is the fuzzy proportion F^* such that $h(x)$ is less than or equal to B? We have

$$\mu_{F^*}(z) = \sup_{t,\, \int^t_{-\infty}\rho(s)\,ds = z} \mu_B(t) = \mu_F(1-z) = \mu_{\mathrm{ant}(F)}(z).$$

See Fig. 7. Note that the complement of the fuzzy interval $C = (B, +\infty)$ is taken as $(-\infty, B)$ which is different from $\bar{C}$. C is a fuzzy set of intervals of the form $[t, +\infty)$ such that $\mu_C([t, +\infty)) = \mu_B(t)$. $C^* = (-\infty, B)$ is a kind of antonym for $(B, +\infty) = C$ since $\mu_{C^*}((-\infty, t]) = \mu_C([t, +\infty))$.

Consider now the proposition "FX are A" where A is the fuzzy set on $T = \mathbb{R}$ of numbers greater than B in the sense that $\forall t \in \mathbb{R}$, $\mu_A(t) = \sup_{x \leqslant t} \mu_B(x)$. Its shape is shown in Fig. 8. A is similar to $(B, +\infty) = C$. When "FX are A" is translated by $\pi(\rho)$, it is semantically equivalent to "$(\mathrm{ant}\,F)X$ are $\bar{A}$." In the alternative model "FX are greater than B" is the same as "$(\mathrm{ant}\,F)X$ are less than B." The fuzzy set that is similar to $(-\infty, B)$ is A^* such that $\mu_{A^*}(t) = \sup_{x \geqslant t} \mu_B(x)$ and not $\bar{A}$!

d. Rules of Inference (Zadeh, 1977a)

The main rules of inference in approximate reasoning are the projection principle, the particularization/conjunction principle, and the entailment principle. Once combined, the first two lead to generalized modus ponens.

α. Projection Principle

Let $\pi_{(X_1, \ldots, X_n)}$ be a possibility distribution over a universe $T_1 \times \cdots \times T_n$. $\pi_{(X_1, \ldots, X_n)}$ is associated with a fuzzy relation F in $T_1 \times \cdots \times T_n$ that defines the fuzzy restriction $F(h_1(X_1), \ldots, h_n(X_n))$ on the values of $h_1(X_1), \ldots, h_n(X_n)$, where h_i denotes an attribute of X_i. $\pi_{(X_1, \ldots, X_n)} = F$ is a translation of a proposition P. Let $s = (i_1, \ldots, i_k)$ be a subsequence of $(1, \ldots, n)$ and $\pi_{X_{(s)}} = \text{proj}[F; T_{i_1} \times \cdots \times T_{i_k}]$ (see II.3.A.a). $\pi_{X_{(s)}}$ is the marginal possibility distribution of $(h_{i_1}(X_{i_1}), \ldots, h_{i_k}(X_{i_k}))$. Let Q be a retranslation of the possibility assignment equation $\pi_{X_{(s)}} = \text{proj}[F; T_{i_1} \times \cdots \times T_{i_k}]$, then the projection principle asserts that Q may be inferred from P.

For instance, $n = 2$, $F = A \times B$ where A means "tall" and B means "fat," from $P =$ "John is tall and fat," we infer "John is tall," provided that $\text{hgt}(A) = \text{hgt}(B)$.

β. Particularization/Conjunction Principle

The particularization of $\pi_{(X_1, \ldots, X_n)}$ is its modification resulting from the stipulation that the possibility distribution $\pi_{X_{(s)}}$ is a fuzzy set G on $T_{i_1} \times \cdots \times T_{i_k}$. The result is a possibility distribution

$$\pi_{(X_1, \ldots, X_n)}\left[\pi_{X_{(s)}} = G\right] = F \cap c(G)$$

where $c(G)$ is the cylindrical extension of G. From P translated in $\pi_{(X_1, \ldots, X_n)} = F$ and Q translated by $\pi_{X_{(s)}} = G$, we can infer R translated by $\pi_{(X_1, \ldots, X_n)} = F \cap c(G)$.

The particularization principle may be viewed as a special case of a somewhat more general principle, which will be referred to as the *conjunction principle*. Specifically, assume that P is translated by $\pi_{(Y_1, \ldots, Y_k, X_{k+1}, \ldots, X_n)} = F$ and Q translated by $\pi_{(Y_1, \ldots, Y_k, Z_{k+1}, \ldots, Z_m)} = G$, then from P and Q we can infer R translated by

$$\pi_{(Y_1, \ldots, Y_k, X_{k+1}, \ldots, X_n, Z_{k+1}, \ldots, Z_m)} = c(F) \cap c(G),$$

i.e., the join of F and G (II.3.A.a).

γ. Entailment Principle

Stated informally, the entailment principle asserts that from any proposition P, we can infer a proposition Q, if the possibility distribution induced by P is contained in the possibility distribution induced by Q. For instance, from $P =$ "X is very large" we can infer $Q =$ "X is large."

δ. Compositional Rule of Inference

The compositional rule of inference consists in the successive application of the particularization/conjunction principle followed by that of the of the particularization/conjunction principle followed by that of the projection principle. Let P be a proposition translated by $\pi_{(X,Y)} = F$ and Q translated by $\pi_{(Y,Z)} = G$. We can infer a proposition R translated by $\pi_{(X,Z)} = F \circ G$ where $\circ$ denotes sup–min composition (see II.3.A.a).

An important special case of the compositional rule of inference is obtained when P and Q are of the form $P = $ "X is A'", $Q = $ "If X is A, then Y is B."

Propositions such as Q are translated by means of type 2 translation rules, after having made the choice of an implication. For instance, Q is translated by $\pi_{(X,Y)} = c(\overline{A}) \cup c(B)$.

From P and Q we can infer the proposition Y is B' where $B' = A' \circ (c(\overline{A}) \cup c(B))$, for instance. Using the results of B.c, we can state the following chains of inclusions:

$$A' \circ (A \Rightarrow B) \supseteq A' \circ (A \leftrightarrow B) \supseteq A' \circ (A \to B)$$

$$\supseteq A' \circ \left(A \overset{3}{\to} B\right) \supseteq A' \circ (A \times B);$$

$$A' \circ (A \Rightarrow B) \supseteq A' \circ \left(A \overset{4}{\to} B\right) \supseteq A' \circ \left(A \overset{2}{\to} B\right) \supseteq A' \circ (A \times B);$$

$$A' \circ \left(A \overset{2}{\to} B\right) \supseteq A' \circ \left(A \overset{1}{\to} B\right).$$

N.B.: In the above chains the implication symbols have been abusively used as set theoretic operators and the cylindrical extensions are omitted. The most valid inferred proposition is "Y is $A' \circ (A \Rightarrow B)$" since it is the fuzziest one!

2. Very recently, Diaz (1978) has proposed another form of translation "If X is A, then Y is B":

$$\pi_{(X,Y)} = \begin{cases} c(\overline{A}) \cup C(B) & \text{if } \mu_A(t) \leqslant \mu_B(t'), \quad (t,t') \in T \times T' \\ c(\overline{A}) \cap C(B) & \text{if } \mu_A(t) > \mu_B(t'), \quad (t,t') \in T \times T'. \end{cases}$$

T and T' are the universes of A and B, respectively.

Schematically, the inference can be pictured as:

$$P = \text{"}X \text{ is } A'\text{"}$$
$$Q = \text{"If } X \text{ is } A, \text{ then } Y \text{ is } B\text{"}$$
$$R = \text{"}Y \text{ is } A' \circ (c(A) \mapsto c(B))\text{"}$$

where $\mapsto$ denotes any of the introduced implications. This inference scheme is called *generalized modus ponens*.

In the classical modus ponens $A' = A$ and the inferred proposition is "Y is B." However, it can be checked here that generally, setting $A' = A$, we get $B' = A \circ (c(A) \mapsto c(B)) \neq B$. For instance, assuming $A, \bar{A}$, and B are normalized fuzzy sets, μ_A and μ_B continuous,[†] we get for

$$\mapsto = \Rightarrow, \qquad \mu_{B'}(t') = \tfrac{1}{2}(1 + \mu_B(t')) \qquad \forall t' \in T'$$

$$\mapsto = \overset{4}{\to}, \qquad \mu_{B'}(t') = \sqrt{\mu_B(t')} \qquad \forall t' \in T';$$

$$\mapsto = \to, \qquad \mu'_B(t') \begin{cases} = \mu_B(t') & \text{if } \mu_B(t') \geq \tfrac{1}{2} \\ = 0.5 & \text{otherwise.} \end{cases}$$

When A is crisp, we recover $B' = B$. However, when A is fuzzy "the implicit part of Q," namely "if X is $\bar{A}$, then Y is unrestricted, overlaps the explicit part, resulting in an interference term which vanishes when A is nonfuzzy" (Bellman and Zadeh, 1977).

Remarks 1 A rather funny particular case of generalized modus ponens is the well-known rule of three. The classical rule is, "If X equals a, if X equals α implies Y equals β, then Y equals $(\beta/\alpha).a$. This rule can be extended using for instance, positive fuzzy numbers, namely $\tilde{a}, \tilde{\alpha}, \tilde{\beta}$. The result of the inference is then "Y equals $\tilde{a} \odot (\tilde{\beta} \odot \tilde{\alpha})$" where "$\odot$" and "$\odot$" denote here extended product and division. Defining

$$\tilde{\alpha} \mapsto \tilde{\beta} \qquad \text{by} \qquad \mu_{\tilde{\alpha} \mapsto \tilde{\beta}}(X, Y) = \mu_{\tilde{\beta} \odot \tilde{\alpha}}(Y/X), \qquad X \neq 0,$$

we have

$$\tilde{a} \odot (\tilde{\beta} \odot \tilde{\alpha}) = \tilde{a} \circ (\tilde{\beta} \mapsto \tilde{\alpha})$$

Thus, generalized modus ponens may be viewed as a generalized interpolation.

2   If A', A, and B are type 2 fuzzy sets, the compositional rule of inference can be readily extended by means of a $\widetilde{\max} \cdot \widetilde{\min}$ (more generally a $\widetilde{\sup} \cdot \widetilde{\min}$) composition (see II.3.F.c).

A generalized modus ponens may involve several conditional propositions such as "If X is A_i, then Y is B_i," $i = 1, n$. The procedure for making inferences is then to aggregate the n rules (for instance, by performing their union) into a binary fuzzy relation R; the inferred proposition is "Y is $A \circ R$."

Some interesting questions, which have not been completely solved yet are: consistency of the rules, nonredundancy of the set of rules, and the converse problem, i.e., determining the rules from knowledge of R (see Tong, 1976). Lastly, the compositional rule of inference has been extended

[†] Hence A and B cannot be crisp!

to possibility-qualified propositions of the form "X is A is 1-possible" (see b, type 4, (iii)) by Sanchez (1978). Consider the inference scheme:

"X is F is 1-possible" translated by $\pi_X = F^+$;

"(X, Y) is G is 1-possible" translated by $\pi_{(X, Y)} = G^+$ then "Y is $F^+ \circ G^+$";

where $\underset{\sim}{F^+}$ is a Φ-fuzzy set, such that $\mu_{F^+}(X) = [\mu_F(X), 1]$, $\circ$ is the $\underset{\sim}{\max \cdot \min}$ composition of Φ-fuzzy relations (see II.3.F.c). Sanchez has shown that $(F \circ G)^+ = F^+ \circ G^+$ so that "Y is $F^+ \circ G^+$" is semantically equivalent to "Y is $F \circ G$ is 1-possible." We have also $F \circ G^+ = (F \circ G)^+$ iff F is normalized, and $F^+ \circ G = (F \circ G)^+$ iff $\text{proj}[G; V] = V$ where V is the universe of Y. We have supposed here $h(X) = X$, i.e., $U = T$.

REFERENCES

Fuzzy Switching Logic

Benlahcen, D., Hirsch, G., and Lamotte, M. (1977). Codage et minimization des fonctions floues dans une algèbre floue. *RAIRO Autom./Syst. Anal. Control (Yellow Ser.)* **11**, 17–31.

Davio, M., and Thayse, A. (1973). Representation of fuzzy functions. *Philips Res. Rep.* **28**, 93–106.

Hughes, J. S., and Kandel, A. (1977). Applications of fuzzy algebra to hazard detection in combinatorial switching circuits. *Int. J. Comput. Inf. Sci.* **6**, No. 1, 71–82.

Kameda, T., and Sadeh, E. (1977). Bounds on the number of fuzzy functions. *Inf. Control* **35**, 139–145.

Kandel, A. (1973a). Comment on an algorithm that generates fuzzy prime implicants by Lee and Chang. *Inf. Control* **22**, 279–282.

Kandel, A. (1973b). On minimization of fuzzy functions. *IEEE Trans. Comput.* **C-22**, 826–832.

Kandel, A. (1973c). Comment on "minimization of fuzzy functions." *IEEE Trans. Comput.* **C-22**, 217.

Kandel, A. (1974a). Application of fuzzy logic to the detection of static hazards in combinational switching systems. *Int. J. Comput. Inf. Sci.* **3**, No. 2, 129–139.

Kandel, A. (1974b). On the properties of fuzzy switching functions. *J. Cybern.* **4**, 119–126.

Kandel, A. (1974c). On the mimimization of uncompletely specified fuzzy functions. *Inf. Control* **26**, 141–153.

Kandel, A. (1974d). Fuzzy representation CNF minimization and their applications to fuzzy transmission structures. *Symp. Multiple-Valued Logic, IEEE, 1974,* 74CH0845-8C, pp. 361–379.

Kandel, A. (1975). Block decomposition of imprecise models. *Asilomar Conf. Circuits, Syst. Comput., 9th* pp. 522–530.

Kandel, A. (1976a). Inexact switching logic. *IEEE Trans. Syst., Man Cybern.* **6**, 215–219.

Kandel, A. (1976b). Fuzzy maps and their applications to the simplification of fuzzy switching functions. *Proc. Int. Symp. Multiple-Valued Logic, 6th, IEEE* 76CH1111-4C.

Kandel, A. (1976c). On the decomposition of fuzzy functions. *IEEE Trans. Comput.* **C-25**, 1124–1130.

Kandel, A. (1977). Comments on comments by Lee—Author's reply. *Inf. Control* **35**, 109–113.

Kaufmann, A. (1975). "Introduction to the Theory of Fuzzy Subsets," Vol. 1, Chap. III, pp. 191–267. Academic Press, New York. (Reference from I, 1975c.)

Lee, E. T. (1977). Comments on two theorems by Kandel. *Inf. Control* **35**, 106–108.

Lee, R. C. T., and Chang, C. L. (1971). Some properties of fuzzy logic. *Inf. Control* **19**, 417–431.

Lee, S. C., and Kandel, A. (1978). "Fuzzy Switching and Automata: Theory and Applications." Crane, Russak, New York.

Malvache, N., and Willaeys, D. (1974). "Représentation et Minimisation de Fonctions Floues." Rep. Cent. Univ. de Valenciennes, France.

Marinos, P. M. (1969). Fuzzy logic and its application to switching systems. *IEEE Trans. Comput.* **18**, 343–348.

Mukaidono, M. (1972). On the B-ternary logical function—A ternary logic considering ambiguity. *Syst.—Comput.—Control* **3**, No. 3, 27–36.

Mukaidono, M. (1975a). On some properties of fuzzy logic. *Syst.—Comput.—Control* **6**, No. 2, 36–43.

Mukaidono, M. (1975b). An algebraic structure of fuzzy functions and its minimal and irredundant form. *Syst.—Comput.—Control* **6**, No. 6, 60–68.

Neff, T. P., and Kandel, A. (1977). Simplification of fuzzy switching functions. *Int. J. Comput. Inf. Sci.* **6**, No. 1, 55–70.

Negoita, C. V., and Ralescu, D. (1975). "Applications of Fuzzy Sets to Systems Analysis," Chap. 3, pp. 65–84. Birkhaeuser, Basel. (Reference from I.)

Negoita, C. V., and Ralescu, D. (1976). Comment on a comment on an algorithm that generates fuzzy implicants by Lee and Chang. *Inf. Control* **30**, 199–201.

Preparata, F. P., and Yeh, R. T. (1972). Continuously valued logic. *J. Comput. Syst. Sci.* **6**, 397–418.

Schwede, G. W., and Kandel, A. (1977). Fuzzy maps. *IEEE Trans. Syst. Man Cybern.* **7**, No. 9, 669–674.

Siy, P., and Chen, C. S. (1972). Minimization of fuzzy functions. *IEEE Trans. Comput.* **21**, 100–102.

Srini, V. P. (1975). Realization of fuzzy forms. *IEEE Trans. Comput.* **24**, 941–943.

General

Albert, P. (1978). The algebra of fuzzy logic. *Int. J. Fuzzy Sets Syst.* **1**, No. 3, 203–230.

Bellman, R. E., and Zadeh, L. A. (1977). Local and fuzzy logics. In "Modern Uses of Multiple-Valued Logic" (J. M. Dunn and G. Epstein, eds.), pp. 103–165. Reidel Publ., Dordrecht, Netherlands.

Chang, C. C. (1964). Infinite valued logic as a basis for set theory. *Proc. Int. Congr. Logic Methodol. Philos. Sci.* (Y. Bar-Hillel, ed.), North Holland, pp. 93–100.

Diaz, M. (1978). Professor Zadeh's Seminar, May 9. Univ. of California, Berkeley.

Dubois, D., and Prade, H. (1978a). Fuzzy logics and fuzzy control. In "Le flou, Mécédonksa" Tech. Rep. CERT-DERA, Toulouse, France.

Dubois, D., and Prade, H. (1978b). Operations in a fuzzy-valued logic. In "Fuzzy Algebra, Analysis, Logics," Tech. Rep. TR-EE 78/13. Purdue Univ., Lafayette, Indiana. *Inf. Control*, **43**, No. 2, 224–240, 1979.

Dubois, D., and Prade, H. (1978c). An alternative fuzzy logic. In "Fuzzy Algebra, Analysis, Logics," Tech. Rep. TR-EE 78/13. Purdue Univ., Lafayette, Indiana.

Gaines, B. R. (1975). Stochastic and fuzzy logics. *Electron. Lett.* **11**, 188–189.

Gaines, B. R. (1976a). Fuzzy reasoning and the logics of uncertainty. *Proc. Int. Symp. Multiple-Valued Logic, 6th, IEEE* pp. 179–188.

Gaines, B. R. (1976b). Foundations of fuzzy reasoning. *Int. J. Man-Mach. Stud.* **8**, 623–668. (Also in "Fuzzy Automata and Decision Processes" (M. M. Gupta, G. N. Saridis, and B. R. Gaines, eds.), pp. 19–75. North-Holland Publ., Amsterdam, 1977.)

Gaines, B. R. (1978). Fuzzy and probability uncertainty logics. *Inf. Control* **38**, No. 2, 154–169.

Giles, R. (1976a). Lukasiewicz logic and fuzzy theory. *Int. J. Man-Mach. Stud.* **8**, 313–327.

Giles, R. (1976b). A logic for subjective belief. In "Foundations of Probability, Statistical Inference and Statistical Theories of Science" (W. Harper and C. A. Hooker, eds.), Vol. 1, pp. 41–72. Reidel Publ., Dordrecht, Netherlands.

Goguen, J. A. (1969). The logic of inexact concepts. *Synthese* **19**, 325–373.

Gottwald, S. (1976). Untersuchungen zur mehrwertigen Mengenlehre. *Math. Nachr.* **72**, 297–303; **74**, 329–336.

Klaua, D. (1973). Zur Arithmetik mehrwertigen Zahlen. *Math. Nachr.* **57**, 275–306.

Lakoff, G. (1973). Hedges: A study in meaning criteria and the logic of fuzzy concepts. *J. Philos. Logic* **2**, 458–508. (Reference from IV.2.)

Lee, R. C. T. (1972). Fuzzy logic and the resolution principle. *J. Assoc. Comput. Mach.* **19**, 109–119.

Machina, K. F. (1976). Truth, belief and vagueness. *J. Philos. Logic* **5**, 47–77.

Maydole, R. E. (1975). Many-valued logic as a basis for set theory. *J. Philos. Logic* **4**, 269–291.

Moisil, G. (1972). La logique des concepts nuancés. In "Essais sur les Logiques Non Chrysippiennes," pp. 157–163. Editions Acad. Repub. Soc. Roum., Bucharest.

Pinkava, V. (1976). "Fuzzification" of binary and finite multivalued logical calculi. *Int. J. Man-Mach. Stud.* **8**, 717–730.

Sanchez, E. (1978). On possibility-qualification in natural languages. *Inf. Sci. (N.Y.)* **15**, 45–76.

Schotch, P. K. (1975). Fuzzy modal logic. *Int. Symp. Multiple-Valued Logic, IEEE* pp. 176–182.

Skala, H. J. (1978). On many-valued logics, fuzzy sets, fuzzy logics and their applications. *Int. J. Fuzzy Syst.* **1**, No. 2, 129–149.

Tong, R. M. (1976). Analysis of fuzzy control algorithms using the relation matrix. *Int. J. Man-Mach. Stud.* **8**, 679–686.

Zadeh, L. A. (1974a). Fuzzy logic and its application to approximate reasoning. *Inf. Process. 74, Proc. IFIP Congr.* **3**, 591–594.

Zadeh, L. A. (1974b). The concept of a linguistic variable and its application to approximate reasoning. In "Learning Systems and Intelligent Robots" (K. S. Fu and J. T. Tou, eds.), pp. 1–10. Plenum, New York.

Zadeh, L. A. (1975). Fuzzy logic and approximate reasoning. (In memory of Grigore Moisil.) *Synthese* **30**, 407–428.

Zadeh, L. A. (1977a). "A Theory of Approximate Reasoning (AR)," Memo UCB/ERL M77/58. Univ. of California, Berkeley. (Also in "Machine Intelligence" (J. E. Hayes, D. Michie, and L. I. Mikulich, eds.), Vol. 9. Elsevier, New York, pp. 149–194, 1979.)

Zadeh, L. A. (1977b). "PRUF: A Meaning Representation Language for Natural Languages," Memo UCB/BRL M77/61. Univ. of California, Berkeley. [Also in *Int. J. Man-Mach. Stud.* **10**, No. 4, 395–460 (1978).] (Reference from IV.2, 1977a.)

Chapter *2*

DYNAMIC FUZZY SYSTEMS

The idea of fuzzy systems appeared very early in the literature of fuzzy sets; it was originated by Zadeh (1965). Research on fuzzy systems seems to have developed in two main directions. The first is rather formal and considers fuzzy systems as a generalization of nondeterministic systems. These have been studied within the same conceptual framework as classical systems. This approach has given birth to a body of abstract results in such fields as minimal realization theory and formal automata theory, sometimes expressed in the setting of category theory. These results are sketched in Sections B and C of this chapter. Section D gives two models of deterministic systems in a fuzzy environment. Section E deals with the practical computation of linear systems whose parameters are fuzzy numbers. It does not seem that the abstract theory of fuzzy systems has been applied yet to the study of real processes. Perhaps this situation is because this formal approach is based on the implicit idea that crisp statements can still be asserted to describe fuzzy behavior. This idea seems to contradict Zadeh's rationale in favor of linguistic models and approximate reasoning. The second direction of research is the linguistic approach to fuzzy systems, in which a fuzzy model is viewed as a linguistic description by means of fuzzy logical propositions. A first extensive outline of the linguistic approach was given by Zadeh (Reference from III.3, 1973). Since then it has been applied to the synthesis of linguistic controllers by Mamdani and Assilian (Reference from IV.4) followed by many others. This chapter is devoted to a formal approach to fuzzy systems. Linguistic aspects are mainly discussed in Part IV and are closely related to the theory of approximate reasoning (Section E of Chapter 1). Various topics, including identification and validation of models, are gathered in Section F. Let us begin this chapter with general considerations about complexity and fuzziness.

A. COMPLEXITY AND FUZZINESS IN SYSTEM THEORY

a. Complex Systems and the Principle of Incompatibility

Zadeh (1972) pointed out that "excessive concern with precision has become a stultifying influence in control and system theory, largely because it tends to focus the research in this field on those, and only those, problems which are susceptible of exact solutions."

Complexity in systems stems from too large a size and/or difficulty in gathering precise information or data to describe their behavior. Precise models of complex systems are often mathematically intractable. Again quoting Zadeh (1972): "The conventional quantitative techniques of system analysis are intrinsically unsuited for dealing with humanistic systems or, for that matter, any system whose complexity is comparable to that of humanistic systems."

The deep reason for this inadequacy can be summarized in what Zadeh called the *principle of incompatibility*: "Stated informally, the essence of this principle is that as the complexity of a system increases, our ability to make precise and yet significant statements about its behavior diminishes until a threshold is reached beyond which precision and significance (or relevance) become almost mutually exclusive characteristics" (see Zadeh, Reference from III.3, 1973). Partial precise information is useless as long as other important aspects of the system cannot be precisely described.

The determination of a satisfying model for a complex process is a matter of approximation. More specifically, when complex systems are considered, there is no longer a sense in which a model must best fit the data. The problem is "that of determining those models that are as good as possible in that no simpler or equally simple model is a better approximation to the data" (Gaines, 1977). Such models are termed *admissible*. Gaines (1977) formulates the general system identification problem as follows. Let B be a set of possible observed behaviors and M be a set of models, Ord_M is the set of all partial order relations on M and $\leqslant$ is a specified, particular order relation. Let f be a mapping from B to Ord_M. $\forall b \in B, f(b)$ is denoted $\leqslant_b$. The relation $\leqslant$ is supposed to rank the models with respect to complexity. Note that there is a set of minimal models rather than a unique minimum. $\leqslant_b$ ranks models with respect to the quality of approximation of behavior b. The admissibility relation $\leqslant_b^*$ on M is defined by

$$\forall m, n \in M, \quad m \leqslant_b^* n \qquad \text{iff} \qquad m \leqslant n \quad \text{and} \quad m \leqslant_b n;$$

i.e., $m \leqslant_b^* n$ iff m is neither a more complex nor a worse approximation of b than n. The admissible models of b are the minimal elements of M in the sense of $\leqslant_b^*$. The complexity of a model depends mainly on its size when its type is given, but this notion remains more or less arbitrary. So are measures of the quality of approximation. Some of these will be hinted at in Section F, which deals with validation of fuzzy models.

b. Fuzzy Systems

A system is viewed here has a set of relations between measurable attributes (i.e., inputs and outputs). The system is considered over a given period during which inputs, outputs, and relations may change. A system will be called fuzzy as soon as inputs or outputs are modeled as fuzzy sets or their interactions are represented by fuzzy relations. Usually, a system is also described in terms of state variables. In a fuzzy system a state can be a fuzzy set. However, the notion of a fuzzy state is quite ambiguous and needs to be clarified. Note that generally a fuzzy system is an approximate representation of a complex process that is not itself necessarily fuzzy. According to Zadeh, the human ability to perceive complex phenomena stems from the use of names of fuzzy sets to summarize information. The notion of probabilistic system corresponds to a different point of view: all the available information at any time is modeled by probability distributions, built from repeated experiments.

A fuzzy system can be described either as a set of fuzzy logical rules or as a set of fuzzy equations. Fuzzy logical rules must be understood as propositions associated with possibility distributions in the sense of 1.E. For instance, "if last input is small, then if last output is large, then current output is medium," where "small" is a fuzzy set on the universe of inputs, and "medium" and "large" are fuzzy sets on the universe of outputs. Such linguistic models will be presented later (see IV.2 and IV.4). Fuzzy equations may provide a representation for systems having fuzzy parameters, fuzzy inputs. Fuzzy constraints or goals can also be taken into account. Note that mathematically there is no essential difference between fuzzy equations and fuzzy logical rules. In both cases results are obtained by sup–min composition of fuzzy relations. However, the composition is sometimes precalculated and thus no longer explicit in the formulation.

Several situations may be encountered from which a fuzzy model can be derived:

there is available linguistic description that reflects a qualitative understanding of the process; a set of fuzzy logical rules can then be built directly;

there are known equations that (at least roughly) describe the behavior of the process, but parameters cannot be precisely identified;

too-complex equations are known to hold for the process and are interpreted in a fuzzy way to build, for instance, a linguistic model;

input–output data are used to estimate fuzzy logical rules of behavior.

Real situations may be hybrid.

B. DISCRETE-TIME FUZZY SYSTEMS

For simplicity, we shall restrict our attention to time-invariant discrete-time systems in which the time t ranges over integers. Time-variant fuzzy systems seem not to have been investigated in the literature to date.

In this section a formulation of state equations of fuzzy systems is given. The usual notions of reachability and observability are presented in the framework of fuzzy systems.

a. State Equations for Fuzzy Systems (Zadeh, 1965, 1971)

Let u_t, y_t, and s_t denote respectively the input, output, and state of a system $\mathbb{S}$ at time t. U, Y, S are respectively the set of possible inputs, outputs, and states. Such a system is said to be deterministic if it is characterized by state equations of the form

$$s_{t+1} = \delta(u_t, s_t), \qquad y_t = \sigma(s_t), \qquad t \in \mathbb{N}.$$

s_0 is called the initial state; δ and σ are functions from $U \times S$ and from S to S and Y, respectively.

$\mathbb{S}$ is said to be nondeterministic if s_{t+1} and/or y_t are not uniquely determined by u_t and s_t. Let S_{t+1} and Y_t be the sets of possible values of s_{t+1} and y_t, respectively, given u_t and s_t. S_{t+1} and Y_t may be understood as binary possibility distributions over S and Y, respectively.

α. Nonfuzzy Inputs

The next step is to assume that S_{t+1} and Y_t are fuzzy sets on S and Y. They can be interpreted as possibility distributions. δ and σ are now fuzzy relations in $S \times U \times S$ and $Y \times S$, respectively. δ is called the *fuzzy transition relation* and σ the *fuzzy output map*. The state equations of the fuzzy system are now

$$\mu_{\tilde{s}_{t+1}}(s_{t+1}) = \sup_{s_t \in s} \min\left(\mu_{\tilde{s}_t}(s_t), \mu_\delta(s_{t+1}, u_t, s_t)\right),$$

$$\mu_{\tilde{y}_t}(y_t) = \sup_{s_t \in S} \min\left(\mu_{\tilde{s}_t}(s_t), \mu_\sigma(y_t, s_t)\right), \tag{1}$$

where $\tilde{s}_t$ is the fuzzy state at time t, $\tilde{y}_t$ the fuzzy output at time t, u_t the nonfuzzy input at time t, and $\tilde{s}_0$ the fuzzy initial state. More compactly, we have

$$\tilde{s}_{t+1} = \tilde{s}_t \circ \delta_{u_t}, \qquad \tilde{y}_t = \tilde{s}_t \circ \sigma,$$

where δ_{u_t} is the fuzzy binary relation of transition between states when the input is u_t.

The fuzzy state at time $t + 1$ can be expressed as a function of the fuzzy state at time $t - 1$ and the input at time $t - 1$ and t:

$$\mu_{\tilde{s}_{t+1}}(s_{t+1}) = \sup_{s_t \in S} \min\left(\sup_{s_{t-1} \in S} \min\left(\mu_{\tilde{s}_{t-1}}(s_{t-1}), \mu_\delta(s_t, u_{t-1}, s_{t-1})\right), \mu_\delta(s_{t+1}, u_t, s_t)\right)$$

$$= \sup_{s_t \in S} \sup_{s_{t-1} \in S} \min\left(\mu_{\tilde{s}_{t-1}}(s_{t-1}), \mu_\delta(s_t, u_{t-1}, s_{t-1}), \mu_\delta(s_{t+1}, u_t, s_t)\right)$$

$$= \sup_{s_{t-1} \in S} \min\left(\mu_{\tilde{s}_{t-1}}(s_{t-1}), \sup_{s_t \in S} \min\left(\mu_\delta(s_t, u_{t-1}, s_{t-1}), \mu_\delta(s_{t+1}, u_t, s_t)\right)\right).$$

Hence $\tilde{s}_{t+1} = \tilde{s}_{t-1} \circ (\delta_{u_t} \circ \delta_{u_{t-1}})$. More generally,

$$\tilde{s}_{t+1} = \tilde{s}_0 \circ (\delta_{u_t} \circ \delta_{u_{t-1}} \circ \cdots \circ \delta_{u_0}) = \tilde{s}_0 \circ \Delta_{\theta_t} \tag{2}$$

where $\tilde{s}_0$ is the fuzzy initial state, $\theta_t = u_0 u_1 \ldots u_t$ is an input string of length $t + 1$, and

$$\mu_\Delta(s_{t+1}, u_0, \ldots, u_t, s_0) = \mu_{\Delta_{\theta_t}}(s_{t+1}, s_0)$$

$$= \sup_{s_1, s_2, \ldots, s_t} \min\left(\mu_\delta(s_1, u_0, s_0), \ldots, \mu_\delta(s_{t+1}, u_t, s_t)\right)$$

The *response function* of the system $\mathfrak{S}$, denoted $\tilde{f}_{\tilde{s}_0}$, is equal to

$$\tilde{y}_{t+1} = \tilde{f}_{\tilde{s}_0}(\theta_t) = \tilde{s}_0 \circ \Delta_{\theta_t} \circ \sigma. \tag{3}$$

The notion of fuzzy state may have two interpretations, which correspond to different representations. First, a fuzzy state can be a possibility distribution over S, i.e., the actual state is one of the elements of S; but since the process behavior is partially unknown, several states of S are possible with a nonzero possibility degree. δ and σ can be then viewed as conditional possibility distributions. For instance, $\mu_\delta(s_{t+1}, u_t, s_t)$ is the possibility for the state to be s_{t+1} at time $t + 1$, knowing that the state and the input at time t are s_t and u_t, respectively. Secondly, S is a set of fuzzy sets on a set Q of possible values for the actual states. Each element of S is a fuzzy cluster of elements of Q; the fuzzy sets belonging to S are an approximate covering of Q in the sense that the union of the fuzzy sets, elements of S, is contained in Q (in the sense of $\subseteq$, see II.1.E.a), but the union of their supports is Q. For instance, Q is $\{0, 1, \ldots, 9, 10\}$, $S = $ {small, medium, large} where

$$\text{small} = 1/0 + 0.9/1 + 0.7/2 + 0.5/3 + 0.3/4,$$

$$\text{medium} = 0.5/2 + 0.7/3 + 0.9/4 + 1/5 + 0.9/6 + 0.7/7 + 0.5/8,$$

$$\text{large} = 0.3/6 + 0.5/7 + 0.7/8 + 0.9/9 + 1/10.$$

The actual state cannot usually be precisely described and thus is represented by a fuzzy set $\tilde{q}$ on Q; however, in the case of a mechanical process, the state can be sometimes precisely measured, i.e., q. The behavior of the process under consideration can be directly described using Q as a state space—which corresponds to the first point of view. In the second point of view we use S, assumed to be built on Q, to describe the process in a more approximate way. Any element q of Q or fuzzy sets $\tilde{q}$ on Q may be expressed as a fuzzy set on S (see II.1.E.c.γ): $\tilde{q} \simeq \tilde{s}(\tilde{q}) = \int_S \text{hgt}(s \cap \tilde{q})/s$. For instance, if $q = 7$, $7 \simeq 0.7/\text{medium} + 0.5/\text{large}$; if $\tilde{q} = 0.6/6 + 1/7 + 0.6/8$, $\tilde{q} \simeq 0.7/\text{medium} + 0.6/\text{large}$. In this example both representations are very close because there is no essential difference between q and $\tilde{q}$ from the approximation point of view. δ_Q and σ_Q, fuzzy relations on $Q \times U \times Q$ and on $Y \times Q$ respectively, may induce δ and σ on $S \times U \times S$ and on $Y \times S$ respectively in the following way:

$$\mu_\delta(s, u, s') = \sup_{q, q'} \min\left(\mu_{\delta_Q}(q, u, q'), \mu_s(q), \mu_{s'}(q') \right), \tag{4}$$

$$\mu_\sigma(y, s) = \sup_q \min\left(\mu_{\sigma_Q}(y, q), \mu_s(q) \right).$$

In some situations δ and σ are directly obtained through a linguistic description using names of fuzzy sets belonging to S, involved in fuzzy conditional propositions (see 1.E). State equations can be established on S, using formulas (4), and they are formally the same as (1). Such a fuzzy model corresponds to an approximate (linguistic) description of a complex system whose equations are possibly unknown. Or when they exist, their precise solution is either quite untractable or inessential.

β. Fuzzy Inputs

In a more general formulation σ may depend on the input and the input may be fuzzy. Then, Eqs. (1) become

$$\mu_{\tilde{s}_{t+1}}(s_{t+1}) = \sup_{\substack{s_t \in S \\ u_t \in U}} \min(\mu_{(\widetilde{su})_t}(s_t, u_t), \mu_\delta(s_{t+1}, u_t, s_t)), \tag{5}$$

$$\mu_{\tilde{y}_t}(y_t) = \sup_{\substack{s_t \in S \\ u_t \in U}} \min(\mu_{(\widetilde{su})_t}(s_t, u_t), \mu_\sigma(y_t, s_t, u_t)),$$

where $(\widetilde{su})_t$ is a fuzzy relation on $S \times U$; $\mu_{(\widetilde{su})_t}(s_t, u_t)$ is the fuzzy counterpart of the joint probability of s_t and u_t. If the input and the state are not interactive, then $\mu_{(\widetilde{su})_t}(\cdot)$ is separable and can be written as $\min(\mu_{\tilde{s}_t}(\cdot), \mu_{\tilde{u}_t}(\cdot))$. (5) becomes then

$$\mu_{\tilde{s}_{t+1}}(s_{t+1}) = \sup_{\substack{s_t \in S \\ u_t \in U}} \min(\mu_{\tilde{s}_t}(s_t), \mu_{\tilde{u}_t}(u_t), \mu_\delta(s_{t+1}, u_t, s_t)), \tag{6}$$

$$\mu_{\tilde{y}_t}(y_t) = \sup_{\substack{s_t \in S \\ u_t \in U}} \min(\mu_{\tilde{s}_t}(s_t), \mu_{\tilde{u}_t}(u_t), \mu_\sigma(y_t, s_t, u_t)).$$

Assume now a string $\tilde{\theta}_t$ of fuzzy inputs; provided that the fuzzy inputs are noninteractive, we have

$$\mu_{\tilde{\theta}_t}(u_0, u_1, \ldots, u_t) = \min(\mu_{\tilde{u}_0}(u_0), \ldots, \mu_{\tilde{u}_t}(u_t));$$

and then, for instance,

$$\mu_{\tilde{s}_{t+1}}(s_{t+1})$$
$$= \sup_{\substack{s_0 \in S \\ u_0, u_1, \ldots, u_t \in U}} \min(\mu_{\tilde{s}_0}(s_0), \mu_{\tilde{u}_0}(u_0), \ldots, \mu_{\tilde{u}_t}(u_t), \mu_\Delta(s_{t+1}, u_0 \cdots u_t, s_0)).$$

N.B.:   A fuzzy system $\mathbb{S}$ will be said to be *memoryless* if the fuzzy set $\tilde{y}_t$ is independent of $\tilde{s}_t$, i.e.,

$$\mu_{\tilde{y}_t}(y_t) = \sup_{u_t \in U} \min(\mu_{\tilde{u}_t}(u_t), \mu_\sigma(y_t, u_t)).$$

In (1978b) Tong has proposed the block diagram

This is the feedbacklike representation of the equations

$$\tilde{e}_t = (\tilde{s}_t \times \tilde{u}_t) \circ \delta_e, \qquad \tilde{s}_{t+1} = (\tilde{s}_t \times \tilde{e}_t) \circ \delta_p.$$

Note that $\tilde{s}_{t+1} = (\tilde{s}_t \times [(\tilde{s}_t \times \tilde{u}_t) \circ \delta_e] \circ \delta_p) \neq (\tilde{s}_t \times \tilde{u}_t) \circ (\delta_e \circ \delta_p)$ because

$\mu_{\tilde{s}_{t+1}}(s_{t+1})$

$$= \sup_{\substack{s_t^* \in S \\ e_t \in U}} \min\left\{\left[\sup_{\substack{s_t \in S \\ u_t \in U}} \min\left(\mu_{\tilde{s}_t}(s_t), \mu_{\tilde{u}_t}(u_t)\mu_{\delta_e}(e_t, u_t, s_t)\right)\right], \mu_{\tilde{s}_t}(s_t^*), \mu_{\delta_p}(s_{t+1}, e_t, s_t^*)\right\}$$

$$= \sup_{\substack{s_t, s_t^* \in S \\ u_t \in U}} \min\left(\mu_{\tilde{s}_t}(s_t^*), \mu_{\tilde{s}_t}(s_t), \mu_{\tilde{u}_t}(u_t), \sup_{e_t \in U} \min\left(\mu_{\delta_e}(e_t, u_t, s_t), \mu_{\delta_p}(s_{t+1}, e_t, s_t^*)\right)\right)$$

$$\neq \sup_{\substack{s_t \in S \\ u_t \in U}} \min\left(\mu_{\tilde{s}_t}(s_t), \mu_{\tilde{u}_t}(u_t), \sup_{e_t \in U} \min\left(\mu_{\delta_e}(e_t, u_t, s_t), \mu_{\delta_p}(s_{t+1}, e_t, s_t)\right)\right),$$

which is the membership function of $(\tilde{s}_t \times \tilde{u}_t) \circ (\delta_e \circ \delta_p)$.

b. Reachability, Observability of Fuzzy Systems

α. Reachability

We consider here the extension of very well-known notions of systems theory to time-invariant discrete-time fuzzy systems. Denoting by U^* the set of finite input strings, a system is classically said to be reachable from s_0 iff $\forall s \in S, \exists t, \exists \theta_t = u_0 u_1 \ldots u_t$ such that $\Delta(\theta_t, s_0) = s$ where $\Delta(\theta_t, s_0) = \delta[u_t, \Delta(\theta_{t-1}, s_0)]$ and $\Delta(u_0, s_0) = \delta(u_0, s_0)$. That is to say, $\Delta(\cdot, s_0)$ is a surjective mapping from U^* to S (see, for instance, Arbib, Zeiger NF 1969).

Negoita and Ralescu (1975) have extended this definition to fuzzy systems with a nonfuzzy input and a nonfuzzy initial state s_0. The fuzzy system S is said to be *reachable* from s_0 iff

$$\forall s \in S, \quad \exists t, \quad \exists \theta_t = u_0 u_1 \ldots u_t \quad \text{such that} \quad \mu_{\Delta_{\theta_t}}(s, s_0) = 1.$$

The α-cuts of a fuzzy system are nondeterministic (nonfuzzy) systems defined by the α-cuts of δ and σ. This definition is consistent whenever sup–min composition and α-cutting commute (II.2.A.b). Then the α-cut of the state at time $t + 1$ can be obtained by the composition of the α-cuts of the fuzzy state at time t and of the transition relation δ.

Thus, the above definition of reachability means that s is reachable from s_0 by the 1-cut of S, which does not take fuzziness into account. Clearly, this definition may be relaxed to any given α-cut of S. More generally, a fuzzy system with a fuzzy initial state $\tilde{s}_0$ and fuzzy inputs will be said to be reachable from $\tilde{s}_0$ iff $\forall \tilde{s} \in \tilde{\mathcal{P}}(S), \exists t, \exists \tilde{\theta}_t, \tilde{\theta}_t = \tilde{u}_0 \tilde{u}_1 \ldots \tilde{u}_t$ where $\tilde{u}_i \in \tilde{\mathcal{P}}(U)$, $i = 0, t$, such that $\tilde{s}_{t+1} = \tilde{s}$ (Negoita and Ralescu, 1975).

This latter definition is very strict because the quantities $\tilde{s}_{t+1}$ and $\tilde{s}$ are by essence ill known. A less strict definition, intuitively more appealing, can be established by weakening the above equality into an inclusion: $\tilde{s}_{t+1} \subseteq \tilde{s}$, or even a weak inclusion (II.1.E.c.α) or an ϵ-inclusion (II.1.E.c.β). A very relaxed definition may use the concept of consistency, replacing $\tilde{s}_{t+1} = \tilde{s}$ by hgt($\tilde{s}_{t+1} \cap \tilde{s}$) $\geqslant \epsilon$ where ϵ is a threshold.

$\tilde{s}$ can be interpreted as a fuzzy goal and the final state must be inside the fuzzy goal. When S is a set of elements s that are already fuzzy sets on Q (see the second point of view at the end of a.α), the goal may be chosen as an element of s^* of S. The reachability condition then may become

$$\forall s' \neq s^*, \quad \mu_{\tilde{s}_{t+1}}(s^*) > \mu_{\tilde{s}_{t+1}}(s')$$

or more rigidly

$$\mu_{\tilde{s}_{t+1}}(s^*) \geqslant \eta \quad \text{and} \quad \forall s' \neq s^*, \quad \mu_{\tilde{s}_{t+1}}(s') < \epsilon \quad \text{where} \quad \eta \geqslant \epsilon.$$

Here $\mu_{\tilde{s}_{t+1}}(s) = \text{hgt}(s \cap \tilde{q}_{t+1})$ where $\tilde{s}_{t+1}$ is the approximation on S of $\tilde{q}_{t+1}$.

N.B.: 1.   Tong (1978b) proposes a definition of reachability, replacing the equality $\tilde{s}_{t+1} = \tilde{s}$ by the equality of their "peaks"; the peak p_A of a fuzzy set A is the nonfuzzy set of elements whose membership value in A is hgt(A). This definition can be questioned since the fuzziness is not really taken into account.

2.   The relaxed definition of reachability, using consistency, is in the same spirit as the notion of fuzzy surjection (II.4.A.a.β). That is to say, a reachable fuzzy system could be one such that the fuzzy mapping $\Delta(\cdot, s_0)$ from U^* to S is fuzzily surjective.

β. Observability

Recall that the response function of a nonfuzzy system is $f_{s_0}(\cdot) = \sigma(\Delta(\cdot, s_0))$ from U^* to Y. A nonfuzzy system is said to be observable iff $\forall s_0, s'_0$, if $s_0 \neq s'_0$, then $f_{s_0}(\cdot) \neq f_{s'_0}(\cdot)$ i.e., the mapping $s \mapsto f_s(\cdot)$ from S to $(U^*)^Y$ is injective. (See, e.g., Arbib and Zeiger, NF 1969.)

Negoita and Ralescu (1975) have extended this definition to a fuzzy system $\tilde{S}$ with a nonfuzzy input and a nonfuzzy initial state s_0: $\tilde{S}$ is said to be *observable* iff

$$\forall s_0, s'_0, \quad \text{if} \quad s_0 \neq s'_0, \quad \text{then} \quad \tilde{f}_{s_0}(\cdot) \neq \tilde{f}_{s'_0}(\cdot)$$

where $\tilde{f}_{s_0}(\cdot)$ is defined by

$$\mu_{\tilde{f}_{s_0}(\theta_t)}(y_t) = \sup_s \min(\mu_{\Delta_{\theta_t}}(s_0, s), \mu_\sigma(y_t, s)).$$

This definition can be readily extended to fuzzy inputs and fuzzy inital state.

Since $f_s(\cdot)$ is basically an ill-known response function, the above definition may once more be considered as far too strict. The concept of a fuzzily injective function (see II.4.A.a.β.) might be useful to define a fuzzy observability for fuzzy systems.

The problem of minimal realization was solved by Negoita and Ralescu (1975) when the inputs and the initial state are nonfuzzy. They use the above definition of observability and the first definition of reachability introduced. Their approach is very similar to the minimal realization of nonfuzzy systems (see Arbib and Zeiger, NF 1969). They use Nerode equivalence of input strings, i.e.,

$$\forall \theta_1, \theta_2 \in U^*, \quad \theta_1 \sim \theta_2 \quad \text{iff} \quad \forall \theta, \quad f(\theta_1 \theta) = f(\theta_2 \theta)$$

where f is a given response function. The set of states S is the quotient space $U^*/\sim$. For more details, see Negoita and Ralescu (1975).

N.B.: The same authors also investigated a very general kind of fuzzy systems where the transition function δ is a function from $\tilde{\mathcal{P}}(S) \times \tilde{\mathcal{P}}(U)$ to $\tilde{\mathcal{P}}(S)$ and the output map σ is a function from $\tilde{\mathcal{P}}(S)$ to $\tilde{\mathcal{P}}(Y)$. The minimal realization theory for these systems is formally equivalent to that of nonfuzzy systems. However, such an approach may appear very rigid for fuzzy systems.

c. Fuzzy Observation, Fuzzy Feedback Control System

Fuzzy systems with feedback control have been scarcely studied in the literature. The only attempt seems to be that of Chang and Zadeh (1972). The authors first introduce the notion of *fuzzy observation*. Let $\tilde{s} \in \tilde{\mathcal{P}}(S)$ be a fuzzy state. An observation of $\tilde{s}$, denoted $\hat{s}$, is any fuzzy set included in $\tilde{s}$ and renormalized; for instance, if $\tilde{s}' \subseteq \tilde{s}$, $\mu_{\hat{s}}(s) = (\mu_{\tilde{s}'}(s)/\text{hgt}(\tilde{s}'))$. An instrument or means of observation is represented by an operator O. $O(\tilde{s})$ represents the set of possible observations of $\tilde{s}$. Let O_1 and O_2 be two observation operators. O_2 is said to be more *definite* than O_1 iff

$$\forall \hat{s}_2 \in O_2(\tilde{s}), \quad \exists \hat{s}_1 \in O_1(\tilde{s}) \quad \text{such that} \quad \hat{s}_2 \subseteq \hat{s}_1.$$

A *fuzzy feedback control system* is composed of a fuzzy relation δ on $S \times U \times S$, an observation operator O, a goal set that is a fuzzy set $\tilde{g}$ on S, a fuzzy control policy η that maps the observed fuzzy state to a fuzzy control $\tilde{u}$, and a fuzzy initial state $\tilde{s}_0$. δ represents the transition of the controlled dynamic system. The equations are

$$\mu_{\tilde{s}_{t+1}}(s_{t+1}) = \sup_{u_t, s_t} \min(\mu_{\tilde{s}_t}(s_t), \mu_{\tilde{u}_t}(u_t), \mu_\delta(s_{t+1}, u_t, s_t)),$$

$$\mu_{\tilde{u}_t}(u_t) = \sup_{s_t} \min(\mu_{\hat{s}_t}(s_t), \mu_\eta(u_t, s_t)), \quad \hat{s}_t \in O(\tilde{s}_t), \quad t \in \mathbb{N}. \tag{7}$$

The goal $\tilde{g}$ is *attainable* iff $\exists \eta, \exists t, \hat{s}_t \subseteq \tilde{g}$. When the control is not fuzzy, the authors proved that given $(\delta_1, O_1, \tilde{g})$ and $(\delta_2, O_2, \tilde{g})$, two control problems such that δ_2 is "finer" than δ_1 and O_2 is more definite than O_1, then if $\tilde{g}$ is attainable in $(\delta_1, O_1, \tilde{g})$, it is attainable in (δ_2, O_2, g). (δ_2 is *finer* than δ_1 means: $\forall u_1, \exists u_2$ such that $\forall s, s'$, $\mu_{\delta_2}(s', u_2, s) \leqslant \mu_{\delta_1}(s', u_1, s)$.) "The power of this feedback concept is demonstrated by showing that a precise goal can be attained with a rather sloppy control and observation concept except that as the goal is approached the observation must be precise" (Chang and Zadeh, 1972).

d. Fuzzy Topological Polysystems

Nazaroff (1973) introduced the notion of fuzzy topological polysystem; it consists in a fuzzification of the topological aspects of the optimal control of dynamical polysystems as contributed by Halkin (NF 1964). Warren (1976) refocused the main conclusions of Nazaroff. This section gives only an outline of the basic notions.

Let E be a set whose elements may be considered as events. A time structure can be imposed on E by assuming a map k from E to $\mathbb{R}$ called the clock; $k(e)$ is the time of occurrence of e. Let r denote a binary relation on $E \times E$ such that $e_1 r e_2$ means "the event e_2 follows the event e_1." r is assumed transitive, reflexive, antisymmetric, and forward (i.e., if $e_1 r e_2$ and $e_1 r e_3$, then $e_2 r e_3$ or $e_3 r e_2$). r is called a strategy, and the set of strategies on E is denoted R.

A fuzzy topological polysystem (Nazaroff, 1973) is a triple $(E, \mathcal{E}, R)$ where $(E, \mathcal{E})$ is a fuzzy topological space (see II.4.E) and R a strategy set such that $\forall e_1, e_2 \in E$, $\forall A \in \mathcal{E}$, $\forall r \in R$ with $e_1 r e_2$ and $e_1 \in \operatorname{supp} A$, $\exists B \in \mathcal{E}$ such that $e_2 \in \operatorname{supp} B$ and $B \subseteq \{e', e r e' \text{ and } e \in \operatorname{supp} A\}$.

A fuzzy dynamical polysystem (Warren, 1976) is a fuzzy topological polysystem $(E, \mathcal{E}, R)$ such that $\forall e_1, e_2, e_3 \in E$, $\forall r_1, r_2 \in R$ with $e_1 r_1 e_2$ and $e_2 r_2 e_3$, $\exists r \in R$ with $e_1 r e_2$ and $e_2 r e_3$. Only such polysystems are considered below.

$\forall e_1, e_2 \in E$, $\forall r \in R$ with $e_1 r e_2$, the set $t(e_1, e_2, r) = \{e \in E, e_1 r e\} \cap \{e \in E, e r e_2\}$ is called the trajectory from the event e_1 to the event e_2 using the strategy r. The reachable set $K(e_1)$ from e_1 is defined by $\forall e_1 \in E$, $K(e_1) = \{e, \exists r \in R, e_1 r e\}$; the reachable set $K(A)$ from the fuzzy set $A \in \mathcal{E}$ is $K(A) = \bigcup_{e_1 \in \operatorname{supp} A} K(e_1)$. Note that $K(e_1), K(A), t(e_1, e_2, r)$ are crisp sets. In this framework Nazaroff (1973) and Warren (1976) give some properties of the trajectory, using the concept of boundary of a fuzzy set (see Warren, Reference from II.4, 1977). A fuzzy control problem for the fuzzy dynamical polysystem is also defined.

Remark Considering fuzzy events $\tilde{e}_1, \tilde{e}_2$ in E and a fuzzy strategy $\tilde{r} \in \tilde{\mathcal{P}}(R)$, one could define a fuzzy trajectory $\tilde{t}(\tilde{e}_1, \tilde{e}_2, \tilde{r})$ with membership function

$$\mu_{\tilde{r}}(e) = \min\Bigg(\sup_{\substack{e_1 \in E \\ r \in R}} \min(\mu_{\tilde{e}_1}(e_1), \mu_r(e_1, e), \mu_{\tilde{r}}(r)),$$

$$\sup_{\substack{e_2 \in F \\ r' \in R}} \min(\mu_{\tilde{e}_2}(e_2), \mu_{r'}(e, e_2), \mu_{\tilde{r}}(r')) \Bigg).$$

The fuzzy reachable set from $\tilde{e}_1$ using nonfuzzy strategies is $\tilde{K}(\tilde{e}_1)$ such that

$$\mu_{\tilde{K}(\tilde{e}_1)}(e) = \sup_{\substack{e_1 \in E \\ r \in R}} \min(\mu_{\tilde{e}_1}(e_1), \mu_r(e_1, e))$$

where r and r' are nonfuzzy binary relations.

e. Concluding Remarks

Classical concepts such as stability have not been extended to fuzzy systems yet (except a recent attempt by Tong (1978b), defining an equilibrium state as a state whose peak does not change within a given period). Moreover, some years ago, Zadeh (1971) evoked a fuzzy theory of aggregates as an open problem. In the theory of aggregates a system is viewed as a collection of input–output pairs; an aggregate is a bundle of input–output pairs satisfying certain conditions and a state is the name of an aggregate (see Zadeh, NF 1969). It is still an open problem for fuzzy systems. On the other hand, the rather rigid approach used for extending reachability and observability to fuzzy systems may not seem intuitively very appealing. A general theory of fuzzy systems perhaps demands more imagination than a straightforward extension of classical concepts of nonfuzzy system theory. Since the theory of approximate reasoning, initiated by Zadeh, radically departs from multivalent logics, a theory of fuzzy systems should perhaps be developed outside of the conceptual framework of classical system theory.

C. FUZZY AUTOMATA

A fuzzy automaton[†] is a fuzzy system in the sense of (1) where the sets U of inputs, S of states, and Y of outputs are finite. The mathematical formulation of a fuzzy automaton with a nonfuzzy initial state and

---

[†] Completely different but related to Poston's work (see II.4.F) are the fuzzy-state automata considered by Dal Cin (1975a, b).

nonfuzzy inputs was first proposed by Wee (1967) and can also be found in Wee and Fu (1969) or Santos (Reference from III.3, 1968). A fuzzy automaton with a fuzzy initial state was first considered by Mizumoto *et al.* (Reference from III.3; 1969) in the framework of language theory (see 3.A.g).

The problem of the reduction of fuzzy automata is investigated by Santos (1972a). For this purpose, the author develops a max–min algebra of real numbers playing a role in the theory of max–min automata similar to that played by linear algebra in the theory of stochastic automata (see Rabin, NF 1963; Paz, Reference from III.3). However, max–min algebra strongly differs from linear algebra, and has scarcely been studied in the literature. Various criteria of irreducibility and minimality are provided (Santos, 1972a).

A general formulation of sequential machines encompassing deterministic, nondeterministic, probabilistic, stochastic, and fuzzy finite machines valued on [0, 1] is proposed in Santos and Wee (1968). Semantic aspects of such machines are discussed at length by Gaines and Kohout (1976).

Valuation sets more general than [0, 1] can be used, especially any ordered semiring R. An R-fuzzy automaton is a complex $(U, S, Y, \delta, \sigma, \tilde{s}_0, R)$ where U, S, Y are sets of inputs, states, and outputs, respectively, δ an R-fuzzy relation on $S \times U \times S$, σ an R-fuzzy relation on $S \times Y$, and $\tilde{s}_0$ an R-fuzzy set (initial state). Max and min operations are replaced by the sum and the product of the semiring (see 3.A.f). Such automata were studied by Wechler and Dimitrov (Reference from III.3) in the framework of language theory. (See also Gaines and Kohout, 1976.) Further, Gaines and Kohout (1976) suggested "possible automata" whose valuation set consists of the semiopen interval]0, 1] and the elements N, E, P, and I, respectively interpreted as "necessary," "eventual," "possible," and "impossible." $\{N, E, P, I\}$, equipped with operations playing the role of max and min for fuzzy automata, is a 4-value Post algebra; but the interaction of]0, 1] with P is inconsistent with a lattice structure. The authors conclude that a more general structure than a distributive lattice is needed for the valuation set, i.e., an ordered semiring. (N.B.: Gaines and Kohout (1976) coined the term "possible," in the sense of possibility distributions[†]—later introduced by Zadeh.)

Other types of automata are max–product automata (Santos, 1972b; see also Santos, Reference from III.3, 1976) where product replaces min and $\mathbb{R}^+$ replaces [0, 1], and fuzzy–fuzzy automata (Mizumoto and Tanaka, Reference from III.3) where [0, 1] is replaced by the set of normalized

[†] Actually, the word *possibilistic* was also introduced by Arbib and Manes (1975a). However, in the late fifties, Shackle (NF 1961) was already discussing a concept of *possibility* much related to Zadeh's approach.

convex fuzzy sets on $[0, 1]$ and max and min by $\widetilde{\max}$ and $\widetilde{\min}$. A distributive lattice structure is preserved in this last case.

Lastly, a very general approach to automata theory was developed by Arbib and Manes (1975a, b) in the framework of category theory. Bobrow and Arbib (NF 1974) had already unified the theories of minimal realization of deterministic automata and linear systems. The study of fuzzy machines in a category is based on the concept of "fuzzy theory" (see II.4.F). The fuzzy transition relation δ and the output map σ become morphisms in an extended category. Because of its very high level of abstraction, the theory of fuzzy machines in a category is beyond the scope of this book. This approach considers fuzziness as a special mathematical property of a system rather than a lack of precise knowledge about the behavior of a complex process.

Remark A generalization of automata and graphs is a Petri net (see e.g., Holt, NF 1971). Fuzzy Petri nets may be worth considering.

D. DETERMINISTIC SYSTEMS IN A FUZZY ENVIRONMENT

This section deals with deterministic systems subject to fuzzily constrained behavior or fuzzy inputs.

a. Deformed Systems (Negoita and Ralescu, 1975)

A deformed system is a complex $((U, A), (S, B), (Y, C), \delta, \sigma, s_0)$ where U, S, Y are the sets of inputs, states, and outputs fuzzily constrained by the fuzzy sets A, B, and C, respectively, δ is a transition function from $U \times S$ to S whose domain is fuzzily constrained by $B \times A$ and its range by B, σ is the output map, i.e., a function from S to Y whose domain is fuzzily restricted by B and its range fuzzily restricted by C, and s_0 is a nonfuzzy initial state. This concept was introduced by Negoita and Ralescu (1974). The state equations of a deformed system are

$$s_{t+1} = \delta(u_t, s_t), \qquad y_t = \sigma(s_t)$$

where δ and σ satisfy

$$\mu_B(s_{t+1}) \geq \min(\mu_A(u_t), \mu_B(s_t)), \qquad \mu_C(y_t) \geq \mu_B(s_t). \tag{8}$$

These two inequalities express the fact that δ and σ are fuzzily constrained functions (see II.4.A.α). The fuzzy set A can be extended to the set U^* of input strings θ in a canonical way:

$$\mu_{A^*}(\theta_t) = \min(\mu_A(u_0), \mu_A(u_1), \ldots \mu_A(u_t))$$

where $\theta_t = u_0 u_1 \ldots u_t$.

The extension Δ of δ from U to U^* is defined by

$$\Delta(\theta_t, s_0) = \delta(u_t, \Delta(\theta_{t-1}, s_0)), \qquad \Delta(u_0, s_0) = \delta(u_0, s_0)$$

It is easy to check that Δ is a fuzzily constrained function from $U^* \times S$ to S, i.e.,

$$\mu_B(\Delta(\theta_t, s_0)) = \mu_B(s_{t+1}) \geqslant \min(\mu_{A^*}(\theta_t), \mu_B(s_0)).$$

It is clear that the response map $\sigma \circ \Delta$ is a fuzzily constrained function. Negoita and Ralescu (1975) have developed a theory of minimal realization for a deformed system; this theory is completely parallel to the same theory for classical nonfuzzy systems.

b. Fuzzy Noise (Sugeno and Terano, 1977)

In this section a representation of a deterministic system subject to fuzzily noised input is derived. The transition function δ of the system is viewed as a mapping from $U \times \Omega \times S$ where Ω is a set of inputs, assumed to be uncontrollable. The state equations of the system are

$$s_{t+1} = \delta(u_t, \omega_t, s_t), \qquad y_t = \sigma(s_t).$$

Ω is assumed to be equipped with a fuzzy measure g (see II.5.A.a), which is assumed to be time-invariant and expresses the fuzzy noise.

At time t the noised input induces on S a fuzzy measure h_t that is recursively defined as follows: let s_0 be the initial state; the fuzzy measure h_0 is such that

$$\forall A \in \mathscr{P}(S), \quad h_0(A) = \mu_A(s_0).$$

Consider now the fuzzy product measure in $S \times \Omega$, denoted $h_t \times g$, such that

$$\forall A \in \mathscr{P}(\Omega \times S), \quad (h_t \times g)(A) = \oint_S\left[\oint_\Omega \mu_A(\omega, s) \circ g(\cdot)\right] \circ h_t(\cdot) \quad (9)$$

$$\forall B \in \mathscr{P}(S), \quad h_{t+1}(B) = (h_t \times g)(A_{u_t})$$

$$\text{where} \quad A_{u_t} = \{(\omega, s), \delta(u_t, \omega, s) \in B\}.$$

g is similar to the probability measure of a noise that disturbs the input. The uncertainty in the knowledge of the state is due to the fuzzy noise, and the corresponding fuzzy measure h_{t+1} is canonically induced from both g and h_t.

It is supposed that

$$\Delta(\theta'\theta'', \omega'\omega'', s) = \Delta\left[\theta'', \omega'', \Delta(\theta', \omega', s)\right]$$

where Δ is defined as in (a), $\theta' = u_t u_{t+1} \ldots u_{t'}$, $\theta'' = u_{t'+1}, \ldots, u_{t''}$, $\omega' = \omega_t \omega_{t+1} \ldots \omega_{t'}$, and $\omega'' = \omega_{t'+1}, \ldots, \omega_{t''}$ with $t < t' < t''$ and $t, t', t'' \in \mathbb{N}$. Denoting by $\sigma_S(B \mid u_t, s)$ the conditional fuzzy measure of the state transition

$$\sigma_S(B \mid u_t, s) = \oint_\Omega \mu_{A_{u_t}}(\omega, s) \circ g(\cdot),$$

it can be shown that

$$h_{t+1}(B) = \oint_S \sigma_S(B \mid u_t u_{t-1} \ldots u_0, s) \circ h_0(\cdot)$$

with

$$\sigma_S(B \mid u_t u_{t-1} \ldots u_0, s) = \oint_S \sigma_S(B \mid u_t u_{t-1} \ldots u_k, s') \circ \sigma_S(\cdot \mid u_{k-1} \ldots u_0, s)$$

$$\sigma_S(B \mid u_0, s) = \oint_\Omega \mu_{A_{u_0}}(\omega, s) \circ g(\cdot).$$

Assume now that g is a possibility measure Π associated with a possibility distribution π on Ω (see II.5.A.a.η). h_t is now a possibility measure associated with a possibility distribution γ_t. Formula (9) becomes

$$(h_t \times \Pi)(A) = \sup_{s \in S} \sup_{\omega \in \Omega} \min(\mu_A(s, \omega), \pi(\omega), \gamma_t(s)).$$

The image of A under δ_{u_t} is $B = \delta_{u_t}(A)$. The possibility distribution γ_{t+1} is obtained by setting $B = \{s_{t+1}\}$:

$$\gamma_{t+1}(s_{t+1}) = \sup_{\substack{s_t, \omega_t \\ s_{t+1} = \delta_{u_t}(\omega_t, s_t)}} \min(\gamma_t(s_t), \pi(\omega_t)),$$

which is nothing but (6) where δ is crisp and the fuzzy part of the input is distinguished from the controllable (nonfuzzy part) using the formal equivalence between possibility distribution and fuzzy sets.

E. LINEAR FUZZY SYSTEMS

We shall call *fuzzy linear systems*, systems defined by linear state equations whose coefficients are fuzzy numbers. The state will be fuzzy. The initial state and the inputs may also be vectors of fuzzy numbers. The state at tine $t + 1$ is given by the equations, in the sense of II.2.B.h.α.,

$$\tilde{s}_{t+1} = \tilde{A} \odot \tilde{s}_t \oplus \tilde{B} \odot \tilde{u}_t \tag{10}$$

where $\tilde{A}$ and $\tilde{B}$ are $n \times n$ and $n \times m$ fuzzy matrices, respectively (see II.2.B.i); $\tilde{s}_{t+1}, \tilde{s}_t$, and $\tilde{u}_t$ are n, n, and m fuzzy vectors respectively; the sum and product of fuzzy matrices can be expressed, using the extended addition and product,

$$\tilde{s}_{t+1}^i = (\tilde{a}_{i1} \odot s_t^1) \oplus \cdots \oplus (\tilde{a}_{in} \odot \tilde{s}_t^n) \oplus (\tilde{b}_{i1} \odot \tilde{u}_t^1) \oplus \cdots \oplus (\tilde{b}_{im} \odot \tilde{u}_t^m).$$

An equation similar to (10) was first hinted at by Negoita and Stefanescu (1974) in a category-theoretic formulation. Jain (1976, 1977) studied fuzzy linear systems in the one-dimensional case. Equation (10) can be expanded as

$$\tilde{s}_1 = \tilde{A} \odot \tilde{s}_0 \oplus \tilde{B} \odot \tilde{u}_0,$$

$$\tilde{s}_2 = \tilde{A} \odot (\tilde{A} \odot \tilde{s}_0 \oplus \tilde{B} \odot \tilde{u}_0) \oplus \tilde{B} \odot \tilde{u}_1,$$

$$\vdots$$

$$\tilde{s}_{t+1} = \tilde{A} \odot (\tilde{A} \odot (\cdots \odot (\tilde{A} \odot \tilde{s}_0 \oplus \tilde{B} \odot \tilde{u}_0) \oplus \cdots) \oplus \tilde{B} \odot \tilde{u}_{t-1}) \oplus \tilde{B} \odot \tilde{u}_t.$$

Generally, the expression of $\tilde{s}_{t+1}$ cannot be reduced (e.g., $\tilde{s}_2 \neq \tilde{A}^2 \odot \tilde{s}_0 \oplus \tilde{A} \odot \tilde{B} \odot \tilde{u}_0 \oplus \tilde{B} \odot \tilde{u}_1$) because of the nondistributivity of $\odot$ over $\oplus$ (see II.2.B.d.β.), which forbids the associativity of the fuzzy matrix product. Especially, we have $\tilde{A} \odot (\tilde{A} \odot \tilde{A}) \neq (\tilde{A} \odot \tilde{A}) \odot \tilde{A}$. However, a sufficient condition that validates the associativity is the positivity of the fuzzy entries of $\tilde{A}$. At any rate, it is always possible to compute $\tilde{s}_{t+1}$ using the above expression provided that we perform the operations recursively.

N.B.: Since (10) is an explicit equation yielding $\tilde{s}_{t+1}$, it is consistent with the extension principle, that is to say it is also a fuzzy equation in the sense of II.2.B.h.δ.

F. OTHER TOPICS RELATED TO FUZZINESS AND SYSTEMS

This last section gives a survey of works dealing with fuzzy identification, validation of fuzzy models, and fuzzy classifications of systems. Research in these domains is only at its initial stages of development. We begin with a remark about fuzzy models.

a. Behavior of Fuzzy Models

Consider a fuzzy model that simulates a complex dynamical system. We suppose that we know only the initial state and the transition function between states in a fuzzy way. The problem is to forecast the future states of the system by means of a fuzzy model. Intuitively, in such models the state at time t' is at least as fuzzily known as state $s_t, t < t'$, unless some external information is provided. Clearly, this is usually true for a fuzzy linear system, for instance, because the result of extended additions or products is always relatively more fuzzy than the least fuzzy of the involved operands. Note that this situation is similar to that of chains of fuzzy inferences in approximate reasoning. An interesting feature of fuzzy models in forecast analysis may be their ability to exhibit their own limit of significance; beyond a certain horizon the forecast becomes too imprecise to be of any use. A periodical restatement of feedback terms may be necessary to limit the increase of fuzziness.

N.B.: 1. Recall that fuzziness is one aspect of imprecision and models the lack of sharp boundaries of tolerance intervals. The level of fuzziness can be evaluated by means of entropy.

2. Thomason (Reference from II.3) showed that the fuzzy state of a finite n-state fuzzy system in free motion (without input) either converges or oscillates with a finite period. Denoting by δ the binary transition matrix and by $\tilde{s}_0$ the initial state, a sufficient convergence condition is $\forall i$, $1 \leqslant i \leqslant n$, $\exists j$, $1 \leqslant j \leqslant n$, such that $s_0^i \leqslant \min(s_0^j, \delta_{ji})$. The convergence occurs in a finite number of states (see II.3.B.b.δ).

b. Identification

The problem of identification of fuzzy systems was recently considered for the first time by R. M. Tong (1978). A fuzzy model is viewed there as a set of fuzzy conditional propositions such as "if ⟨last input⟩ is small and if ⟨last output⟩ is large, then ⟨current output⟩ is medium." Those propositions are called rules. Tong proposes indices of quality for the assessment of such models so as to compare them with respect to a set of data (i.e., input–output pairs). The identification method is called "logical examination." A class of models is characterized by the structure of the rules, which corresponds to a data pattern. For the above example, the data pattern is (u_{t-1}, y_{t-1}, y_t). The logical examination technique is then to match each data pattern that can be built out of the data set with all possible rules that can be defined for the class of models. When the consistency between a data pattern and a rule is high enough, the rule is kept as part of the model unless a significant data pattern is found contradicting the rule.

Identification of fuzzy models must not be confused with fuzzy identification of models. (Gaines, 1977) has proposed a fuzzification of the general identification problem formulated in Section A.a. The behavior of the process to be identified is assumed not to be observed precisely, but is instead a fuzzy restriction on the set of possible behaviors B, i.e., a mapping μ from B to $[0, 1]$. The mapping can clearly be extended in the usual way to M_b, the nonfuzzy set of admissible models that describe the nonfuzzy behavior b: the mapping μ^* from M, the set of models, to $[0, 1]$, such that

$$\forall m \in M, \quad \mu^*(m) = \sup_b \min(\mu_{M_b}(m), \mu(b))$$

defines the fuzzy admissible subset of models induced by the fuzzy behavior. The author illustrates his approach on an example of fuzzy identification of a stochastic automaton. Note that "this simple extension does not take into account the relative degrees of approximation of the same models to differing behaviors."

c. Validation of Models

This aspect of identification was considered by Chang (1977) for the validation of economic models. Given a structural equation $y = F(x, \beta)$ where x is a vector of exogenous variables, y a vector of endogenous variables, and β a vector of parameters, the problem of determining the parameters β from economic data is called the estimation problem. The author considers β as a fuzzy set B constructed as follows

$$\mu_B(\beta) = \exp\left[(N - 1)(1 - C(\beta)/C(\beta_0))\right]$$

where $C(\beta) = \sum_{i=1}^{N}[y_i - F(x_i, \beta)]'W[y_i - F(x_i, \beta)]$, $\{(x_i, y_i), i = 1, N\}$ is a set of data, W a matrix of weights, $C(\beta)$ the cost associated with a forecast error, and β_0 a value of β minimizing $C(\beta)$.

N.B.:   This is not the only way of defining B.

Given B and a fuzzy real vector $\tilde{x}$, the model $y = F(x, \beta)$ induces a fuzzy set A of possible values for y:

$$\mu_A(y) = \sup_{y = F(x, \beta)} \min(\mu_{\tilde{x}}(x), \mu_B(\beta)).$$

$\mu_B(\beta)$ may be viewed as an evaluation of the validity of the parameter value β with respect to the set of data. When m different forecasting models are available, the author suggests a way of combining their results. Let A_i be the fuzzy result of model i that is assumed to have a reliability $r_i \in [0, 1]$. The consensus of the m models is given by

$$A = \bigcap_{i=1, m} A_i^{r_i} \quad \text{where} \quad \mu_{A_i^{r_i}}(x) = \left[\mu_{A_i}(x)\right]^{r_i}.$$

When hgt(A) is close to 1, the m models have a consensus that is likely to be reliable; if hgt(A) is close to 0, no consensus can be reached among the forecasts.

Moreover, Yager (1978) recently outlined a linguistic approach for the validation of fuzzy models with respect to a set of fuzzy data. Formally, the fuzzy model is described by the equation $\tilde{y} = F(\tilde{x})$ where $\tilde{x}$ and $\tilde{y}$ are fuzzy sets of $\mathbb{R}$. The set of data is made up of pairs (A, B) of real fuzzy sets where A is the input and B the output of the process under consideration. The problem is to compare $F(A)$ with the observation B taken as a reference. The author suggests the use of a truth qualification rule (see 1.E.b), i.e., find $\tau \in \tilde{\mathscr{P}}([0, 1])$ such that the fuzzy proposition "y is $F(A)$ is τ" is semantically equivalent to the fuzzy proposition "y is B." τ linguistically measures the compatibility of the fuzzy model with the pair of fuzzy data (A, B).

d. Fuzzy Classes of Systems

The basic idea is that the class of nonfuzzy systems that are approximately equivalent to a given (type of) system from the point of view of their behaviors is a fuzzy class of systems, for instance, the class of systems that are approximately linear. This idea of fuzzy classification of systems was first hinted at by Zadeh (1965). Saridis (1975) applied it to the classification of nonlinear systems according to their nonlinearities. Pattern recognition methods are first used to build crisp classes. "Generally this approach does not answer the question of complete identification of the nonlinearities involved within one class." To distinguish between the nonlinearities belonging to a single class, membership values in this class are defined for each nonlinearity. One of these is considered as a reference with a membership value 1. The membership value of each nonlinearity is calculated by comparing the coefficients of its polynomial series expansion to that of the reference nonlinearity. This technique of classification is similar to those used in fuzzy pattern classification (IV.6).

REFERENCES

Arbib, M. A., and Manes, E. G. (1975a). A category-theoretic approach to systems in a fuzzy world. *Synthese* **30**, 381–406.

Arbib, M. A., and Manes, E. G. (1975b). Fuzzy machines in a category. *J. Aust. Math. Soc.* **13**, 169–210.

Chang, S. S. L. (1977). Application of fuzzy set theory to economics. *Kybernetes* **6**, 203–207.

Chang, S. S. L., and Zadeh, L. A. (1972). On fuzzy mapping and control. *IEEE Trans. Syst. Man. Cybern.* **2**, No. 1, 30–34.

Dal Cin, M. (1975a). Fuzzy-state automata: Their stability and fault tolerance. *Int. J. Comput. Inf. Sci.* **4**, No. 1, 63–80.

Dal Cin, M. (1975b). Modification tolerance of fuzzy-state automata. *Int. J. Comput. Inf. Sci.* **4**, No. 1, 81–93.

Gaines, B. R. (1977). Sequential fuzzy system identification. *Proc. IEEE Conf. Decision Control, New Orleans* pp. 1309–1314. [Also in *Int. J. Fuzzy Sets Syst.* **2**, No. 1, 15–24 (1979).]

Gaines, B. R., and Kohout, L. J. (1976). The logic of automata. *Int. J. Gen. Syst.* **2**, 191–208.

Ishikawa, A., and Mieno, H. (1976). Feedforward control system and fuzzy entropy. Working Paper No. 76-12. Grad. School Bus. Adm. Rutgers, The State University, New Brunswick, New Jersey.

Jain, R. (1976). Outline of an approach for the analysis of fuzzy systems. *Int. J. Control* **23**, No. 5, 627–640.

Jain, R. (1977). Analysis of fuzzy systems. In "Fuzzy Automata and Decision Processes" (M. M. Gupta, G. N. Saridis, and B. R. Gaines, eds.), pp. 251–268. North-Holland Publ., Amsterdam.

Maarschalk, C. G. D. (1977). Exact and fuzzy concepts surimposed to the G.S.T. (General Systems Theory): A meta-theory. In "Modern Trends in Cybernetics and Systems" (J. Rose and C. Bilciu, eds.), Vol. 2, pp. 95–102. Springer-Verlag, Berlin and New York.

Meseguer, J., and Sols, I. (1974). Automata in semi-module categories. *Proc. Int. Symp. Category Theory Appl. Comput. Control, 1st* pp. 196–202.

Nazaroff, G. J. (1973). Fuzzy topological polysystems. *J. Math. Anal. Appl.* **41**, 478–485.

Negoita, C. V., and Kelemen, M. (1977). On the internal model principle. *Proc. IEEE Conf. Decision Control, New Orleans* pp. 1343–1344.

Negoita, C. V., and Ralescu, D. A. (1974). Fuzzy systems and artificial intelligence. *Kybernetes* **3**, 173–178.

Negoita, C. V., and Ralescu, D. A. (1975). "Application of Fuzzy Sets to System Analysis." Chap. 4. Birkaeuser, Basel. (Reference from I.)

Negoita, C. V., and Ralescu, D. A. (1977). Some results in fuzzy systems theory. In "Modern Trends in Cybernetics and Systems" (J. Rose and C. Bilciu, eds.), Vol. 2, pp. 87–93. Springer-Verlag, Berlin and New York.

Negoita, C. V., Ralescu, D. A., and Ratiu, T. (1977). Relations on monoids and realization theory for dynamic systems. In "Modern Trends in Cybernetics and Systems" (J. Rose and C. Bilciu, eds.), Vol. 2, pp. 337–350. Springer-Verlag, Berlin and New York.

Negoita, C. V., and Stefanescu, A. C. (1975). On the state equation of fuzzy systems. *Kybernetes* **4**, 213–214.

Pollatschek, M. A. (1977). Hierarchical systems and fuzzy set theory. *Kybernetes* **6**, 147–152.

Santos, E. S. (1968). Maximin automata. *Inf. Control* **13**, 363–377.

Santos, E. S. (1972a). On reductions of maximin machines. *J. Math. Anal. Appl.* **40**, 60–78.

Santos, E. S. (1972b). Max-product machines. *J. Math. Anal. Appl.* **37**, 677–686.

Santos, E. S. (1973). Fuzzy sequential functions. *J. Cybern.* **3**, No. 3, 15–31.

Santos, E. S., and Wee, W. G. (1968). General formulation of sequential machines. *Inf. Control* **12**, 5–10.

Saridis, G. N. (1975). Fuzzy notions in nonlinear system classification. *J. Cybern.* **4**, No. 2, 67–82.

Sugeno, M., and Terano, T. (1977). Analytical representation of fuzzy systems. In "Fuzzy Automata and Decision Processes" (M. M. Gupta, G. H. Saridis, and B. R. Gaines, eds.), pp. 177–189. North-Holland Publ., Amsterdam.

Tong, R. M. (1978a). Synthesis of fuzzy models for industrial process—Some recent results. *Int. J. Gen. Syst.* **4**, 143–162.

Tong, R. M. (1978b). Analysis and control of fuzzy systems using finite discrete relations. *Int. J. Control* **27**, No. 3, 431–440.

Warren, R. H. (1976). Optimality in fuzzy topological polysystems. *J. Math. Anal. Appl.* **54**, 309–315.

Wee, W. G. (1967). "On Generalization of Adaptive Algorithms and Applications of the Fuzzy Sets Concept to Pattern Classification," Ph.D. Thesis, TR-EE 67/7. Purdue Univ., Lafayette, Indiana.

Wee, W. G., and Fu, K. S. (1969). A formulation of fuzzy automata and its application as a model of learning systems. *IEEE Trans. Syst. Sci. Cybern.* **5**, 215–223. (Reference from IV.5.)

Yager, R. R. (1978). Validation of fuzzy-linguistic models. *J. Cybern.* **8**, 17–30.

Zadeh, L. A. (1965). Fuzzy sets and systems. *Proc. Symp. Syst. Theory, Polytech. Inst. Brooklyn* pp. 29–37.

Zadeh, L. A. (1971). Toward a theory of fuzzy systems. In "Aspects of Network and System Theory" (R. E. Kalman and N. De Claris, eds.), pp. 469–490.

Zadeh, L. A. (1972). A rationale for fuzzy control. *J. Dyn. Syst., Meas. Control* **34**, 3–4. (Reference from III.3.)

Chapter **3**

FUZZY LANGUAGES—
FUZZY ALGORITHMS

This chapter is divided into two distinct parts. The first deals with the application of fuzzy set theory to formal languages. Many papers have been published on this topic. These are mainly interested in studying the properties of fuzzy grammars and the recognition capabilities of fuzzy automata. However, other models have been developed where the original max and min operators and the valuation set [0, 1] were more or less given up and other structures were investigated. These models share little with the initial motivations and purposes of fuzzy set theory. On the contrary, in order to reduce the gap between formal languages and natural language, Zadeh has proposed an alternative approach where the semantic aspects are no longer neglected.

The second part is devoted to fuzzy algorithms. A clear distinction is made between usual algorithms extended to deal with fuzzy data and algorithms that are approximate descriptions of complex actions or procedures, yielding fuzzy or nonfuzzy results. Both formal and semantic aspects are discussed.

A. FUZZY LANGUAGES AND FUZZY GRAMMARS

"The precision of formal languages contrasts rather sharply with the imprecision of natural languages. To reduce the gap between them, it is natural to introduce randomness into the structure of formal languages, thus leading to the concept of stochastic languages" (see, e.g., Fu and

Huang, NF 1972). "Another possibility lies in the introduction of fuzziness" (Lee and Zadeh, 1969).

This section gives a survey of fuzzy *formal* languages and grammars and more general formal models that encompass them. The reader is assumed familiar with the theory of formal languages (see, e.g., Hopcroft and Ullman, NF 1969).

a. Fuzzy Languages   (Lee and Zadeh, 1969)

Let V_T be a finite set called an *alphabet*. We denote by V_T^* the set of finite strings constructed by concatenation of elements of V_T, including the null string Λ. V_T^* is a free monoid over V_T. A language is a subset of V_T^*. Very naturally, a *fuzzy formal language* is a fuzzy set $\mathcal{L}$ on V_T^*, i.e.,

$$\mathcal{L} = \sum_{x \in V_T^*} \mu_\mathcal{L}(x)/x$$

with $\mu_\mathcal{L}$ a function from V_T^* to $[0, 1]$. $\mu_\mathcal{L}(x)$ is the degree of membership of x in $\mathcal{L}$ and can be interpreted as a degree of properness of the string x, valuating to what extent it is suitable to use it.

Union and intersection of fuzzy languages can be defined as usual

$$\mathcal{L}_1 \cup \mathcal{L}_2: \quad \mu_{\mathcal{L}_1 \cup \mathcal{L}_2}(x) = \max(\mu_{\mathcal{L}_1}(x), \mu_{\mathcal{L}_2}(x)), \qquad \forall x \in V_T^*,$$

$$\mathcal{L}_1 \cap \mathcal{L}_2: \quad \mu_{\mathcal{L}_1 \cap \mathcal{L}_2}(x) = \min(\mu_{\mathcal{L}_1}(x), \mu_{\mathcal{L}_2}(x)), \qquad \forall x \in V_T^*.$$

And the complement $\bar{\mathcal{L}}$ of $\mathcal{L}$ has membership function $1 - \mu_\mathcal{L}$.

A specific operation between languages is *concatenation*: any string x in V_T^* is the concatenation of a prefix string u and a suffix string $v: x = uv$. According to the extension principle, the concatenation $\mathcal{L}_1\mathcal{L}_2$ of two fuzzy languages $\mathcal{L}_1$ and $\mathcal{L}_2$ is defined by

$$\mu_{\mathcal{L}_1\mathcal{L}_2}(x) = \sup_{x = uv} \min(\mu_{\mathcal{L}_1}(u), \mu_{\mathcal{L}_2}(v)).$$

The concatenation of fuzzy languages is associative. Denoting by $\mathcal{L}^n$ the concatenation of $\mathcal{L}$ n times, the Kleene closure of $\mathcal{L}$ is $\hat{\mathcal{L}} = \{\Lambda\} \cup \mathcal{L} \cup \mathcal{L}^2 \cup \mathcal{L}^3 \cup \cdots \cup \mathcal{L}^n \cup \cdots$. Note that $\forall x \in V_T^*$, if $x = a_1 a_2 \cdots a_k$, $a_i \in V_T$, $i = 1, k$, then

$$\mu_{\hat{\mathcal{L}}}(x) = \sup_{i = 1, k} \left[\sup_{\substack{x = u_1 u_2 \cdots u_i \\ \forall j,\, u_j \in V_T^*}} \left[\min_{j = 1, i} \mu_\mathcal{L}(u_j) \right] \right];$$

k is the length of x, denoted $l(x)$.

The following property holds (Negoita and Ralescu, 1975): $\mathcal{L} = \hat{\mathcal{L}}$ iff $\mu_\mathcal{L}(\Lambda) = 1$ and $\mu_\mathcal{L}(uv) \geqslant \min(\mu_\mathcal{L}(u), \mu_\mathcal{L}(v))$ $\forall u, v \in V_T^*$.

A fuzzy language that is its own Kleene closure is said to be closed, and it is obviously a fuzzy monoid in V_T^*, in the sense of Rosenfeld (Reference from II.1).

N.B.:   The idea of valuating the strings of a language is not new. In a probabilistic context Rabin (NF 1963) already used weighted languages, but the semantics were different: the weight of a string reflected a frequency of occurrence. At about the same time Chomsky and Schützenberger (NF 1970) assigned integer values to strings in order to model structural ambiguity.

b.   Fuzzy Grammars   (Lee and Zadeh, 1969)

"Informally, a *fuzzy grammar* may be viewed as a set of rules for generating the elements of a fuzzy set" (Lee and Zadeh, 1969). More precisely, a fuzzy grammar is a quadruple $G = (V_N, V_T, P, s)$ where: V_T is a set of terminals or alphabet; V_N is a set of nonterminals ($V_N \cap V_T = \emptyset$), i.e., labels of certain fuzzy sets on V_T^* called fuzzy syntactic categories; P is a finite set of rules called productions; and $s \in V_N$ is the initial symbol, i.e., the label of the syntactic category "string." The elements of P are expressions of the form $\alpha \overset{\rho}{\to} \beta$; $\rho \in [0, 1]$ where α and β are strings in $(V_T \cup V_N)^*$. ρ is the grade of membership of β given α. The symbol * always indicates a free monoid X^* over the set X. ρ also expresses a degree of properness of the rule $\alpha \to \beta$.

Let $\alpha_1, \ldots, \alpha_m$ be strings in $(V_T \cup V_N)^*$, and $\alpha_1 \overset{\rho_2}{\to} \alpha_2, \ldots, \alpha_{m-1} \overset{\rho_m}{\to} \alpha_m$ be productions. Then α_m is said to be *derivable* from α_1 in G, more briefly $\alpha_1 \Rightarrow_G \alpha_m$. The expression $\alpha_1 \overset{\rho_2}{\to} \alpha_2, \ldots, \overset{\rho_m}{\to} \alpha_m$ will be referred to as a derivation chain from α_1 to α_m.

A fuzzy grammar G generates a fuzzy language $\mathcal{L}(G)$ in the following manner. A string x of V_T^* is said to be in $\mathcal{L}(G)$ iff x is derivable from s. The grade of membership $\mu_G(x)$ of x in $\mathcal{L}(G)$ is

$$\mu_G(x) = \sup \min(\mu(s \to \alpha_1), \mu(\alpha_1 \to \alpha_2), \ldots, \mu(\alpha_m \to x)) > 0 \qquad (1)$$

where $\mu(\alpha_i \to \alpha_{i+1})$ is the nonnull ρ_{i+1} such that

$$\left(\alpha_i \overset{\rho_{i+1}}{\to} \alpha_{i+1} \right) \in P \;\forall i = 0, m,$$

if $\alpha_0 = s$ and $\alpha_{m+1} = x$.

The supremum is taken over all derivation chains from s to x. Note that "x is in $\mathcal{L}(G)$" means $x \in \text{supp} \, \mathcal{L}(G)$. The degree of properness of a derivation chain is that of its least proper link, and $\mu_G(x)$ is calculated on the "best" chain. Two fuzzy grammars G_1 and G_2 are said to be equivalent iff $\forall x \in V_T^*$, $\mu_{G_1}(x) = \mu_{G_2}(x)$.

Example $V_N = \{s, A, B\}$; $V_T = \{a, b\}$;

$$P = \left\{ \begin{array}{llll} s \xrightarrow{0.3} a, & s \xrightarrow{0.5} aA, & s \xrightarrow{0.2} bA, & A \xrightarrow{0.5} b \\ s \xrightarrow{0.3} as, & s \xrightarrow{0.7} aB, & s \xrightarrow{0.1} b, & B \xrightarrow{0.4} b \end{array} \right\}.$$

Consider the string ab. There are three derivation chains:

$$s \xrightarrow{0.7} aB \xrightarrow{0.4} ab, \qquad s \xrightarrow{0.5} aA \xrightarrow{0.5} ab, \qquad s \xrightarrow{0.3} as \xrightarrow{0.1} ab,$$

and $\mu_G(ab) = \max(\min(0.7, 0.4), \min(0.5, 0.5), \min(0.3, 0.1)) = 0.5$.

Paralleling the standard classification of ordinary grammars, we can distinguish four types of fuzzy grammars:

Type 0 grammar. The allowed productions are of the general form $\alpha \xrightarrow{\rho} \beta$, $\rho > 0$, $\alpha, \beta \in (V_T \cup V_N)^*$.

Type 1 grammar (context-sensitive). The productions are of the form $\alpha_1 A \alpha_2 \xrightarrow{\rho} \alpha_1 \beta \alpha_2$, $\rho > 0$, $\alpha_1, \alpha_2, \beta$ are in $(V_T \cup V_N)^*$, A in V_N, $\beta \neq \Lambda$. $s \xrightarrow{1} \Lambda$ is also allowed.

Type 2 grammar (context-free). The allowable productions are now $A \xrightarrow{\rho} \beta$, $\rho > 0$, $A \in V_N$, $\beta \in (V_N \cup V_T)^*$, $\beta \neq \Lambda$, and $s \xrightarrow{1} \Lambda$.

Type 3 grammar (regular). The allowable productions are $A \xrightarrow{\rho} aB$ or $A \xrightarrow{\rho} a$, $\rho > 0$, where $a \in V_T$; $A, B \in V_N$, and $s \xrightarrow{1} \Lambda$.

In the above example the grammar G is regular. If G is of type i, $\mathcal{L}(G)$ is said to be of type i.

A grammar is said to be recursive iff there is an algorithm that computes $\mu_G(x)$. Lee and Zadeh (1969) showed that a fuzzy context-sensitive grammar was recursive. The proof uses loop-free derivation chains; the set of such chains, over which the supremum is taken in (1), can be further restricted to those of bounded length l_0 which depends on $l(x)$ and $|V_T \cup V_N|$. The number of loop-free chains is finite because a production of a context-sensitive grammar is noncontracting, i.e., $\forall (\alpha_i \xrightarrow{\rho} \alpha_j)$, $l(\alpha_j) \geqslant l(\alpha_i)$: and thus the search process is finite. Note that type 2 and 3 grammars are recursive too since they are particular cases of context-sensitive grammars.

Another kind of extension of classical results to fuzzy grammar is Chomsky and Greibach normal forms for context-free grammars (Lee and Zadeh 1969); let G be a fuzzy context-free grammar:

G is equivalent to a fuzzy grammar whose productions are of the form $A \xrightarrow{\rho} BC$ or $A \xrightarrow{\rho} a$ where $A, B, C \in V_N$, $a \in V_T$ (Chomsky);

G is equivalent to a fuzzy grammar whose productions are of the form $A \overset{\rho}{\rightarrow} a\alpha$, $a \in V_T$, $\alpha \in V_N^*$, $A \in V_N$ (Greibach).

The canonical forms are obtained as for nonfuzzy grammars, provided we add formula (1) for valuating derivation chains. For a detailed proof, the reader is referred to Lee and Zadeh (1969).

c. Cut-Point Languages

Let $\mathcal{L}(G)$ be a fuzzy language and G a grammar that generates $\mathcal{L}(G)$. Several nonfuzzy languages can be generated from $\mathcal{L}(G)$. For instance,

$$\mathcal{L}(G, \lambda) = \{ x \in V_T^* \mid \mu_G(x) > \lambda \},$$

$$\mathcal{L}(G, \geqslant, \lambda) = \{ x \in V_T^* \mid \mu_G(x) \geqslant \lambda \},$$

$$\mathcal{L}(G, =, \lambda) = \{ x \in V_T^* \mid \mu_G(x) = \lambda \},$$

$$\mathcal{L}(G, \lambda_1, \lambda_2) = \{ x \in V_T^* \mid \lambda_1 < \mu_G(x) \leqslant \lambda_2 \},$$

where $\lambda, \lambda_1, \lambda_2$ are thresholds that belong to $[0, 1]$. These languages are called *cut-point* languages. Note that since P is finite, the image of V_T^* through μ_G, $\mu_G(V_T^*) \subseteq [0, 1]$, is also finite because we use only max and min operators to valuate strings. Let $\mu_G(V_T^*)$ be $\{0, \mu_1, \mu_2, \ldots, \mu_p\}$ where p is at most the number of rules in P. Thus the number of cut-point languages of each kind is finite, and depends on the number of distinct production valuations.

Mizumoto *et al.* (1970) have proven the following properties:

(i) $\forall \lambda$, $\forall i = 0, 3$, if G is a fuzzy grammar of type i, then $\mathcal{L}(G, \lambda)$ is of type i.

(ii) $\forall \lambda$, for $i = 0$ and 2, if G is a fuzzy grammar of type i, then $\mathcal{L}(G, \lambda_1, \lambda_2)$ and $L(G, =, \lambda)$ may not be of type i. For $i = 3, \mathcal{L}(G, \lambda_1, \lambda_2)$ and $\mathcal{L}(G, =, \lambda)$ are of type 3. For $i = 1$, the result is unknown.

(iii) $\forall \lambda$, $i = 0, 3$, if G is a fuzzy grammar of type i, then $\mathcal{L}(G, \geqslant, \lambda)$ is of type i.

Proof: (i) and (iii) stem from the fact that only rules of production $\alpha \overset{\rho}{\rightarrow} \beta$ such that $\rho > \lambda$ (resp. $\rho \geqslant \lambda$) are used to build $\mathcal{L}(G, \lambda)$ (resp. $\mathcal{L}(G, \geqslant, \lambda)$. Moreover, $\mathcal{L}(G, \lambda_1, \lambda_2) = \mathcal{L}(G, \lambda_1) - \mathcal{L}(G, \lambda_2)$; $\mathcal{L}(G, =, \lambda) = \mathcal{L}(G, \geqslant, \lambda) - \mathcal{L}(G, \lambda)$, and the sets of all languages of type 0 and 2 are not closed under subtraction. For $i = 1$, the result is unknown. For $i = 3$, the closure property holds.

d. **Fuzzy Syntax Directed Translations** (Thomason, 1974)

A translation T of a language $\mathcal{L}_1$ in V_T^* into a language $\mathcal{L}_2$ in W_T^* where V_T, W_T are alphabets is a fuzzy relation on $V_T^* \times W_T^*$ such that dom $T = \mathcal{L}_1$ and ran $T = \mathcal{L}_2$, where dom T and ran T are respectively the domain and the range of T (see II.3.B.a). $\mu_T(x, y)$ is the grade of properness of translating x by y, $x \in \text{supp}\,\mathcal{L}_1$, $y \in \text{supp}\,\mathcal{L}_2$.

An efficient model in formal language translation theory is that of a syntax-directed translation scheme (SDTS). A fuzzy STDS is a 5-tuple $\mathcal{T} = (V_N, V_T, W_T, s, D)$ where V_N is a set of nonterminals, V_T and W_T are alphabets, s an initial symbol and D a set of double productions $A \overset{\rho}{\to} \alpha, \beta$ with $A \in V_N, (\alpha, \beta) \in (V_T \cup V_N)^* \times (W_T \cup V_N)^*$, and $\rho > 0$ valuates the translation of α into β.

N.B.: α and β are assumed to contain the same nonterminal elements, but not necessarily to have the same length.

Obviously, when a string is generated in V_T^*, another string is generated in W_T^* by means of a double derivation chain. A fuzzy STDS builds the translation relation T.

Example $V_N = \{s, A, B\}$, $V_T = \{a, b\}$, $W_T = \{c, d, e\}$

$$\text{Productions:}\ \begin{cases} s \overset{0.3}{\to} as, cs; & s \overset{0.3}{\to} bs, es; & s \overset{0.5}{\to} aA, dA \\ s \overset{0.7}{\to} aB, dB; & s \overset{0.2}{\to} bA, cA; & s \overset{0.1}{\to} b, e \\ A \overset{0.5}{\to} b, c; & B \overset{0.4}{\to} b, c \end{cases}.$$

Consider $ab \in V_T^*$. There are two derivation chains that translate ab into dc:

$$s \overset{0.7}{\to} aB, dB \overset{0.4}{\to} ab, dc \quad \text{and} \quad s \overset{0.5}{\to} aA, dA \overset{0.5}{\to} ab, dc.$$

According to (1), $\mu_T(ab, dc) = \max(\min(0.7, 0.4), \min(0.5, 0.5)) = 0.5$. Another possible translation of ab is ce since $(s \overset{0.3}{\to} as, cs \overset{0.1}{\to} ab, ce)$ and $\mu_T(ab, dc) = \min(0.3, 0.1) = 0.1$.

N.B.: Note that a regular STDS, as the one of the example, always translates a string into one or several strings of the same length, that of the string to be translated.

e. *N*-fold Fuzzy Grammars (Mizumoto *et al.*, 1970, 1973a)

Ordinary fuzzy grammars have one major drawback that prevents them from being a convenient tool for the modeling of natural language: the grade of properness of productions is always context-free, which is not realistic.

Instead of increasing V_N or P in order to cope with this context-dependency, it is possible to define a grammar where grades of properness of productions depend on productions that have been previously used.

An N-fold fuzzy grammar is a 6-tuple $(V_T, V_N, s, P, J, \mathcal{R}) = G$ where V_T is a set of terminals, V_N of nonterminals, and s an initial symbol. J is a set of labels for production rules $(J \subseteq \mathbb{N})$; P a set of production rules $\alpha \underset{l}{\rightarrow} \beta$ where $l \in J$; and $\mathcal{R}$ a set of $N + 1$ fuzzy relations denoted R_i, $i = 1, N + 1$. R_i is an i-ary relation on J, $i = 2, N + 1$, such that $\mu_{R_i}(l_1, \ldots, l_i)$ is the degree of properness of using production l_i when productions $l_1, \ldots, l_{i-1}$ have already been used successively just before l_i in a derivation chain.

For $i = 1$, $\mu_{R_1}(l)$ valuates productions of the form $s \underset{l}{\rightarrow} \alpha$, where l belongs to a subset J_s of J, which labels such productions.

Consider the derivation chain

$$s \overset{\rho_1}{\underset{l_1}{\rightarrow}} \alpha_1 \overset{\rho_2}{\underset{l_2}{\rightarrow}} \alpha_2 \rightarrow \cdots \overset{\rho_p}{\underset{l_p}{\rightarrow}} \alpha_p.$$

We have $\rho_1 = \mu_{R_1}(l_1)$, $\rho_2 = \mu_{R_2}(l_1, l_2)$, $\ldots$, $\rho_{N+1} = \mu_{R_{N+1}}(l_1, \ldots, l_{N+1})$; and for $i > N + 1$, $\rho_i = \mu_{R_{N+1}}(l_{i-N}, \ldots, l_i)$. Hence, the grade of properness of a production depends on the N previously used productions.

N.B.:   A 0-fold fuzzy grammar is an ordinary fuzzy grammar.

Example (Mizumoto *et al.*, 1970) We give an example of a 1-fold fuzzy grammar:

$$V_N = (A, B, C, s), \quad V_T = \{a, b, c\} \quad J_s = \{1\} \quad \mu_{R_1}(1) = 0.9.$$

R_2 is defined by

	1	2	3	4	5	6	7	8	9	10
(1) $s \rightarrow ABC$ (		0.7			0.8			0.9		)
(2) $A \rightarrow aA$ (		0.7								)
(3) $B \rightarrow bB$ (			0.7							)
(4) $C \rightarrow cC$ (	0.7							0.7		)
(5) $A \rightarrow aAa$ (						0.8				)
(6) $B \rightarrow bBb$ (							0.8			)
(7) $C \rightarrow cCc$ (					0.8			0.8		)
(8) $A \rightarrow a$ (									0.9	)
(9) $B \rightarrow b$ (										0.9)
(10) $C \rightarrow c$ (										)

In the list of productions each blank space indicates that the value of ρ at that space is a numerical value within the range $[0, 0.65]$. Consider the generation of the sequence $a^3b^3c^3$:

(i) $s \xrightarrow[1]{0.9} ABC \xrightarrow[2]{0.7} aABC \xrightarrow[3]{0.7} aAbBC \xrightarrow[4]{0.7} aAbBcC \xrightarrow[2]{0.7} a^2AbBcC$

$\xrightarrow[3]{0.7} a^2Ab^2BcC \xrightarrow[4]{0.7} a^2Ab^2Bc^2C \xrightarrow[8]{0.7} a^3b^2Bc^2C \xrightarrow[9]{0.9} a^3b^3c^2C \xrightarrow[10]{0.9} a^3b^3c^3.$

(ii) $s \xrightarrow[1]{0.9} ABC \xrightarrow[5]{0.8} aAaBC \xrightarrow[6]{0.8} aAabBbC \xrightarrow[7]{0.8} aAabBbcCc$

$\xrightarrow[8]{0.8} a^3bBbcCc \xrightarrow[9]{0.9} a^3b^3cCc \xrightarrow[10]{0.9} a^3b^3c^3.$

The grades of properness of (i) and (ii) are respectively 0.7 and 0.8. Other derivations are possible. It could be checked that $\mu_G(a^3b^3c^3) = 0.8$. Moreover, the cut-point languages of G are

$$\mathcal{L}(G, 0.95) = \varnothing; \quad L(G, 0.85) = \{a, b, c\};$$

$$\mathcal{L}(G, 0.75) = \{a^{2n-1}b^{2n-1}c^{2n-1} \mid n > 0, n \in \mathbb{N}\};$$

$$\mathcal{L}(G, 0.65) = \{a^h b^h c^h \mid h > 0, h \in \mathbb{N}\};$$

$$\mathcal{L}(G, 0) = \{a^p b^q c^r \mid p, q, r \in \mathbb{N} \quad \text{and} \quad pqr \neq 0\}.$$

Note that G has a context-free structure, and $\mathcal{L}(G, 0.75)$, $\mathcal{L}(G, 0.65)$ are context-sensitive languages. Hence, type 2 N-fold fuzzy grammars can generate type 1 fuzzy languages, i.e., N-fold fuzzy grammars are more powerful than ordinary fuzzy grammars.

Mizumoto et al. (1973a) showed that given a regular N-fold fuzzy grammar, it is always possible to build an $(N + 1)$-fold fuzzy grammar and an $(N - 1)$-fold fuzzy grammar ($N \geqslant 1$) that are equivalent to the initial grammar. Hence, a regular N-fold fuzzy grammar is able to generate only regular fuzzy languages and is a useless notion.

N.B.:   Another way of reducing the gap between formal and natural languages by modifying a fuzzy grammar was suggested by Kandel (1974) who constrains derivations through a control language. This is a set of strings that encode allowed derivation chains in a deterministic fashion. Although this approach seems much more rigid than that of Mizumoto et al. (1973a) the determination of the control language looks more straightforward than that of the relations R_i, from the point of view of grammatical inference.

f. Other Kinds of Grammars

Other kinds of fuzzy grammars have been considered in the literature. Santos (1975a, b) has studied the so-called max–product grammars, which are fuzzy grammars such that the valuation set of productions is $\mathbb{R}^+$. Moreover, for the evaluation of a derivation chain, the product replaces min in (1). Max–product grammars also arise naturally from the application of maximal interpretation of stochastic grammars (Fu and Huang, NF 1972). Santos (1975a) proved that context-free max–product grammars generate a set of fuzzy languages that contains the set of fuzzy context-free languages as a proper subset, without being the set of fuzzy context-sensitive languages. Regular max–product grammars were investigated in Santos (1975b). They are more general than the max–min ones, in that they generate nonregular languages. But some context-free and stochastic languages are not obtained.

DePalma and Yau (1975) introduced *fractionally fuzzy grammars*. A string is derived in the same manner as in the case of a fuzzy grammar. However, the membership of a string is given by

$$\mu_G(x) = \sup_k \left[\frac{\sum\limits_{i=1}^{n_k} g(l_k^i)}{\sum\limits_{i=1}^{n_k} h(l_k^i)} \right] \in [0,1]$$

with k = index of a derivation chain leading to x; n_k = length of the kth derivation chain; g and h functions from J to $\mathbb{R}$, where J labels the productions, and $h(l) \geqslant g(l)$ $\forall l \in J$; and the convention $0/0 = 0$. Fractionally fuzzy grammars were used by both authors instead of fuzzy grammars in order to reduce the combinatorial aspect of parsing in a pattern recognition procedure (see IV.6.A.b).

Fuzzy tree grammars were also considered by Inagaki and Fukumura (1975). A tree grammar (see Brainard, NF 1969) is a 5-tuple (V_N, V_T, r, P, s) where V_N, V_T are sets of nonterminals and terminals, respec ively. r is a mapping $V_T \cup V_N \to \mathbb{N}$ ranking V_T and V_N, P is a finite set of productions $\Phi \to \Psi$ with Φ and Ψ being trees over $(V_N \cup V_T, r)$. r is used for encoding the trees. Inagaki and Fukumura use production rules valuated on [0, 1].

Another way of increasing the generative power of a fuzzy grammar is to take a lattice as a valuation set. A general formulation of formal lattice-valued grammars was proposed by Mizumoto *et al.* (1972).

An L-fuzzy language over an alphabet V_T is an L-fuzzy set (see II.1.G.a) on V_T^*. Union and intersection are defined using the operators sup and inf of the lattice L, as in II.1.G.a. The concatenation of two L-fuzzy languages $\mathcal{L}_1$ and $\mathcal{L}_2$ is very similar to that of ordinary fuzzy languages:

$$\forall x \in V_T^*, \quad \mu_{\mathcal{L}_1 \mathcal{L}_2}(x) = \sup_{\substack{u \in V_T^*, v \in V_T^* \\ x = uv}} \inf(\mu_{\mathcal{L}_1}(u), \mu_{\mathcal{L}_2}(v)).$$

Mizumoto *et al.* suggest that sup and inf may be exchanged. We then obtain an inf–sup concatenation. Kleene closure is also defined very similarly, as above in a.

An *L*-fuzzy grammar is a fuzzy grammar where productions are valued in a lattice *L*, also called a *weighting space* (see Kim *et al.*, 1974). The properties of *L*-fuzzy grammars have been studied by several authors for particular types of lattices. Here we only state some results without proofs:

L is a Boolean finite lattice B (Mizumoto *et al.*, 1975a):

the class of cut-point languages $\mathfrak{L}(G,\lambda), \lambda \in B$, generated by context-free *B*-fuzzy grammars properly contains the class of context-free languages.

any cut-point language $\mathfrak{L}(G,\lambda), \lambda \in B$, generated by regular *B*-fuzzy grammars is a regular language;

regular *B*-fuzzy languages are a closed set under union, intersection, sup–inf concatenation, and Kleene closure; however, the complement (in the sense of Brown, Reference from II.1; see II.1.G.a) of a regular *B*-fuzzy language is generally not a regular *B*-fuzzy language, but can be generated through inf–sup concatenation of regular *B*-fuzzy languages.

context-free *B*-fuzzy languages are a closed set under union, sup–inf concatenation, and Kleene closure, but not under intersection and complementation.

L is an ordered semiring R (Wechler, 1975a):

a semiring $(R, +, \cdot)$ is an algebraic structure equipped with two operations $+$ and $\cdot$, such that:

$(R, +)$ is a semigroup (or monoid), i.e., $\forall r_1, r_2 \in R_1 \; r_1 + r_2 \in R$, $+$ is associative, $\exists 0 \in R$; $r_1 + 0 = 0 + r_1 = r_1$ for any $r_1 \in R$; 0 is an identity;

$(R - \{0\}, \cdot)$ is a semigroup; the identity is 1;

0 is a zero for $\cdot$, i.e., $\forall r \in R, 0 \cdot r = r \cdot 0 = 0$;

$\cdot$ is distributive over $+$.

A semiring *R* is said to be an ordered semiring $(R, \leqslant)$ iff $\leqslant$ is a partial ordering and:

$\forall r_1, r_2, r_3 \in R$ if $r_1 \leqslant r_2$, then $r_1 + r_3 \leqslant r_2 + r_3$ and $r_3 + r_1 \leqslant r_3 + r_2$;

$\forall r_1, r_2 \in R, \forall r_3 \neq 0$ such that $0 \leqslant r_3$; if $r_1 \leqslant r_2$, then $r_1 \cdot r_3 \leqslant r_2 \cdot r_3$ and $r_3 \cdot r_1 \leqslant r_3 \cdot r_2$.

Examples Examples include $[0, 1]$ with the usual ordering, $+ = \max$, $\cdot = \min$, $\mathbb{N}$, the set of positive integers, $\mathbb{Z}$ the set of integers, $\mathbb{R}$ that of real numbers, any complete distributive lattice equipped with their usual operations.

The weight of a string generated by an R-fuzzy grammar, where R is an ordered semiring is the sum (+) of the weights of all the derivation chains from the initial symbol to the string. The weight of a derivation chain is the product of the weights of the productions involved in the chain. The partial ordering $\leqslant$ is used for the definition of cut-point languages.

Wechler (1975a) gives the following theorem: Let G be a regular R-fuzzy grammar. Then $\mathcal{L}(G,r)$ is a regular language for every $r \in R$ if R is a finite ordered semiring or a complete distributive lattice or $\mathbb{N}$.

N.B.:   The idea of using an ordered semiring to weight the productions of a grammar and to valuate strings in a language is not new. Chomsky and Schützenberger (NF 1970) first suggested it in 1963. Each string x generated by a grammar G was associated with a positive integer n which was the number of possible derivation chains existing from the initial symbol to x. n measured the grade of ambiguity of x with respect to G. Both authors studied the algebraic properties of such "$\mathbb{N}$-fuzzy languages" generated by context-free grammars. They introduced a representation of the language which was a formal power series $\sum_{x \in V_T^*} \langle w, x \rangle x$ where w is the weighting function and $\langle w, x \rangle = w(x) \in \mathbb{N}$; this notation is very similar to that introduced by Zadeh (Reference from II.2, 1972; see II.1.A) for representing fuzzy sets.

Other examples of formal grammars with weights can be found in Mizumoto et al. (1973b).

In conclusion, it seems rather easy to enhance the generative power of fuzzy context-free grammars by modifying the weighting space; however, very often, L-fuzzy grammars, when regular, generate only regular L-fuzzy languages for usual L. Modifying also the valuation rules of derivation chains and strings looks more efficient, as shown by the properties of max–product fuzzy grammars.

g. Languages and Automata

α. Generation of a Fuzzy Language by a Fuzzy Automaton

Fuzzy automata have been introduced in the previous chapter as models of fuzzy systems. Here they will be considered as acceptors of fuzzy languages.

Let $\mathcal{Q} = (U, S, Y, \tilde{s}_0, \delta, \sigma)$ be the fuzzy automaton already introduced in Chapter 2. Recall that:

U is a finite set of inputs, $U = \{a_1, \ldots, a_m\}$;
S is a finite set of states, $S = \{q_1, \ldots, q_n\}$;
Y is a finite set of outputs, $Y = \{y_1, \ldots, y_p\}$;
$\tilde{s}_0$ is a fuzzy set on X, the fuzzy initial state;

δ is a fuzzy ternary relation on $S \times U \times S$, made up of m transition relations $\{\delta_u\}_{u \in U}$ for the states;

σ is a fuzzy relation on $S \times Y$, i.e., the output map.

When a nonfuzzy input u is processed by the automaton, the output can be symbolically written $\tilde{y} = \tilde{s}_0 \circ \delta_u \circ \sigma$ where $\circ$ is the (associative) composition of binary relations. Once a string of inputs $\theta = u_1 u_2 \cdots u_k$ has been processed by the automaton, the fuzzy output is

$$\tilde{y} = \tilde{s}_0 \circ \left(\delta_{u_1} \circ \delta_{u_2} \circ \cdots \circ \delta_{u_k} \right) \circ \sigma = \tilde{s}_0 \circ \Delta_\theta \circ \sigma.$$

Denote by Δ_θ the result of the composition $\delta_{u_1} \circ \cdots \circ \delta_{u_k}$. Δ_Λ is the identity relation.

From now on we consider automata whose output set is the singleton $\{y\}$. What such automata compute are membership values, i.e., $\tilde{y} = f_{\mathcal{Q}}(\theta)/y$ using Zadeh's notation, where $f_{\mathcal{Q}}$ is the response function of the automaton. For simplicity, we shall write $f_{\mathcal{Q}}(\theta) = \tilde{s}_0 \circ \Delta_\theta \circ \sigma$.

Note that the image of U^* (the free monoid over U containing all the finite strings of inputs) under $f_{\mathcal{Q}}$ is a fuzzy language $\mathcal{L}(\mathcal{Q})$. Hence, fuzzy automata can recognize fuzzy languages.

β. *Structural Properties of Automata-Generated Fuzzy Languages*

Structural properties of fuzzy languages accepted by fuzzy automata were studied by Mizumoto *et al.* (1969) and by Santos (1968, 1969b).

First, there is a closure property under $\cup$ and $\cap$ (Mizumoto *et al.*, 1969).

(1) If $\mathcal{Q}_1$ and $\mathcal{Q}_2$ are fuzzy automata, $\exists \mathcal{Q} = \mathcal{Q}_1 \coprod \mathcal{Q}_2$ such that

$$\mathcal{L}(\mathcal{Q}) = \mathcal{L}(\mathcal{Q}_1) \cup \mathcal{L}(\mathcal{Q}_2).$$

A is defined by $U = U_1 \cup U_2, S = S_1 \cup S_2, Y = Y_1 \cup Y_2$. Using matrix notations (see I.3.B.b.γ);

$$\delta = \begin{pmatrix} \delta_1 & 0 \\ 0 & \delta_2 \end{pmatrix}, \qquad \tilde{s}_0 = (\tilde{s}_{0_1}, \tilde{s}_{0_2}), \qquad \sigma = \begin{pmatrix} \sigma_1 \\ \sigma_2 \end{pmatrix}.$$

where the subscript 1 or 2 denotes a component of $\mathcal{Q}_1$ or $\mathcal{Q}_2$, respectively.

(2) If $\mathcal{Q}_1$ and $\mathcal{Q}_2$ are fuzzy automata, $\exists \mathcal{Q} = \mathcal{Q}_1 \otimes \mathcal{Q}_2$ such that

$$\mathcal{L}(\mathcal{Q}) = \mathcal{L}(\mathcal{Q}_1) \cap \mathcal{L}(\mathcal{Q}_2).$$

It is defined by $U = U_1 \times U_2, S = S_1 \times S_2, Y = Y_1 \times Y_2, \delta = c(\delta_1) \cap c(\delta_2) = \delta_1 \times \delta_2; \tilde{s}_0 = \tilde{s}_{0_1} \times \tilde{s}_{0_2}; \sigma = \sigma_1 \times \sigma_2. \delta, \tilde{s}_0, \sigma$ are the joins (II.3.A.a) of δ_1 and δ_2, $\tilde{s}_{0_1}$ and $\tilde{s}_{0_2}$, σ_1 and σ_2 respectively.

Now, a property relating fuzzy languages and their complement is given.

(3) $\forall \mathcal{C}$, $\exists \bar{\mathcal{C}}$ that recognizes $\overline{\mathcal{L}(\mathcal{C})}$ and $\bar{\mathcal{C}}$ is a "min–max automaton," i.e.,

$$1 - f_{\mathcal{C}}(\theta) = \tilde{s}_0 \circ \Delta_\theta \circ \sigma = \overline{\tilde{s}_0} \; \overline{\circ} \; \overline{\Delta_\theta} \; \overline{\circ} \; \overline{\sigma} = f_{\bar{\mathcal{C}}}(\theta).$$

$\bar{\mathcal{C}}$ is a "min–max automaton" because $\overline{\circ}$, min–max composition is used instead of $\circ$.

Lastly, the following propositions were proved by Santos (1977):

(4)   If $\mathcal{C}_1$ and $\mathcal{C}_2$ are fuzzy automata, $\exists \mathcal{C} = \mathcal{C}_1 \circ \mathcal{C}_2$ such that

$$\mathcal{L}(\mathcal{C}) = \mathcal{L}(\mathcal{C}_1)\mathcal{L}(\mathcal{C}_2) \qquad \text{(concatenation)}.$$

(5)   If $\mathcal{C}$ is a fuzzy automaton, then $\exists \hat{\mathcal{C}}$ such that

$$\mathcal{L}(\hat{\mathcal{C}}) = \hat{\mathcal{L}}(\mathcal{C}) \qquad \text{(Kleene closure)}$$

(6)   If $\mathcal{C}_1$ is a fuzzy automaton, then $\exists \mathcal{C}$ such that $\mathcal{L}(\mathcal{C}) = a \hat{\ } \mathcal{L}(\mathcal{C}_1)$, defined as

$$f_{\mathcal{C}}(\theta) = \min(a, f_{\mathcal{C}_1}(\theta)), \qquad \forall \theta \in U^*, \qquad \text{and} \qquad a \in [0,1].$$

$\mathcal{C}$ is the same as $\mathcal{C}_1$ except its output map σ is defined by $\forall q_j \in S$, $\mu_\sigma(q_j, y) = \min(a, \mu_{\sigma_1}(q_j, y))$ where σ_1 is the output map of $\mathcal{C}_1$.

γ.   Fuzzy Automata and Regular Grammars

In this section we consider a restricted type of fuzzy automaton whose initial state is not fuzzy, and σ is a classical function from F to $\{y\}$, F being a nonfuzzy subset of states, called "final states," i.e., $\mu_\sigma(s_j, y) = 1$ iff $s_j \in F$.

Two automata $\mathcal{C}_1$ and $\mathcal{C}_2$ are said to be equivalent iff $\mathcal{L}(\mathcal{C}_1) = \mathcal{L}(\mathcal{C}_2)$.

Any fuzzy automaton as defined in α is equivalent to a restricted fuzzy automaton. For consider $\mathcal{C} = (U, S, Y, \tilde{s}_0, \delta, \sigma)$. $\mathcal{C}' = (U, S \cup \{\hat{s}, y\}, Y', \hat{s}, \delta', \sigma')$ is equivalent to $\mathcal{C}$ provided that

$\hat{s}$ is an artificial state added to S, and the initial state of $\mathcal{C}'$;

δ' is defined as follows: $\mu_{\delta'}(a, s, s') = \mu_\delta(a, s, s') \; \forall s, s' \in S, \; \forall a \in U$;

$$\mu_{\delta'}(a, \hat{s}, s) = \max_{s' \in S} \min(\mu_{\tilde{s}_0}(s'), \mu_\delta(a, s', s)) \qquad \forall s \in S;$$

$$\mu_{\delta'}(a, s, \hat{s}) = 0 \qquad \text{for} \quad s \in S \cup \{\hat{s}, y\};$$

$$\mu_{\delta'}(a, s, y) = \max_{s' \in S} \min(\mu_\delta(a, s, s'), \mu_\sigma(s', y)) \qquad \forall s \in S;$$

$$\mu_{\delta'}(a, y, s) = 0 \qquad \forall s \in S \quad \text{or} \quad s = \hat{s};$$

$$\mu_{\delta'}(a, \hat{s}, y) = \max_{s \in S} \min(\mu_{\delta'}(a, \hat{s}, s), \mu_\sigma(s, y)).$$

The final state of $\mathcal{Q}'$ is y. $Y' = \{y'\}$ where Y' is artificially introduced and σ' is such that $\mu_{\sigma'}(s, y') = 0 \; \forall s \in S \cup \{\hat{s}\}$ and 1 for $s = y$. Then, given an input string $\theta = u_1 u_2 \cdots u_k$, $u_i \in U$,

$$f_{\mathcal{Q}}(\theta)$$

$$= \max_{s_0, s_1, \ldots, s_k \in S} \min\left(\mu_{\tilde{s}_0}(s_0), \mu_\delta(u_1, s_0, s_1), \ldots, \mu_\delta(u_k, s_{k-1}, s_k), \mu_\sigma(s_k, y) \right)$$

$$= \max_{s_1, \ldots, s_{k-1} \in S} \min\left(\mu_{\delta'}(u_1, \hat{s}, s_1), \mu_{\delta'}(u_2, s_1, s_2) \cdots \mu_{\delta'}(u_k, s_{k-1}, y) \right)$$

$$= f_{\mathcal{Q}'}(\theta)$$

because $\hat{s}$ is not fuzzy and $\mu_{\sigma'}(y, y') = 1$. When $k = 1$, we use the expression for $\mu_{\delta'}(a, \hat{s}, y)$.

Santos (1968) showed that the capacity of a fuzzy automaton as an acceptor is equal to that of a nonfuzzy automaton. More specifically, for a given regular fuzzy grammar G, there exists a fuzzy automaton $\mathcal{Q}$ such that $\mathcal{L}(G) = \mathcal{L}(\mathcal{Q})$ and conversely. The proof given here is similar to Mizumoto et al.'s (1970):

(i) Let $G = (V_T, V_N, s, P, J)$ be a regular fuzzy grammar. The corresponding fuzzy automaton $\mathcal{Q} = \{ U, S, Y, s_0, \delta, F)$ where the initial state s_0 is nonfuzzy and F is the set of final states is defined by

$$U = V_T, \qquad S = J \cup \{s\}, \qquad Y = \{y\} \quad \text{(any singleton)},$$

$$s_0 = s, \qquad F = \left\{ l \in J, \left(A \xrightarrow{\rho}{}_l a \right) \in P \right\},$$

$\forall s_i, s_j \in S, \; \forall a \in U, \; \mu_\delta(s_i, a, s_j) = \rho_j$ iff (s_i is the index of a production $A \to bB$ and s_j that of the production $B \xrightarrow{\rho_j} a$ or $B \xrightarrow{\rho_j} aC$) or ($s_i = s$ and s_j is the index of the production $s \xrightarrow{\rho_j} aA$); $\mu_\delta(s_i, a, s_j) = 0$ otherwise. It is easy to verify that any sequence of transitions of $\mathcal{Q}$ from the initial state to a final state has a nonzero membership value iff the corresponding input string is generated by G.

(ii) Let $\mathcal{Q} = (U, S, Y, s_0, \delta, F)$ be a fuzzy restricted automaton. It is possible to consider only such an automaton since a general one can always be put into a restricted form. The equivalent fuzzy grammar is defined by

$$V_T = U, \qquad V_N = S - F, \qquad s = s_0,$$

P contains productions $s_i \xrightarrow{\rho_j} a s_j$ when $\mu_\delta(s_i, a, s_j) = \rho_j > 0$ and $s_i \xrightarrow{\rho_j} a$ when $\mu_\delta(s_i, a, s_j) = \rho_j > 0$, and $s_j \in F$ where s_i can be the initial state s_0. Q.E.D.

Hence, fuzzy automata generate regular fuzzy languages.

δ.   *Other Properties of Fuzzy Automata*

The following property gives a sufficient condition for a fuzzy automaton to generate a fuzzy language that is its own Kleene closure.

Let $\mathcal{C} = (U, S, Y, \tilde{s}_0, \delta, \sigma)$ be a fuzzy automaton such that (1) $\forall a \in U, \forall s_i \in S$, $\mu_\delta(s_i, a, s_i) = 1$; (2) $\tilde{s}_0 \circ \sigma = 1$; then $\mathcal{L}(\mathcal{C})$ is a closed language (i.e., $\hat{\mathcal{L}}(\mathcal{C}) = \mathcal{L}(\mathcal{C})$) (Negoita, Ralescu 1975, Reference from I).

Proof: $f_{\mathcal{C}}(\Lambda) = \tilde{s}_0 \circ \Delta_\Lambda \circ \sigma = \tilde{s}_0 \circ \sigma = 1$. Since δ_a is reflexive, Δ_θ is also reflexive for each $\theta \in U^*$. Hence, $\Delta_{\theta\theta'} = \Delta_\theta \circ \Delta_{\theta'} \supseteq \Delta_\theta$, and we have

$$f_{\mathcal{C}}(\theta\theta') = \tilde{s}_0 \circ \Delta_{\theta\theta'} \circ \sigma \geqslant \tilde{s}_0 \circ \Delta_{\theta'} \circ \sigma = f_{\mathcal{C}}(\theta') \geqslant \min(f_{\mathcal{C}}(\theta), f_{\mathcal{C}}(\theta')).$$

Q.E.D.

Now, we consider a property of fuzzy automata related to cut-point languages. Let $\mathcal{L}(\mathcal{C}, \alpha)$ be the cut-point language $\{\theta \in U^*, f_{\mathcal{C}}(\theta) > \alpha\}$ of $\mathcal{L}(\mathcal{C})$; $\forall \alpha_0 \in [0, 1], \exists \mathcal{C}_0, \mathcal{L}(\mathcal{C}, \alpha) = \mathcal{L}(\mathcal{C}_0, \alpha_0)$ (for a proof see Mizumoto, et al. 1969).

h.   Other Recognition Devices

Let us quote Santos (1976): "The model of fuzzy automata obtained [using max and min operators] is not an interesting model when we view it as a recognition device for fuzzy languages. . . . Most of the results obtained in this manner are trivial extensions of existing ones." This remark points out the need for more powerful devices: the related literature is briefly surveyed.

We must however mention first a paper by Thomason (1974) in which a fuzzy transducer automaton is constructed. This machine is equivalent to the fuzzy STDS (see d) and is still a max–min machine.

Paz (1967) deals with the problem of the approximate recognition of fuzzy languages and their cut-points by means of deterministic and probabilistic automata.

Nasu and Honda (1968) define a probabilistic event as a fuzzy language accepted by a finite probabilistic automaton. The set of probabilistic events is a proper subset of the set of fuzzy languages. In this paper it is shown that the set of probabilistic events is closed under transposition and convex combination. Less strong results for the fuzzy union and intersection of probabilistic events are given.

The recognition capabilities of max-product automata (see 2.C) were investigated by Santos (1975a) and compared to those of probabilistic and max–min automata. Santos (1976) also proved that the union of all cut-point languages of the forms $\mathcal{L}(\mathcal{C}, \lambda), \mathcal{L}(\mathcal{C}, \geqslant, \lambda), \mathcal{L}(\mathcal{C}, =, \lambda)$ over a given alphabet contains the set of regular languages over this alphabet as a proper subset.

L-fuzzy automata were studied by various authors: Wechler and Dimitrov (1974) when L is an ordered semiring; Mizumoto *et al.* (1975a) when L is a boolean lattice; Mizumoto and Tanaka (1976) when L is the set of normalized convex fuzzy sets of $[0, 1]$; and Santos (1977) when L is a linearly ordered semigroup. These automata have the same recognition capabilities as max–min automata in the sense that L-fuzzy automata recognize regular L-fuzzy languages whose cut-points of the form $\mathfrak{L}(\mathcal{C}, \lambda)$ are regular.

Wechler (1975b) also considered languages recognized by time-variant L-fuzzy automata where L is an ordered semiring.

Lastly, Honda and Nasu (1975) and Honda *et al.* (1977) present general results on recognition of L-fuzzy languages where L is a lattice with a minimum element, by L-fuzzy Turing machines (see the next section for $L = [0, 1]$), L-fuzzy linear bounded automata, L-fuzzy push-down automata, and L-fuzzy automata.

Remark   Augmented transition networks (Woods, NF 1970), which are related to transformational grammars, do not seem to have been fuzzified yet, although this may be worth considering. The stochastic version has already been studied by Chou and Fu (NF 1975).

i. Nonformal Fuzzy Languages   (Zadeh, 1972)

As may be seen above, it is quite easy to generalize much of the theory of formal languages to fuzzy sets of strings. However, as Zadeh (1972) points out, "the resulting theory still falls far short of providing an adequate model for the syntax of natural language." This is because "it fails to reflect the primary function of a language as a system of correspondences between strings of words and sets of objects or constructs which are described by these strings."

In order to explicitly take into account these correspondences—which are fuzzy by essence in natural language—Zadeh (1972) gives the following broader definition of a fuzzy language.

A *fuzzy language* $\mathfrak{L}$ is a quadruple (U, T, E, N) in which:

U is a universe of discourse, i.e., a set of objects, actions, relations, concepts, . . . ;

T, the term set, is a fuzzy set of terms that serve as names of fuzzy subsets of U;

E, an embedding set for T, is a collection of symbols and their combinations from which the terms are drawn, i.e., T is a fuzzy subset on E;

N, the naming relation, is a fuzzy relation on $(\operatorname{supp} T) \times U$.

The grade of membership $\mu_T(t)$ of the term t may be viewed as the degree of well formedness or grammaticality of t. $\mu_N(t, u)$ is interpreted as the degree to which the term t fits the element $u \in U$.

When T and U are sets with a small number of elements, it is easy to define μ_N and μ_T by tabulation. However, generally, both supp T and U are infinite sets and "the characterization of T and N requires that they be endowed with a structure allowing the computation of μ_T and μ_N." Hence, there is introduced the notion of a *structured fuzzy language* which is a quadruple (U, S_T, E, S_N) where U and E are defined as above, S_T is a set of rules, called the syntactic rules of £, which provide an algorithm for computing μ_T, and S_N is a set of rules, called the semantic rules of £, which provide an algorithm for computing μ_N.

Obviously, a formal fuzzy language is a particular case of the fuzzy language defined above; more specifically, in a formal fuzzy language, only T and E are considered. Further, when T is nonfuzzy, the domain dom(N) (see II.3.B.a) of the fuzzy relation N may be viewed as a fuzzy formal language.

Semantic aspects of fuzzy languages are studied in Part IV, Chapter 2.

B. FUZZY ALGORITHMS

According to Zadeh (1973), "a fuzzy algorithm is an ordered set of fuzzy instructions which upon execution yield an approximate solution to a specified problem." The idea of fuzzy algorithm was first introduced by Zadeh (1968). Such a definition subsumes most human action and thinking: people use fuzzy algorithms when they drive a car, search for an object, untie a knot, cook food (the recipe of a scrumptious chocolate fudge was given under the form of a fuzzy flowchart in Zadeh, 1973), recognize patterns, or make a decision. Since fuzzy algorithms can face a range of slightly different situations, they summarize information in a concise, although approximate manner.

a. Fuzzy Instructions   (Zadeh, 1973)

The instructions in a fuzzy algorithm belong to one of three categories:

(1) *Assignment statements*: a possibly fuzzy value is assigned to a variable. For instance:

"x equals *approximately* 5."
"y is not *small* and not very *large*."

(2) *Fuzzy conditional statements*: a possibly fuzzy value is assigned to a variable or an action is executed, provided that a fuzzy condition holds. For instance:

"If x is *large*, then y is *small* else y is not *small*."
"If x is *much smaller* than 8, then stop."

(3) *Unconditional action statements*: a possibly fuzzy mathematical operation or an action is executed:

"Decrease x *slightly*."
"Multiply x by itself *a few* times."

An instruction is thus said to be fuzzy as soon as the name of a fuzzy set appears in it, and blurs somewhat its execution ("write small" is fuzzy, but "write 'small'" is not). Note that in a fuzzy algorithm, not all the instructions are necessarily fuzzy. Note that fuzzy instructions of type 1 and 2 are very similar to fuzzy propositions in approximate reasoning (see 1.E).

b. Formal Algorithmic Machines

α. Fuzzy Algorithms as Fuzzy Systems

As pointed out by Zadeh (1968, 1972b), the notion of fuzzy algorithm is closely related to that of fuzzy system. We may view a fuzzy algorithm as a fuzzy system whose equations are $\forall t \in \mathbb{N}$;

$$\mu_{\tilde{s}_{t+1}}(s_{t+1}) = \sup_{u_t, s_t} \min\left(\mu_{\tilde{s}_t}(s_t), \mu_{\tilde{u}_t}(u_t), \mu_\delta(s_{t+1}, u_t, s_t) \right),$$

$$\mu_{\tilde{u}_{t+1}}(u_{t+1}) = \sup_{s_{t+1}} \min\left(\mu_{\tilde{s}_{t+1}}(s_{t+1}), \mu_\sigma(u_{t+1}, s_{t+1}) \right), \tag{2}$$

where $\tilde{s}_t$ is a fuzzy state of the algorithm at time t, $\tilde{u}_t$ a fuzzy input (representing a fuzzy instruction) at time t, $\tilde{s}_{t+1}$ the result of the execution of the fuzzy instruction $\tilde{u}_t$, δ expresses the dependence of $\tilde{s}_{t+1}$ on $\tilde{s}_t$ and $\tilde{u}_t$, and σ the dependence of the fuzzy instruction at time t on the fuzzy state at time t. $\tilde{s}_0$ is the initial state.

Formulas (2) correspond to the complete execution of the fuzzy instruction $\tilde{u}_t$: the first equation changes the state, the second is a fuzzy branching. A fuzzy instruction is viewed here as a fuzzy set of instructions executed in parallel.

β. Fuzzy Turing Machines

Algorithms may be thought of in terms of Turing machines. Thus, a natural way to formalize the concept of fuzzy algorithm is via the concept of fuzzy Turing machines. A brief discussion of fuzzy Turing machines was first given by Zadeh (1968), and a detailed formulation can be found in Santos (1970).

A *fuzzy Turing machine* is a complex $Z = (A, B, S, \delta, \tilde{s}_0)$ where A is the printing alphabet, B an auxiliary alphabet that contains special symbols like "blank" (b), S a set of internal states, δ a transition fuzzy relation on $S \times U \times V \times S$ with $U = A \cup B$ and $V = U \cup \{+, -, 0\}$; $\tilde{s}_0$ is the initial state. $+, -, 0$ mean respectively a move of one step to the "right," a move of one step to the "left," and the termination of the algorithmic procedure. It is assumed that

$$\forall u \in U, \quad \mu_\delta(s, u, 0, s') = 0 \quad \text{if} \quad s' \neq s.$$

An instantaneous expression α of Z is a finite sequence (possibly empty) of elements of $U \cup S$ such that α contains only one element of S; α is of the form $\theta su\tau$ with θ and τ elements of U^* (the set of strings made of elements of U), $u \in U$, and $s \in S$. The state s is said to point at u in the instantaneous expression α. The transition between two instantaneous expressions α and β is expressed by the membership value $\mu_Z(\alpha, \beta)$ which is equal to:

$$\mu_\delta(s, u, u', s') \quad \text{if} \quad \alpha = \theta su\tau \quad \text{and} \quad \beta = \theta s' u' \tau$$

(u' has been written in place of u and the new state s' points at the same place);

$$\mu_\delta(s, u, +, s') \quad \text{if} \quad \begin{cases} \alpha = \theta suu'\tau & \text{and} \quad \beta = \theta us'u'\tau \\ \text{or} \quad \alpha = \theta su & \text{and} \quad \beta = \theta us'b \end{cases}$$

(the new state has "moved" one step to the right);

$$\mu_\delta(s, u, -, s') \quad \text{if} \quad \begin{cases} \alpha = \theta u'su\tau & \text{and} \quad \beta = \theta su'u\tau \\ \text{or} \quad \alpha = su\tau & \text{and} \quad \beta = s'bu\tau \end{cases}$$

(the new state has "moved" one step to the left);

$$\mu_\delta(s, u, 0, s) \quad \text{if} \quad \alpha = \beta \quad \text{(end of computation);}$$

$$0 \qquad \text{otherwise.}$$

A computation of Z with input $x \in U^*$ and output $y \in U^*$ is a finite sequence $\alpha_0, \alpha_1, \ldots, \alpha_n$ of instantaneous expressions where $\alpha_0 = s_0 x$ and $\alpha_n = \theta s \tau$ with $y = \theta \tau$. The membership value of the computation is

$$\min(\mu_{\tilde{s}_0}(s_0), \mu_Z(\alpha_0, \alpha_1), \ldots, \mu_Z(\alpha_{n-1}, \alpha_n), \mu_\delta(s, u, 0, s))$$

where u is the symbol immediately to the right of s in α_n. Denoting by $\mu_Z^i(x/y)$ the membership value of a computation of y from x, the possibility of computing y from x is $\mu_Z(x/y) = \sup_i \mu_Z^i(x/y)$.

γ. *Fuzzy Markov Algorithm*

Santos (1970) also proved the equivalence between a fuzzy Turing machine and a fuzzy version of a Markov algorithm. Here we give a simpler definition of a fuzzy Markov algorithm according to Zadeh (1972a).

Let U be an alphabet. A fuzzy Markov algorithm FM is made of a linearly ordered set of production rules P_i, $i = 1, n$, of the form

$$P_i: \quad \alpha_i \to \mu_1/\beta_1 + \cdots + \mu_k/\beta_k \qquad i = 1, n-1,$$
$$P_n: \quad \Lambda \to 1/\Lambda,$$

where α_i, β_j are elements of U^* and Λ denotes the empty string. we omitted for the sake of simplicity the subscript i in the right-hand part of the rule i. The input is a finite support fuzzy set of strings:

$$\mathcal{L}_0 = \mu(\theta_1)/\theta_1 + \cdots + \mu(\theta_m)/\theta_m.$$

First, we use the rule $FM(\mathcal{L}_0) = \mu(\theta_1)/FM(\theta_1) + \cdots + \mu(\theta_m)/FM(\theta_m)$. To compute each $FM(\theta_\rho)$, we find the smallest i such that production P_i can be applied to θ_ρ (i.e., α_i occurs as a substring of θ_ρ). When there is ambiguity about how to apply P_i, it is the leftmost occurrence of α_i that is replaced by $\mu_1/\beta_1 + \cdots + \mu_k/\beta_k$: if $\theta_\rho = \gamma\alpha_i\tau$ with $\gamma, \tau \in U^*$, then $FM(\theta_\rho) = \mu_1/\gamma\beta_1\tau + \cdots + \mu_k/\gamma\beta_k\tau$.

Moreover, we define

$$\mu(\theta_\rho)/FM(\theta_\rho) = \min(\mu(\theta_\rho), \mu_1)/\gamma\beta_1\tau + \cdots + \min(\mu(\theta_\rho), \mu_k)/\gamma\beta_k\tau.$$

When the application of P_i to θ_ρ gives a string $\gamma\beta_j\tau$ that is the terminating part of θ_ρ, the string $\gamma\beta_j\tau$ is considered as dead and no rule will be applied to it any longer. A string is also dead when only rule P_n can be applied to it. The procedure is iterated until only dead strings remain. The fuzzy language made up of all the dead strings is the result.

"The execution of a fuzzy Markov algorithm may be likened to a birth-and-death process in which the operation FM on a string θ_ρ gives rise to the birth of new strings . . . and the death of others. . . . As in a birth and death process the population of 'live' strings can grow explosively . . . . This rather interesting aspect of fuzzy Markov algorithm is not present in conventional Markov algorithms" (Zadeh, 1972a).

δ. *Fuzzy Programs* (Santos, 1976, 1977)

A mathematical formulation of fuzzy programs was introduced by Santos, general enough to emcompass all existing formulations (those of S. K. Chang, 1972; Jakubowski and Kasprzak, 1973; Tanaka and Mizumoto, 1975). The main appeal of this formulation is that it is closer to the

intuitive representation of a fuzzy algorithm as a set of fuzzy instructions.
An instruction is a string of one of the following forms:

Start: go to L;
L: do F; go to L';
L: if P, then go to $(L1, L2, \ldots, LN)$;
L: halt;

where N is a positive integer, $L, L', L1, \ldots, LN$ belong to a set $\mathcal{L}_a$ of
labels, F belongs to $\mathcal{F}$ (the set of function or operation symbols), and P
belongs to $\mathcal{P}_N$ (the set of N-valued predicates or test symbols). The four
types of instructions are called respectively "start," "operation," "test,"
and "halt" instructions.

A program is a finite set φ of instructions containing exactly one start
instruction and no two instructions in φ have the same label. In a test
instruction P is a function valued in $\{1, 2, \ldots, N\}$; and if the value of P is
K, then the next instruction to be executed has label LK.

Let $(R, +, \cdot, \leqslant)$ be an ordered semiring (see A.f). An R-machine is a
complex $(U, M, Y, I, \delta, \tau, \sigma, \mathcal{F}, \mathcal{P})$ such that U, M, and Y are respectively the
input, memory, and output sets. I is an R-fuzzy relation on $U \times M$, δ an
R-fuzzy relation on $M \times \mathcal{F} \times M$, τ an R-fuzzy relation on $\mathcal{P} \times M \times \mathbb{N}$
with $\mathcal{P} = \bigcup_N \mathcal{P}_N$, and σ an R-fuzzy relation on $M \times Y$. Moreover, denoting by τ_P the conditioned fuzzy relation obtained from τ by fixing P, for
every $P \in \mathcal{P}$, there exists a positive integer N such that τ_P is an R-fuzzy
relation on $M \times \{1, 2, \ldots, N\}$.

Let $P \in \mathcal{P}_N$. If $\mu_{\tau_P}(m, K) = \mu_{\tau_P}(m', K)$ for all m and m' belonging to M
and for all $K \in \{1, \ldots, N\}$, then P is said to be unconditional, and so is
any instruction containing P.

Let φ be a program and M be an R-machine. $\mathfrak{M}$ is said to admit φ iff:

(1)   for every operation instruction of the form "L: do F; go to L',"
$F \in \mathcal{F}$;
(2)   for every test instruction of the form "if P, then go to
$(L1, \ldots, LN)$," $P \in \mathcal{P}_N$.

We write $(L', m_1) \overset{r}{\to} (L, m_2)$, where m_1 and m_2 are two elements of M, iff
there is an instruction in φ of the form "L': do F; go to L" and
$\mu_\delta(m_1, F, m_2) = r$. We write $(L', m) \overset{r}{\to} (L'', m)$, where $m \in M$, iff there is an
instruction in φ of the form "L': if P, then go to $(L1, \ldots, LN)$" such that
$L'' = LK$ for some $K \in \{1, \ldots, N\}$ and $r = \mu_{\tau_P}(m, K)$.

A computation by φ on an R-machine is a finite sequence u, L_0, m_0,
$L_1, \ldots, L_n, m_n, y$ where $u \in U, y \in Y$, in which L_0 is a label contained in
the start instruction of φ and L_N is the label of some halt instruction in φ.
The membership value associated with the computation is an element

$r \in R$ such that $r = r_0 \cdot r_1 \cdot \cdots \cdot r_{n+1}$ where $r_0 = \mu_I(u, m_0)$, $r_{n+1} = \mu_\sigma(m_n, y)$, and for every other i: $(L_{i-1}, m_{i-1}) \xrightarrow{r_i} (L_i, m_i)$. The computation of y from u is feasible iff $r \neq 0$. The possibility of computing y from u with φ is valuated by the sum (in the sense of R) of all the membership values r of the computations of y from u. Note that the result is an R-fuzzy set of Y.

"The above discussion gives a precise formulation of the concept of max–min programs, probabilistic programs, max–product programs, non-deterministic programs, deterministic programs and other types of fuzzy programs." (Santos, 1977). Santos (1977) also showed that fuzzy programs and fuzzy Turing machines are closely related. The fuzzy machines executing fuzzy programs introduced in Chang (1972), Jakubowski and Kasprzak (1973), and Tanaka and Mizumoto (1975) are less general because they are based on deterministic, nondeterministic, or R-fuzzy automata, respectively, and fuzzy instructions in the three cases.

Note that a fuzzy program is viewed here as a fuzzy set of nonfuzzy programs in the sense that fuzzy instructions are fuzzy sets of instructions.

ε. *Execution of Fuzzy Programs*

Obviously, the execution of fuzzy programs in the sense of δ is equivalent to the parallel execution of a possibly nonfinite number of deterministic programs. Practically however, for each fuzzy instruction, a deterministic instruction is chosen, which is assumed to be the best interpretation of the fuzzy instruction, and actually executed. If a fuzzy instruction is reached for which there is no deterministic instruction capable of performing a feasible computation ($r = 0$ or at least r falls below a given threshold), then a backtracking process must be initiated. The necessity for backtracking stems from the fact that choosing locally the best interpretation of fuzzy instructions in sequence does not ensure the optimality of the global computation.

A fuzzy test instruction is interpreted by selecting the label of another instruction. Tanaka and Mizumoto (1975) hinted at three ways of selection:

(i) "fuzzy" selection, i.e., choose LK such that $\mu_{\tau_p}(m, K)$ is the greatest of all $\mu_{\tau_p}(m, J)$ for $J = 1, N$;

(ii) "probabilistic" selection, i.e., a random choice of LK, according to the probability values

$$\mu_{\tau_p}(m, J) \Big/ \sum_J \mu_{\tau_p}(m, J) \qquad \forall J;$$

(iii) "nondeterministic" selection, i.e., choose any among the LJ's such that $\mu_{\tau_p}(m, LJ) \geq \alpha$, for a given threshold α.

The operation instructions are interpreted by selecting a deterministic state transition, according to the value of $\mu_g, (m_1, F, m_2)$, with one of the above selection methods. In Tanaka and Mizumoto (1975) this approach for executing fuzzy programs was exemplified with computer simulation of human driver behavior (the "driver" follows a fuzzy routing plan) and of character generation. Here, the action expressed by a fuzzy instruction yields a nonfuzzy result, and thus the action itself is nonfuzzy. In Tanaka and Mizumoto's fuzzy programs it is assumed that when an action fails (because it is unfeasible), it is always possible to backtrack and modify a previous action. Obviously, the backtracking assumption is not always realistic.

"The key to success of a fuzzy algorithm is fuzzy feedback that is a mechanism for (a) observing—not necessarily precisely—the result of execution of a fuzzy instruction; and (b) executing a new instruction based on the result or results of preceding instructions." (Zadeh, 1968). The existence of a feedback allows slight modifications of the result of a fuzzy instruction owing to (fuzzy) tests that evaluate the quality of this result. The fuzzy feedback control loop of the algorithm improves its robustness and suppresses part of the backtracking.

c. Algorithms for Computing Fuzzy Sets

Such algorithms are made of fuzzy instructions that are executed in a deterministic way through logico-algebraic combinations of fuzzy sets. The result consists in fuzzy sets computed from intermediate fuzzy results and fuzzy sets appearing in the fuzzy instructions.

α. Fuzzy Assignment Statements

A fuzzy value is assigned to a variable of the algorithm. For instance, "x equals approximately 5" is interpreted here as $x \leftarrow \tilde{5}$ where $\tilde{5}$ is a fuzzy number whose mean value is 5. Note that in fuzzy algorithms in the sense of b this instruction is executed by stating $x \leftarrow a$ where a is nonfuzzy and chosen by one of the described selection methods above, using $\mu_{\tilde{5}}(a)$.

β. Fuzzy Unconditional Action Statements

An operation is performed on fuzzy sets. For instance, "decrease x slightly" may be interpreted as follows: $x \leftarrow x \odot (1 \ominus \tilde{\epsilon})$ where $\tilde{\epsilon}$ is a positive fuzzy number close to 0. Such an operation was referred to as an "operation" in the previous section, where it was executed by choosing a nonfuzzy ϵ: $x \leftarrow x(1 - \epsilon)$ using $\mu_{\tilde{\epsilon}}(\epsilon)$ for guiding the choice. Another example is "multiply x by itself a few times," with $few = 1/1 + 0.8/2 + 0.6/3 + 0.4/4$ and x assumed nonfuzzy; we then obtain $x \leftarrow 1/x^2 +$

$0.8/x^3 + 0.6/x^4 + 0.4x^5$. Nevertheless, if *few* is modeled by a continuous positive fuzzy number $\tilde{\lambda}$, it is easy to compute at once $x^{\tilde{\lambda}}$ (see II.2.B.d: $\mu_{x^{\tilde{\lambda}}}(y) = \mu_{\tilde{\lambda}}(\ln y/\ln x)$) instead of computing the power of x several times, actually as many times as the number of integer elements in supp$\tilde{\lambda}$. It is always possible to a posteriori extract from $x^{\tilde{\lambda}}$ the membership values corresponding to integer exponents. (Here, the performance of the extended operation commutes with the discretization of the support; this is not always true—see II.2.A.b.) More generally, extended operations are very appealing in the execution of fuzzy unconditional statements that are allowed to have a fuzzy result because parallel computation can be avoided.

γ. *Fuzzy Conditional Statements* (C. L. Chang, 1975)

We consider here fuzzy instructions of the form "if $P(x_1, \ldots, x_n)$, then go to L else go to L'" where P is an n-ary fuzzy predicate and L and L' are labels of fuzzy instructions. C. L. Chang (1975) has proposed the following interpretation of a fuzzy conditional statement when the values of $x_1, \ldots, x_n$ are allowed to be fuzzy. In ordinary nonfuzzy algorithms, a nonfuzzy condition described by a predicate is checked on the n-tuples of nonfuzzy values and only the "then part" of the statement is executed when the condition holds. Otherwise, it is the negation of the condition that holds and it is the "else part" that is executed. When $x_1, \ldots, x_n$ are fuzzy or/and the predicate is fuzzy, none of the complementary conditions may completely hold and both branchings need to be done. The fuzzy instructions L and L' will be executed by using different fuzzy values of the variables, i.e., those that fit respectively P and $\neg P$.

More specifically, let $\tilde{u}_1, \ldots, \tilde{u}_n$ be the fuzzy values of $x_1, \ldots, x_n$ before the execution and R the n-ary fuzzy relation associated with P; the degree to which P holds for $\tilde{u}_1, \ldots, \tilde{u}_n$ is

$$\mu_{\tilde{R}}(\tilde{u}_1, \ldots, \tilde{u}_n) = \sup_{u_1, \ldots, u_n} \min(\mu_{\tilde{u}_1}(u_1), \ldots, \mu_{\tilde{u}_n}(u_n), \mu_R(u_1, \ldots, u_n));$$

this is actually a consistency degree (see II.3.F.d). The fuzzy value $\tilde{u}_i^*$ of x_i that best fits P with respect to the values $\tilde{u}_1, \ldots, \tilde{u}_{i-1}, \tilde{u}_{i+1}, \ldots, \tilde{u}_n$ is such that

$$\mu_{\tilde{u}_i^*}(u_i) = \sup_{j \neq i} \min(\mu_{\tilde{u}_j}(u_j), \mu_R(u_1, \ldots, u_n)).$$

The value of x_i before executing instruction L will be $\tilde{u}_i \cap \tilde{u}_i^*, i = 1, n$. The degree to which $\neg P$ holds for $\tilde{u}_1, \ldots, \tilde{u}_n$ is $\mu_{\bar{\tilde{R}}}(\tilde{u}_1, \ldots, \tilde{u}_n)$ where $\bar{R}$ is the complement of R. The value of x_i before executing instruction L' will be $\tilde{u}_i' \cap \tilde{u}_i$ where $\tilde{u}_i'$ is computed as $\tilde{u}_i^*$, replacing R by $\bar{R}$ in the formula expressing $\tilde{u}_i^*$. Note that generally $\mu_{\tilde{R}}(\tilde{u}_1, \ldots, \tilde{u}_n) \neq 1 - \mu_{\tilde{\tilde{R}}}(\tilde{u}_1, \ldots, \tilde{u}_n)$, except if the values of the x_i are not fuzzy. Moreover, $\forall i = 1, n$,

$$\mu_{\tilde{R}}(\tilde{u}_1, \ldots, \tilde{u}_n) = \mathrm{hgt}(\tilde{u}_i \cap \tilde{u}_i^*) \qquad \text{and} \qquad \mu_{\tilde{\tilde{R}}}(\tilde{u}_i, \ldots, \tilde{u}_n) = \mathrm{hgt}(\tilde{u}_i \cap \tilde{u}_i^{\cdot}).$$

The values of the program variables may become nonnormalized after such a fuzzy branching: their heights indicate the validity of the result. The existence of fuzzy conditional statements entails parallel computations and may cause some programs to loop. A computation may be stopped when output variables of a fuzzy instruction have values whose height is too low. Some examples of fuzzy branching instructions are given below. First consider:

If x is small, then go to $L1$ else go to $L2$;
$L1$: y is large; go to L;
$L2$: y is not large; got to L;
L: output y.

Let $\tilde{s}, \tilde{l}$, and $\tilde{u}$ represent "small," "large," and the initial value of x. The "then part" yields the result $x \leftarrow \tilde{u} \cap \tilde{s}$ and $y \leftarrow \tilde{l}$; the "else part" gives $x \leftarrow \tilde{u} \cap \bar{\tilde{s}}$ and $y \leftarrow \bar{\tilde{l}}$. Note that we obtain a fuzzy set of fuzzy sets $\mathrm{hgt}(\tilde{u} \cap \tilde{s})/\tilde{l} + \mathrm{hgt}(\tilde{u} \cap \bar{\tilde{s}})/\bar{\tilde{l}}$. Adamo (Reference from IV.2, 1978b) has proposed a method to reduce the complexity of the result: $y \leftarrow \tilde{v}$ where $\mu_{\tilde{v}}(l) = \max(\min(\mathrm{hgt}(\tilde{u} \cap \tilde{s}), \mu_{\tilde{l}}(l)), \min(\mathrm{hgt}(\tilde{u} \cap \bar{\tilde{s}}), \mu_{\bar{\tilde{l}}}(l)))$. Instead of keeping x unchanged, as Adamo did, we may prefer $x \leftarrow (\tilde{u} \cap \tilde{s}) \cup (\tilde{u} \cap \bar{\tilde{s}}) = \tilde{u} \cap (\tilde{s} \cup \bar{\tilde{s}})$.

Consider now the following example (C. L. Chang, 1975):

$L0$: if x is approximately equal to y, go to $L1$ else go to $L2$;
$L1$: $y \leftarrow x + y$; go to $L0$;
$L2$: $z \leftarrow x + y$.

Let $\tilde{u}, \tilde{v}$ be respectively the initial value of x and y and R be the fuzzy relation "approximately equal." The result of the first step is for $L1$, $x \leftarrow \tilde{u} \cap (\tilde{v} \circ R)$ and $y \leftarrow [\tilde{u} \cap (\tilde{v} \circ R)] \oplus [\tilde{v} \cap (\tilde{u} \circ R)]$, and for $L2$ we have $z \leftarrow [\tilde{u} \cap (\tilde{v} \circ \bar{R})] \oplus [\tilde{v} \cap (\tilde{u} \circ \bar{R})]$. Stopping may occur when the height of the value of z is maximal, otherwise the program may loop even when x and y are nonzero.

Remark Fuzzy algorithms may be pictured by fuzzy flowcharts. For instance, the flowchart of the last example is shown in Fig. 1 where λ and μ belong to $[0, 1]$ and valuate the validity of the branchings (symbolically, $\lambda = \mathrm{hgt}(x \cap (y \circ \bar{R})) = \mathrm{hgt}(y \cap (x \circ \bar{R}))$ and $\mu = \mathrm{hgt}(x \cap (y \circ R)) = \mathrm{hgt}(y \cap (x \circ R)))$. When one of the validity degrees is very low, a natural approximation leads to canceling the corresponding branching. More generally, we may keep only the most valid branching according to the "rule of the preponderant alternative" (Zadeh, 1973). Besides, we may

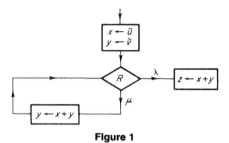

Figure 1

think of linguistic validity degrees for the branchings (see Zadeh, Reference from II.2, 1975); they could be defined as compatibility values (see II.2.A.e.β), instead of consistency values as above.

δ. Inference Statements

Some fuzzy conditional statements can be interpreted as "fuzzy inference statements," which are particular cases of fuzzy unconditional action statements. Consider, for instance, the fuzzy conditional statement: "If x is P, then go to $L1$ else go to $L2$" with $L1$: $y \leftarrow Q$ and $L2$: $y \leftarrow R$, where P, Q, R are fuzzy on the universes of x, y, and y, respectively. Viewed as interpolation, the three fuzzy instructions above can be translated into one fuzzy instruction:

$$y \leftarrow x \circ \left[(P \rightarrow Q) \cup (\neg P \rightarrow R) \right]$$

by analogy with approximate reasoning (see 1.E), provided that after the execution of $L1$ and $L2$, the same fuzzy instruction follows. In the above statement $\rightarrow$ denotes any implication operator considered in 1.E.c.δ. However, when min is used for valuating $\rightarrow$, the above formula is nothing but Adamo's interpretation (Adamo, Reference from IV.2 1978b) of the original three fuzzy instructions (see c.γ).

Proof: Let $\tilde{u}$ be the initial value of x and $\tilde{v}$ be the output value of y. We have

$$\mu_{\tilde{v}}(v) = \sup_{u} \min\left(\mu_{\tilde{u}}(u), \max\left[\min(\mu_P(u), \mu_Q(v)), \min(\mu_{\bar{P}}(u), \mu_R(v)) \right] \right)$$

$$= \sup_{u} \max\left(\min(\mu_{\tilde{u}}(u), \mu_P(u), \mu_Q(v)), \min(\mu_{\tilde{u}}(u), \mu_{\bar{P}}(u), \mu_R(v)) \right)$$

$$= \max\left[\min\left(\sup_{u} \min(\mu_{\tilde{u}}(u), \mu_P(u)), \mu_Q(v) \right), \right.$$

$$\left. \min\left(\sup_{u} \min(\mu_{\tilde{u}}(u), \mu_{\bar{P}}(u)), \mu_R(v) \right) \right]$$

$$= \max\left[\min(\text{hgt}(\tilde{u} \cap P), \mu_Q(v)), \min\left(\text{hgt}(\tilde{u} \cap \bar{P}), \mu_R(v)\right) \right]. \quad \text{Q.E.D.}$$

d. Fuzzification of Algorithms: A Warning

Given two nonfuzzy algorithms that calculate the same quantities using different flowcharts, it may happen that their straightforward fuzzifications (i.e., adapting them to fuzzy data or parameters) do not any longer yield the same result. This point is illustrated in the following example.

Consider the very simple scheduling problem involving one task composed of a known sequence of n elementary operations i; the processing time of operation i is p_i. Let r_0 be the earliest starting operating time and d_{n+1} be the due date of the task. It is assumed that $d_{n+1} - r_0 \geqslant \sum_{i=1}^{n} p_i$ (i.e., the problem is feasible). We want to find an algorithm for computing the earliest starting time r_i, the latest ending time d_i, and the slack time e_i of each operation i (in the nonfuzzy case all the e_i are equal to $d_{n+1} - r_0 - \sum_{i=1}^{n} p_i$). Two possible algorithms are:

(1)

$$r_i = r_0 + \sum_{j=1}^{i-1} p_j, \qquad i = 1, n;$$

$$d_i = d_{n+1} - \sum_{j=i+1}^{n} p_j, \qquad i = 1, n;$$

$$e_i = d_i - r_i - p_i, \qquad i = 1, n.$$

(2)

$$r_{i+1} = r_0 + \sum_{j=1}^{i} p_j, \qquad i = 0, n;$$

$$d_i = d_{n+1} - \sum_{j=i+1}^{n} p_j, \qquad i = 1, n;$$

$$e_i = d_{i-1} + p_i - r_{i+1}, \qquad i = 1, n.$$

Note that d_{i-1} and r_{i+1} are respectively the latest starting time and the earliest ending time of operation i.

When the processing times are fuzzy and now denoted $\tilde{p}_i$, the fuzzy earliest starting times and latest ending times determined using both algorithms are the same. But the fuzzy slack times are different. For the first algorithm, we have

$$\tilde{e}_{i,1} = \tilde{d}_i \ominus \tilde{r}_i \ominus \tilde{p}_i$$

$$= \tilde{d}_{n+1} \ominus \tilde{p}_{i+1} \ominus \cdots \ominus \tilde{p}_n \ominus \tilde{r}_0 \ominus \tilde{p}_1 \ominus \cdots \ominus \tilde{p}_{i-1} \ominus \tilde{p}_i$$

$$= \tilde{d}_{n+1} \ominus \tilde{r}_0 \ominus (\tilde{p}_1 \oplus \cdots \oplus \tilde{p}_n),$$

$$\tilde{e}_{i,2} = \tilde{d}_{n+1} \ominus \tilde{p}_i \ominus \tilde{p}_{i+1} \ominus \cdots \ominus \tilde{p}_n \oplus \tilde{p}_i \ominus \tilde{r}_0 \ominus \tilde{p}_1 \cdots \ominus \tilde{p}_i$$

$$= \tilde{d}_{n+1} \ominus \tilde{r}_0 \ominus (\tilde{p}_1 \oplus \cdots \oplus \tilde{p}_n) \ominus \tilde{p}_i \oplus \tilde{p}_i \neq \tilde{e}_{i,1}$$

(for the use of $\oplus$ and $\ominus$ see II.2.B.d.β). Specifically, $\tilde{e}_{i,2}$ depends on i, but not $\tilde{e}_{i,1}$.

Although the above example is somewhat artificial, the same situation may occur in more realistic algorithms without necessarily being easy to detect. Obviously, here the first fuzzified algorithm is the right one—the second generates redundant fuzziness; but sometimes deciding which is the right algorithm may be more tricky.

When using a fuzzified algorithm, we must make sure that the structure of the mathematical expressions involved reflects the direct logical chain of inferences that gave birth to the algorithm. The transformation of a fuzzified algorithm into a more efficent version must be performed very carefully because some classical mathematical manipulations are no longer authorized with fuzzy quantities.

e. Conclusions

The preceding sections have demonstrated that a fuzzy algorithm (i.e., a set of fuzzy instructions) can be viewed

(1) as a family of nonfuzzy algorithms that are executed in parallel (Santos' approach see b.δ.) unless only one of them is chosen for execution (Tanaka and Mizumoto's approach, see b.ϵ.);

(2) as a single algorithm that processes fuzzy data in a deterministic fashion.

From the semantic point of view, another dichotomy exists regarding the intended purpose of a fuzzy algorithm. In that respect, two general kinds of algorithms that are fuzzy exist. The first consists in ordinary algorithms that realize an implementation of fuzzy models. The aim of such algorithms is to deduce the fuzziness of the outputs knowing that of the inputs and/or of the model. In other words, what is obtained are possibility distributions on the actual nonfuzzy output values of the modeled process. However, when a human subject is presented with a fuzzy instruction, the action he performs will not be fuzzy. Thus, another kind of algorithms exists; these algorithms are fuzzy descriptions of nonfuzzy actions, in the sense of b.ϵ. The result is always a sequence of precise actions. This type of algorithm can be considered as involving some decision process, while the other one bears a forecasting purpose. Combinations of both types can be imagined. For instance, when the next action to be performed is conditioned by an observation, and the observed situation realizes a trade-off between two prototypes of situations, an interpolation (in the sense of approximate reasoning) can be performed in order to generate a fuzzy instruction that is more suitable than the precalculated ones. Then a best interpretation can be determined.

Four classes of fuzzy algorithms were described by Zadeh (1973), each corresponding to a particular type of application:

(i) *Fuzzy definitional algorithm* this is "a finite set of possibly fuzzy instructions which define a fuzzy set in terms of other fuzzy sets or constitute a procedure for computing the grade of membership of any element of the universe in the set under definition"—for instance, the concept of "oval". Used as an identificational device, the algorithm yields a nonfuzzy result.

(ii) *fuzzy generational algorithms*: this serves to generate rather than define a fuzzy set—for instance, generation of "handwritten" characters. Note that the fuzzy generation of a character is different from the generation of fuzzy characters.

(iii) *fuzzy relational and behavioral algorithm*: this serves to describe relations between fuzzy variables. "A relational algorithm which is used for the specific purpose of approximate descriptions of the behaviour of the system will be referred to as a fuzzy behavioral algorithm." Usually such an algorithm will yield fuzzy output; but if it is embedded in a feedback control system, these outputs have to be defuzzified.

(iv) *fuzzy decisional algorithm*: this provides an approximate description of a strategy or decision rule. Obviously, the result of such an algorithm cannot be fuzzy when a strategy is actually applied.

Fuzzy algorithms could be used for the solution of difficult mathematical programming problems. Note that a heuristic method is a nonfuzzy approximation of fuzzy algorithms, expressed in an ordinary programming language. In the field of combinatorial problems the only possible solution methods are often of implicit enumeration type, for instance, branch and bound methods. Finding a solution is equivalent to finding a path in a solution tree. A method is efficient if it builds an optimal path very quickly. The key factors in a branch and bound method are the choice of a separation variable at a given node of the solution tree, the choice of the following node to explore, and the computation of good bounds on the value of the criterion for the unexplored nodes. Some approximate rules are known for determining a good strategy for these choices from the structural features of the data. These rules are very often fuzzy and could be implemented in the framework of a fuzzy decisional algorithm to analyze the set of data at each node of the solution tree.

REFERENCES

Languages and Grammars

DePalma, G. F., and Yau, S. S. (1975). Fractionally fuzzy grammars with application to pattern recognition. *In* "Fuzzy Sets and Their Applications to Cognitive and Decision

Processes" (L. A. Zadeh, K. S. Fu, K. Tanaka, and M. Shimura, eds.), pp. 329–351. Academic Press, New York.

Hirokawa, S., and Miyano, S. (1978). A note on the regularity of fuzzy languages. *Mem. Fac. Sci., Kyushu Univ., Ser. A* **32**, No. 1, 61–66.

Honda, N., and Nasu, M. (1975). Recognition of fuzzy languages. *In* "Fuzzy Sets and Their Applications to Cognitive and Decision Processes" (L. A. Zadeh, K. S. Fu, K. Tanaka, and M. Shimura, eds.), pp. 279–299. Academic Press, New York.

Honda, N., Nasu, M., and Hirose, S. (1977). *F*-recognition of fuzzy languages. *In* "Fuzzy Automata and Decision Processes" (M. M. Gupta, G. N. Saridis, and B. R. Gaines, eds.), pp. 149–168. North-Holland Publ., Amsterdam.

Inagaki, Y., and Fukumura, T. (1975). On the description of fuzzy meaning of context-free languages. *In* "Fuzzy Sets and Their Applications to Cognitive and Decision Processes" (L. A. Zadeh, K. S. Fu, K. Tanaka, and M. Shimura, eds.), pp. 301–328. Academic Press, New York.

Kandel, A. (1974). Codes over languages. *IEEE Trans. Syst., Man Cybern.* **4**, No. 1, 135–138.

Kaufmann, A. (1975). "Introduction à la Théorie des Sous-Ensembles Flous. Vol. 2: Applications à la Linguistique, à la Logique et à la Sémantique." Masson, Paris. (Reference from I, 1975a.)

Kim, H. H., Mizumoto, M., Toyoda, J., and Tanaka, K. (1974). Lattice grammars. *Syst.—Comput.—Controls* **5**, No. 3, 1–9.

Lee, E. T., and Zadeh, L. A. (1969). Note on fuzzy languages. *Inf. Sci.* **1**, 421–434.

Mizumoto, M., and Tanaka, K. (1976). Fuzzy–fuzzy automata. *Kybernetes* **5**, 107–112.

Mizumoto, M., Toyoda, J., and Tanaka, K. (1969). Some considerations on fuzzy automata. *J. Comput. Syst. Sci.* **3**, 409–422.

Mizumoto, M., Toyoda, J., and Tanaka, K. (1970). Fuzzy languages. *Syst.—Comput.—Controls* **1**, 36–43.

Mizumoto, M., Toyoda, J., and Tanaka, K. (1972). General formulation of formal grammars. *Inf. Sci.* **4**, 87–100.

Mizumoto, M., Toyoda, J., and Tanaka, K. (1973a). *N*-fold fuzzy grammars. *Inf. Sci.* **5**, 25–43.

Mizumoto, M., Toyoda, J., and Tanaka, K. (1973b). Examples of formal grammars with weights. *Inf. Process Lett.* **2**, 74–78.

Mizumoto, M., Toyoda, J., and Tanaka, K. (1975a). B-fuzzy grammars. *Int. J. Comput. Math., Sect. A* **4**, 343–368.

Mizumoto, M., Toyoda, J., and Tanaka, K. (1975b). Various kinds of automata with weights. *J. Comput. Syst. Sci.* **10**, 219–236.

Nasu, M., and Honda, N. (1968). Fuzzy events realized by probabilistic automata. *Inf. Control* **12**, 284–303.

Negoita, C. V., and Ralescu, D. A. (1975). "Applications of Fuzzy Sets Theory to Systems Analysis," Chap. 5. Birkhaeuser, Basel. (Reference from I.)

Paz, A. (1967). Fuzzy star functions, probabilistic automata and their approximation by non probabilistic automata. *J. Comput. Syst. Sci.* **1**, 371–390.

Santos, E. S. (1968). Maximin Automata. *Inf. Control* **13**, 363–377.

Santos, E. S. (1969a). Maximin sequential chains. *J. Math. Anal. Appl.* **26**, 28–38.

Santos, E. S. (1969b). Maximin sequential-like machines and chains. *Math. Syst. Theory* **3**, 300–309.

Santos, E. S. (1974). Context-free fuzzy languages. *Inf. Control* **26**, 1–11.

Santos, E. S. (1975a). Realization of fuzzy languages by probabilistic, max-product and maximin automata. *Inf. Sci.* **8**, 39–53.

Santos, E. S. (1975b). Max-product grammars and languages. . *Inf. Sci.* **9**, 1–23.

Santos, E. S. (1976). Fuzzy automata and languages. *Inf. Sci.* **10**, 193–197.

Santos, E. S. (1977). Regular fuzzy expressions. *In* "Fuzzy Automata and Decision Processes" (M. M. Gupta, G. N. Saridis, and B. R. Gaines, eds.), pp. 169–175. North-Holland Publ., Amsterdam.

Thomason, M. G. (1974). Fuzzy syntax-directed translations. *J. Cybern.* **4**, No. 1, 87–94.

Thomason, M. G., and Marinos, P. N. (1974). Deterministic acceptors of regular fuzzy languages. *IEEE Trans. Syst., Man Cybern.* **4**, 228–230.

Wechler, W. (1975a). *R*-fuzzy grammars. *In* "Mathematical Foundations of Computer Science" (J. Bečvář, ed.), Lecture Notes in Computer Science, Vol. 32, pp. 450–456. Springer-Verlag, Berlin and New York.

Wechler, W. (1975b). *R*-fuzzy automata with a time-variant structure. *In* "Mathematical Foundations of Computer Science" (A. Blikle, ed.), Lecture Notes in Computer Science, Vol. 28, pp. 73–76. Springer-Verlag, Berlin and New York.

Wechler, W. (1977). The concept of fuzziness in the theory of automata. *In* "Modern Trends in Cybernetics and Systems" (J. Rose and C. Bilciu, eds.), Vol. 2, pp. 79–86. Springer-Verlag, Berlin and New York.

Wechler, W., and Dimitrov, V. (1974). *R*-fuzzy automata. *Inf. Process., 74, Proc. IFIP Congr.* pp. 657–660.

Zadeh, L. A. (1972). Fuzzy languages and their relations to human and machine intelligence. *Man Comput. Proc. Int. Conf., Bordeaux, 1970* pp. 130–165. S. Karger, Basel.

Fuzzy Algorithms

Chang, C. L. (1975). Interpretation and execution of fuzzy programs. *In* "Fuzzy Sets and Their Applications to Cognitive and Decision Processes" (L. A. Zadeh, K. S. Fu, K. Tanaka, and M. Shimura, eds.), pp. 191–218. Academic Press, New York.

Chang, S. K. (1972). On the execution of fuzzy programs using finite state machines. *IEEE Trans. Comput.* **21**, 241–253.

Chang, S. S. L. (1977). On fuzzy algorithm and mapping. *In* "Fuzzy Automata and Decision Processes" (M. M. Gupta, G. N. Saridis, and B. R. Gaines, eds.), pp. 191–196. North-Holland Publ., Amsterdam.

Hinde, C. J. (1977). Algorithms embedded in fuzzy sets. *Int. Comput. Symp.* (E. Morlet, and D. Ribbens, eds.), pp. 381–387. North-Holland Publ., Amsterdam.

Jakubowski, R., and Kasprzak, A. (1973). Application of fuzzy programs to the design of machining technology. *Bull. Pol. Acad. Sci.* **21**, 17–22.

Kaufmann, A. (1975). Introduction à la Théorie des Sous-Ensembles Flous. Vol. 3: Applications à la Classification et la Reconnaissance des Formes, aux Automates et aux Systèmes, aux Choix des Critères. Masson, Paris. (Reference from I, 1975b.)

Santos, E. S. (1970). Fuzzy algorithms. *Inf. Control* **17**, 326–339.

Santos, E. S. (1976). Fuzzy and probabilistic programs. *Inf. Sci.* **10**, 331–345.

Santos, E. S. (1977). Fuzzy and probabilistic programs. *In* "Fuzzy Automata and Decision Processes" (M. M. Gupta, G. N. Saridis, and B. R. Gaines, eds.), pp. 133–147. North-Holland Publ., Amsterdam.

Tanaka, K., and Mizumoto, M. (1975). Fuzzy programs and their execution. *In* "Fuzzy Sets and Their Applications to Cognitive and Decision Processes" (L. A. Zadeh, K. S. Fu, K. Tanaka, and M. Shimura, eds.), pp. 41–76. Academic Press, New York.

Zadeh, L. A. (1968). Fuzzy algorithms. *Inf. Control* **12**, 94–102.

Zadeh, L. A. (1971). Towards fuzziness in computer systems—Fuzzy algorithms and languages. *In* "Architecture and Design of Digital Computers" (G. Boulaye, ed.), pp. 9–18. Dunod, Paris.

Zadeh, L. A. (1972a). "On Fuzzy Algorithms," Memo UCB/ERL M-325. Univ. of California, Berkeley.

Zadeh, L. A. (1972b). A rationale for fuzzy control. *J. Dyn. Syst. Meas. Control* **94**, 3–4.

Zadeh, L. A. (1973). Outline of a new approach to the analysis of complex systems and decision processes. *IEEE Trans. Syst. Man Cybern.* **3**, 28–44.

Zadeh, L. A. (1974). A new approach to system analysis. *In* "Man and Computer" (M. Marois, ed.), pp. 55–94. North-Holland Publ., Amsterdam.

Chapter 4

FUZZY MODELS FOR
OPERATIONS RESEARCH

This brief chapter is intended to present the state of the art concerning the application of fuzzy sets to theoretical operations research. The conceptual framework for optimization in a fuzzy environment is first reviewed and particularized to fuzzy linear programming. In the second part some existing definitions of fuzzy graph theory are stated. Lastly, some very well-known shortest-path algorithms are extended to graphs where distances between vertices are fuzzy.

A. OPTIMIZATION IN A FUZZY ENVIRONMENT

Optimization models in operations research assume that the data are precisely known, that constraints delimit a crisp set of feasible decisions, and that criteria are well defined and easy to formalize. However, in the real world such assumptions are only approximately true. This section presents the existing conceptual framework for optimization in a fuzzy environment. The linear case is then studied more particularly.

a. General Formulation

Let X be a set of alternatives that contains the solution of a given multicriteria optimization problem. Bellman and Zadeh (1970) pointed out that in a fuzzy environment goals and constraints formally have the same nature and can be represented by fuzzy sets on X. Let C_i be the fuzzy domain delimited by the ith constraint ($i = 1, m$) and G_j the fuzzy domain associated with the jth goal ($j = 1, n$). G_j is, for instance, the optimizing set of an objective function g_j, from X to $\mathbb{R}$ (see II.4.B.a). When goals and constraints have the same importance, Bellman and Zadeh (1970) called a *fuzzy decision* the fuzzy set D on X

$$D = \left(\bigcap_{i=1, m} C_i \right) \cap \left(\bigcap_{j=1, n} G_j \right), \tag{1}$$

that is,

$$\forall x \in X, \quad \mu_D(x) = \min\left[\min_{i=1, m} \mu_{C_i}(x), \min_{j=1, n} \mu_{G_j}(x) \right]$$

A fuzzy decision is pictured in Fig. 1 and corresponds to a constraint "x should be substantially greater than x_0" and an objective function g whose optimizing set is G.

The final decision x_f can be chosen in the set $M_f = \{ x_f, \mu_D(x_f) \geqslant \mu_D(x), \forall x \in X \}$. M_f is called the *maximal decision set*.

When criteria and constraints have unequal importance, membership functions can be weighted by x-dependent coefficients α_i and β_j such that

$$\forall x \in X, \quad \sum_{i=1}^{m} \alpha_i(x) + \sum_{j=1}^{n} \beta_j(x) = 1,$$

and we have according to Bellman and Zadeh (1970)

$$\mu_D(x) = \sum_{i=1}^{m} \alpha_i(x)\mu_{C_i}(x) + \sum_{j=1}^{n} \beta_j(x)\mu_{G_j}(x). \tag{2}$$

Note that D satisfies the property (see II.1.E.b)

$$\left(\bigcap_{i=1, m} C_i \right) \cap \left(\bigcap_{j=1, n} G_j \right) \subseteq D \subseteq \left(\bigcup_{i=1, m} C_i \right) \cup \left(\bigcup_{j=1, n} G_j \right).$$

However, other aggregation patterns for the μ_{C_i} and μ_{G_j} may be worth considering (see IV.4).

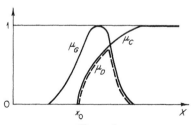

Figure 1

When criteria and constraints refer to different sets X and Y, respectively, and there is some causal link between X and Y, a fuzzy decision can still be constructed. X is, for instance, a set of causes constrained by $C_i, i = 1, m$, and Y a set of effects on which is defined a set of fuzzy goals $G_j, j = 1, n$. Let R be a fuzzy relation on $X \times Y$, the fuzzy decision D can be defined on X by aggregation of the fuzzy domains C_i and the fuzzy goals $G_j \circ R^{-1}$ induced from the G_j.

The definition of an optimal decision by maximizing μ_D (in the sense of formula (1)) is not always satisfactory, expecially when $\mu_D(x_f)$ is very low. It indicates that goals and constraints are more or less contradictory, and thus x_f cannot be a good solution. For such a situation Asai et al. (1975) have proposed the following approach: choose an alternative that better satisfies the constraints and substitute an attainable short-range goal for the nonattainable original one. More specifically, we must find a pair (x_C, x_G) where x_C is a short-range optimal decision and x_G a short-range estimated goal and (x_C, x_G) maximizes

$$\mu_D(x, x') = \min(\mu_C(x), \mu_G(x'), \mu_R(x, x')). \qquad (3)$$

C and G are the fuzzy constrained domain and the fuzzy goal (we take $m = n = 1$ for simplicity), and R expresses a fuzzy tolerance on the discrepancy between the immediate optimal decision x_C and the fuzzy goal G; x_G is the most reasonable objective because it is a trade-off between a feasible decision and G. Note that when R is the identity, (3) gives (1). Asai et al. (1975) discussed the choice of R and found that a likeness relation ($\bar{R}$ is a distance, see II.3.C.c) was most suitable with respect to some natural intuitive assumptions. The authors generalized their approach to the N-period case where N short-range decisions must be chosen together with N short-range goals and μ_C and μ_G may be time-dependent (see Asai et al., 1975). Some definitions pertaining to time-dependency in fuzzy set theory in the scope of planning may be found in Lientz (1972).

b. Fuzzy Linear Programming

We deal now with the very common (in the literature!) situation when constraints and criteria are linear functions of a set of variables.

α. Soft Constraints

We start with the problem

$$\begin{aligned} \text{minimize} \quad & Z = gx \\ \text{subject to} \quad & Ax \leqslant b, \quad x \geqslant 0, \end{aligned} \qquad (4)$$

where g is a vector of coefficients of the objective function, b a vector of constraints, and A the matrix of coefficients of the constraints. The fuzzy version of this problem is (Zimmermann, 1976, 1978):

$$gx \lesssim Z_0, \qquad Ax \lesssim b, \qquad x \geqslant 0. \tag{5}$$

The symbol $\lesssim$ denotes a relaxed version of $\leqslant$ and assumes the existence of a vector μ of membership functions μ_i, $i = 0, m$, defined as follows: Let a_{ij} and b_i be the coefficients of A and b, respectively; then, for $i = 0, m$,

$$\mu_i\left(\sum_{j=1}^{n} a_{ij}x_j\right) = \begin{cases} 1 & \text{for } \sum_{j=1}^{n} a_{ij}x_j \leqslant b_i, \\ 1 - \dfrac{1}{d_i}\left(\sum_{j=1}^{n} a_{ij}x_j - b_i\right) & \text{for } b_i \leqslant \sum_{j=1}^{n} a_{ij}x_j \leqslant b_i + d_i, \\ 0 & \text{for } \sum_{j=1}^{n} a_{ij}x_j > b_i + d_i \end{cases}$$

(see Fig. 2) with $b_0 = Z_0$, $a_{0j} = g_j$; d_i is a subjectively chosen constant expressing a limit of the admissible violation of the constraint i. Z_0 is a constant to be determined.

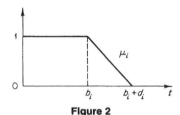

Figure 2

The fuzzy decision of the problem (5) is D such that

$$\mu_D(x) = \min_i \mu_i\left(\sum_{j=1}^{n} a_{ij}x_j\right).$$

The maximization of μ_D is equivalent to the linear program

maximize x_{n+1}

subject to $\quad x_{n+1} \leqslant \mu_i\left(\sum_{j=1}^{n} a_{ij}x_j\right), \qquad i = 0, m, \tag{6}$

$$x_{n+1} \geqslant 0.$$

The constant $Z_0 + d_0$ is determined by solving the above problem (6) without the constraint $i = 0$; let $\underline{x}$ be its solution, we state $Z_0 + d_0 = g\underline{x}$ and Z_0 is defined as the optimal value of the objective function in problem (4) where b_i is replaced by $b_i + d_i$, $\forall i$ (see Sommer and Pollatschek, 1976).

N.B.:   1.   Constraints such as $Ax \geq b$ or $Ax = b$ can be softened in a similar way (see Sommer and Pollatschek, 1976).

2.   The same problem was also studied by Negoita (1976) and Sularia (1977), using the same approach.

3.   Zimmermann (1978) solved the multicriteria linear programming problem in the same way.

4.   In the same paper (Zimmermann, 1978) he defines a fuzzy decision, replacing min by product and compares the two formulations. Sommer and Pollatschek (1976) use arithmetic mean instead of min.

5.   Linear programming with soft constraints is very related to sensitivity analysis in linear programming.

β. Fuzzy Constraints with Fuzzy Coefficients

What happens to a linear constraint $A_i x = b_i$ when the coefficents a_{ij} and b_i become fuzzy numbers? $\tilde{A}_i x$ can be calculated by means of extended addition $\oplus$ (see II.2.B.d.β). The symbol $=$ can be understood in two different ways:

first, as a strict equality between $\tilde{A}_i x$ and $\tilde{b}_i$ (equality of the membership functions); this equality can be weakened into an inclusion $\tilde{A}_i x \subseteq \tilde{b}_i$ which also reduces to equality in the nonfuzzy case; the fuzziness of $\tilde{b}_i$ is interpreted as a maximum tolerance for the fuzziness of $\tilde{A}_i x$;

secondly, as an approximate equality between $\tilde{A}_i x$ and $\tilde{b}_i$ in the sense of II.1.E.c.

Both points of view will be successively investigated. We assume here that the variables are positive ($x \geq 0$) and $\tilde{a}_{ij}$ and $\tilde{b}_i$ are L-R fuzzy numbers (see II.2.B.e.α). See also Dubois and Prade (1978b).

(i)   *Tolerance Constraints*

Consider the system of linear fuzzy constraints

$$\tilde{A}_i x \subseteq \tilde{b}_i, \qquad i = 1, m.$$

Since the coefficients are L-R fuzzy numbers, we can write symbolically

$$A_i = \left(A_i, \underline{A_i}, \overline{A_i} \right)_{LR}$$

where $A_i, \underline{A_i}$, and $\overline{A_i}$ are vectors of mean values and left and right spreads. Since the $\overline{x_j}$ are positive, the system is equivalent to

$$A_i x = b_i, \qquad \underline{A_i} x \leq \underline{b_i}, \qquad \overline{A_i} x \leq \overline{b_i}, \qquad i = 1, m, \qquad x \geq 0.$$

which is an ordinary linear system of equalities and inequalities. According to the value of m and the number n of variables involved, it may or not have solutions.

Owing to this result, the "robust programming problem" (Negoita, 1976)

$$\text{maximize} \quad gx$$
$$\text{subject to} \quad \tilde{A}_i x \subseteq \tilde{b}_i, \quad i = 1, m,$$
$$x \geq 0$$

can be turned into a classical linear programming problem having $3m$ constraints. This approach seems more tractable than that of Negoita (1976).

(ii) *Approximate Equality Constraints*

Let $\tilde{a}$ and $\tilde{b}$ be two fuzzy numbers with $\tilde{a} = (a, \underline{a}, \overline{a})_{LR}$ and $\tilde{b} = (b, \underline{b}, \overline{b})_{RL}$. Recall that in II.2.B.g $\tilde{a}$ is said to be greater than $\tilde{b}$, denoted $\tilde{a} \geq \tilde{b}$, as soon as $a - b \geq \underline{a} + \overline{b}$. $\tilde{a}$ is said to be approximately equal to $\tilde{b}$ iff neither $\tilde{a} \geq \tilde{b}$ nor $\tilde{b} \geq \tilde{a}$ holds.

A system of approximate equalities in the above sense can be considered, i.e.,

$$\tilde{A}_i x \simeq \tilde{b}_i, \quad i = 1, m,$$

where $\tilde{A}_i$ is a vector of L-R fuzzy numbers, $\tilde{b}_i$ an R-L fuzzy number, and $\simeq$ denotes approximate equality. This fuzzy system is equivalent to the nonfuzzy one

$$b_i - A_i x \leq \underline{b}_i + \overline{A}_i x \quad \text{when} \quad 0 \leq b_i - A_i x,$$
$$A_i x - b_i \leq \overline{b}_i + \underline{A}_i x \quad \text{when} \quad 0 \leq A_i x - b_i.$$

The above approach assumes the existence of an equality threshold (see II.2.B.g).

An alternative approach can be, as in a, to define the constraint domain associated with an approximate equality $\tilde{A}_i x \simeq \tilde{b}_i$ by the membership function μ_i such that $\mu_i(x) = \text{hgt}(\tilde{A}_i x \cap \tilde{b}_i)$. More specifically,

$$\mu_i(x) = \begin{cases} R\left[\dfrac{b_i - A_i x}{\underline{b}_i + \overline{A}_i x} \right] & \text{if} \quad b_i - A_i x \geq 0, \\[4mm] L\left[\dfrac{A_i x - b_i}{\overline{b}_i + \underline{A}_i x} \right] & \text{if} \quad A_i x - b_i \geq 0. \end{cases}$$

The problem of finding x_f, maximizing $\min_{i=1, m} \mu_i(x)$, the optimal decision with respect to the m fuzzy constraints, can be thus reduced to a nonlinear program (see Dubois and Prade, 1978b).

N.B.: 1. The approach in (ii) can be extended to fuzzy linear objective functions and to linear approximate inequality constraints. Moreover, (i) and (ii) can be generalized to fuzzy variables.

2. Systems of linear equations with interval-valued coefficients were already considered in Hansen (NF 1969). (See also Jahn, 1974.)

B. FUZZY GRAPHS

Graph theory plays an important role in the modeling of structures, especially in operations research. Fuzzy graphs may be helpful for representing soft or ill-defined structures, for instance, in humanistic systems. A graph is traditionally a pair $G = (V, E)$ where V is a finite set of vertices and E a nonfuzzy relation on $V \times V$, i.e., a set of ordered pairs of vertices; these pairs are the edges of G. A detailed exposition of graph theory and related algorithms can be found in Roy (NF 1969–1970).

a. Fuzzy Vertices, Fuzzy Edges

A fuzzy graph $\tilde{G}$ is a pair $(\tilde{V}, \tilde{E})$ where $\tilde{V}$ is a fuzzy set on V and $\tilde{E}$ is a fuzzy relation on $V \times V$ such that

$$\mu_{\tilde{E}}(v, v') \leqslant \min(\mu_{\tilde{V}}(v), \mu_{\tilde{V}}(v')).$$

This definition is from Rosenfeld (1975).

The above inequality expresses that the strength of the link between two vertices cannot exceed the degree of "importance" or of "existence" of the vertices. In other words $\tilde{E}$ is a fuzzy relation on $\tilde{V} \times \tilde{V}$ in the sense that dom($\tilde{E}$) and ran($\tilde{E}$) (see II.3.B.a) are contained in $\tilde{V}$. However, in some situations it may be desirable to relax this inequality.

Classical concepts and definitions pertaining to graphs have been extended to fuzzy graphs:

A *path* whose length is n in a fuzzy graph is a sequence of distinct vertices $v_0, v_1, \ldots, v_n$ such that $\mu_{\tilde{E}}(v_{i-1}, v_i) > 0$ $\forall i = 1, n$. The strength of the path is $\min_{i=1,n} \mu_{\tilde{E}}(v_{i-1}, v_i)$ and $\mu_{\tilde{V}}(v_0)$ if $n = 0$. A *strongest path* joining two vertices v_0 and v_n has a strength $\mu_{\hat{\tilde{E}}}(v_0, v_n)$ where $\hat{\tilde{E}}$ is the transitive closure of $\tilde{E}$ (see II.3.B.c.α) (Rosenfeld 1975).

In an ordinary graph the distance between two vertices is the length of the shortest path linking them. A set U of vertices is called a cluster of order k iff:

(i) $\forall v, v' \in U, d(v, v') \leqslant k$,
(ii) $\forall v \notin U, \exists v' \in U, d(v, v') > k$,

where $d(v, v')$ denotes the distance between v and v'. When $k = 1$, a k-cluster is called a clique, i.e., a maximum complete subgraph. In a fuzzy graph a nonfuzzy subset U of V is called a *fuzzy cluster of order k* if

$$\min_{\substack{v \in U \\ v' \in U}} \mu_{\tilde{E}^k}(v, v') > \max_{v \notin U} \min_{v' \in U} \mu_{\tilde{E}^k}(v, v')$$

where $\tilde{E}^k$ is the kth power of $\tilde{E}$ (see II.3.B.b.δ) (Rosenfeld, 1975).

The following definitions assume that the set of vertices is not fuzzy and $\tilde{E}$ is symmetrical ($\mu_{\tilde{E}}(v, v') = \mu_{\tilde{E}}(v', v)$). The *degree* of a vertex v is $\mathrm{dg}(v) = \sum_{v' \neq v} \mu_{\tilde{E}}(v, v')$. The minimum degree of G is $\delta(G) = \min_v \mathrm{dg}(v)$.

G is said to be λ-*degree connected* (Yeh and Bang, 1975) iff:

(i) $\forall v, v' \in V$, $\mu_{\tilde{E}}(v, v') \neq 0$ (if $v \neq v'$),

(ii) $\delta(G) \geqslant \lambda$.

A λ-*degree component* of G is a maximal λ-degree connected subgraph of G. For any $\lambda > 0$, the λ-degree components of a fuzzy graph are disjoint (Yeh and Bang, 1975).

Yeh and Bang (1975) have given an algorithm for the determination of λ-degree components of a finite symmetric fuzzy graph. Moreover, they defined other kinds of connectivity and provided the corresponding decomposition algorithms (Yeh and Bang, 1975). Other definitions related to fuzzy graphs can be found in Rosenfeld (1975) (bridge, cut-node, forest, tree, . . .) and in Halpern (1975) (set-adjacency measures).

b. Shortest-Path Algorithms for Fuzzily Weighted Graphs (Dubois and Prade, 1978a)

In this section we consider fuzzily weighted graphs: i.e., for instance, to each edge is assigned a positive weight that represents the "length" of the edge. Shortest-path algorithms have a common feature: they require only additions and comparisons. It is thus easy to fit these algorithms to fuzzy weights, thanks to the extended addition $\oplus$ and subtraction $\ominus$ together with the $\widetilde{\max}$ and $\widetilde{\min}$ operators.

α. Floyd's Algorithm

As an example, let us first focus our attention on Floyd's algorithm (Floyd, NF 1962) for symmetric connected graphs. Let $G = (V, E)$ be such a nonfuzzy graph. Let w_{ij} be the weight of the edge (v_i, v_j) belonging to E. Let l_{ij} be a value assigned to each pair of vertices (v_i, v_j). At the beginning of the procedure we set $l_{ij} = w_{ij}$ for $(v_i, v_j) \in E$, $l_{ij} = \infty$ for $(v_i, v_j) \notin E$ and $i \neq j$, and $l_{ii} = 0$ $\forall v_i \in V$. The procedure consists in modifying the l_{ij}, replacing l_{ij} by $\min(l_{ij}, l_{ik} + l_{kj})$ for v_k ranging over V, v_i ranging over $V - \{v_k\}$ and v_j ranging over V for a fixed v_i. At the end of the procedure l_{ij} is the length of a shortest path between v_i and v_j. Obviously, this procedure can be readily extended to deal with fuzzy weights $\tilde{w}_{ij}$ ($\tilde{w}_{ij}$ is assumed to be a positive fuzzy number). At the beginning, set $\tilde{l}_{ij} = \tilde{w}_{ij}$ for $(v_i, v_j) \in E$ and as above otherwise. Replace $\tilde{l}_{ij}$ by $\widetilde{\min}(\tilde{l}_{ij}, \tilde{l}_{ik} \oplus \tilde{l}_{kj})$ where k, i, j vary as indicated above.

The length of a shortest path between v_i and v_j may also be defined as $l_{ij} = \min_{k \in \mathcal{P}(i, j)} l_k$ where $\mathcal{P}(i, j)$ is the set of all paths between i and j and l_k the length of path k. l_{ij} is an increasing function of the w_{kl}. The result of Floyd's algorithm is also an increasing function of the w_{kl}. Both functions coincide for nonfuzzy w_{kl}. Hence, the extensions of these functions to fuzzy numbers $\tilde{w}_{kl}$ also coincide (as a consequence of Theorem 1 for n-ary operations see II.2.B.a). Thus, the fuzzy Floyd algorithm does calculate the fuzzy shortest distances between vertices.

Although the fuzzy shortest distance $\tilde{l}_{ij}$ between v_i and v_j is still obtained, a shortest path (or a set of shortest paths) whose length is $\tilde{l}_{ij}$ does not necessarily exist any longer. The identity $\tilde{l}_{ij} = \widetilde{\min}_{k \in \mathcal{P}(i, j)} \tilde{l}_k$ is valid, but because the $\widetilde{\min}$ of several fuzzy numbers does not necessarily yield one of those numbers, it is possible that no path has fuzzy length $\tilde{l}_{ij}$. A criticity value of path k can be hgt($\tilde{l}_k \cap \tilde{l}_{ij}$).

β. Ford's Algorithm

Another example is Ford's algorithm (see, e.g., Roy, NF 1969–1970) applied to a connected directed graph without loop where the vertices are weighted. Let p_i be the positive weight of v_i. The vertices represent, for instance, a set of tasks and E the precedence constraints between the tasks. v_i is assumed to have a fuzzy weight, i.e., the ill-known processing time $\tilde{p}_i$ of the task v_i. Let $P(i)$ and $S(i)$ be respectively the set of vertices immediately preceding v_i and the set of vertices that immediately follow v_i. The classical formulas giving the earliest starting time r_i and the latest ending time d_i of the task v_i become

$$\tilde{r}_i = \widetilde{\max_{v_j \in P(i)}} (\tilde{r}_j \oplus \tilde{p}_j), \qquad \tilde{d}_i = \widetilde{\min_{v_j \in S(i)}} (\tilde{d}_j \ominus \tilde{p}_j),$$

where $\tilde{r}_i$ and $\tilde{d}_i$ are respectively the fuzzy earliest starting time and the fuzzy latest ending time of task v_i. The fuzzy earliest starting time (resp. latest ending time) of the tasks without predecessor (resp. successor), which are used to initialize the procedure, may be also fuzzy. It is easy to see that the above fuzzification of Ford's algorithm is valid in the sense of 3.B.d. Note that, owing to the use of L-R fuzzy numbers (see II.2.B.e), the fuzzy versions of the Floyd and Ford algorithms do not require much computation.

The case of a fuzzily weighted fuzzy graphs is considered in Dubois and Prade (1978a).

REFERENCES

Asai, K., Tanaka, H., and Okuda, T. (1975). Decision-making and its goal in a fuzzy environment. *In* "Fuzzy Sets and Their Applications to Cognitive and Decision Processes" (L. A. Zadeh, K. S. Fu, K. Tanaka, and M. Shimura, eds.), pp. 257–277. Academic Press, New York.

Bellman, R. E., and Zadeh, L. A. (1970). Decision-making in a fuzzy environment. *Manage. Sci.* **17**, No. 4, B141–B164.

Dubois, D., and Prade, H. (1978a). Algorithmes de plus courts chemins pour traiter des données floues. *RAIRO Oper. Res.* **12**, No. 2, 213–227.

Dubois, D., and Prade, H. (1978b). Systems of linear fuzzy constraints. *In* "Fuzzy Algebra, Analysis, Logics," Tech. Rep. TR-EE 78/13. Purdue Univ., Lafayette, Indiana. [*Int. J. Fuzzy Sets Syst.* **3**, No. 1, 37–48, 1980.

Halpern, J. (1975). Set-adjacency measures in fuzzy graphs. *J. Cybern.* **5**, No. 4, 77–87.

Jahn, K. U. (1974). Eine Theorie der Gleichungesysteme mit Intervall-koeffizienten. *Z. Angew. Math. Mech.* **54**, 405–412.

Lientz. B. P. (1972). On time-dependent fuzzy sets. *Inf. Sci.* **4**, 367–376.

Negoita, C. V. (1976). Fuzziness in management. *ORSA/TIMS, Miami.*

Negoita, C. V., Minoiu, S., and Stan, E. (1976). On considering imprecision in dynamic linear programming. *Econ. Comput. Econ. Cybern. Stud. Res.* No. 3, 83–96.

Negoita, C. V., and Sularia, M. (1976). On fuzzy programming and tolerances in planning. *Econ. Comput. Econ. Cybern. Stud. Res.* No. 1, 3–14.

Negoita, C. V., Flondor, P. and Sularia, M. (1977). On fuzzy environment in optimization problems. *In* "Modern Trends in Cybernetics and Systems" (J. Rose and C. Bilciu, eds.), Vol. 2, pp. 475–486. Springer-Verlag, Berlin and New York.

Negoita, C. V., and Ralescu. D. A. (1977). On fuzzy optimization. *Kybernetes* **6**, No. 3, 193–196.

Rosenfeld, A. (1975). Fuzzy graphs. *In* "Fuzzy Sets and Their Applications to Cognitive and Decision Processes" (L. A. Zadeh, K. S. Fu, K. Tanaka, and M. Shimura, eds.), pp. 77–95. Academic Press, New York.

Sommer, G., and Pollatschek, M. A. (1976). "A Fuzzy Programming Approach to an Air Pollution Regulation Problem," Working Pap., No. 76/01. Inst. Wirtschaftswiss. R.W.T. H., Aachen.

Sularia, M. (1977). On fuzzy programming in planning. *Kybernetes* **6**, 229–230.

Yeh, R. T., and Bang, S. Y. (1975). Fuzzy relations, fuzzy graphs, and their applications to clustering analysis. *In* "Fuzzy Sets and Their Applications to Cognitive and Decision Processes" (L. A. Zadeh, K. S. Fu, K. Tanaka, and M. Shimura, eds.), pp. 125–149. Academic Press, New York.

Zimmermann, H.-J. (1976). Description and optimization of fuzzy systems. *Int. J. Gen. Syst.* **2**, 209–215.

Zimmermann, H.-J. (1978). Fuzzy programming and LP with several objective functions. *Int. J. Fuzzy Sets Syst.* **1**, No. 1, 45–55.

REFERENCES

Part **IV**
_____

SYSTEMS-ORIENTED
FUZZY TOPICS

In the field of systems science there are many common situations that are pervaded by fuzziness. Classical models and methods dealing with these situations must thus be revised in order to take this basic aspect into account. Most work concerned with this is still at an early stage of development, and in most problems no general methodology is yet available. However, several fuzzy approaches seem worth considering and some have already yielded promising results.

After a first chapter devoted to the estimation of membership functions, nine system-oriented topics are surveyed from a fuzzy-set-theoretic point of view: knowledge representation and natural language, decision-making, control, learning, pattern classification, diagnosis, structural identification, games, and catastrophe theory. These chapters have unequal length according to the respective states of the art.

Chapter *1*

WHERE DO "THEY" COME FROM?

A very widespread question about fuzzy set theory is, From what kind of data and how can membership functions actually be derived? Answering this question is very important for practical applications. Another problem is to check whether the choices of fuzzy set-theoretic operators have an experimental basis.

A. INFORMAL PRELIMINARY DISCUSSION

The membership function is supposed to be a good model of the way people perceive categories. Experiments made by psychologists showed a distinction between central members of a category and peripheral members. If subjects have to respond true or false to questions of the form Does x belong to such a category?, the response time is shorter if x is a central member (i.e., a good example of the category) than if it is a peripheral one (i.e., a not very good example of the category); see Lakoff (1973) for more details. The existence of classes of central and peripheral elements in a category reminds us of *flou* sets (see II.1.G.c) which seem to thus have an intuitive basis.

Clearly, category membership is not always a yes-or-no matter, but rather a matter of degree. However, Lakoff (1973) pointed out that some speakers seem to turn relative judgments of category membership into absolute judgments by assigning the member in question to the category in which it has the highest degree of membership. Since category membership is a matter of degree, the question naturally arises as to what determines the degree of membership for each element. One common hypothesis is that there is a prototype for each category, and the degree of membership for each item is directly related to the similarity of the item to the prototype. But, as indicated by Oden (1977a), the prototype may be an ideal element that does not lie in the category (the corresponding fuzzy set will not be normalized) or, on the contrary, the category may have multiple and noncomparable prototypes.

Although fuzzy set theory is capable of dealing with degrees of set membership, the membership function is not a primitive concept from a psychological point of view. A membership value is generally not absolutely defined; take for example the concept of tallness; how one perceives other people's tallness may depend upon what one's height is. Undoubtedly, the membership function itself is fuzzy; as soon as it has a good shape, it can be considered a satisfactory approximation. According to Lakoff (1973), the membership function is perceived more like a continuum than a discrete set of membership values, although it may be sampled for practical purposes. The choice of continuous set-theoretic operators is consistent with fuzzy knowledge of membership functions: a slight modification of the membership values does not drastically affect the rough shape of the result of a set operation. To take into account the imprecision of membership functions, we may think of using type 2 fuzzy sets (II.1.G. d), probabilistic sets (II.1.G.e), tolerance classes of fuzzy sets (II.3.F.e), or level 2 fuzzy sets (II.2.C.a). Estimating the membership function of such higher order fuzzy sets is certainly more difficult than in the case of ordinary fuzzy sets, but the parameters of higher order fuzzy sets tolerate less-precise estimation. On the whole, ordinary membership functions will be sufficient for an approximate quantitative representation of this intrinsically qualitative notion, that is gradual category membership.

Remark In the framework of experiments on human height, MacVicar-Whelan (1977) noticed that the location of the boundary of a fuzzy set such as "tall" seems "to be equiprobable within some range of values."

B. PRACTICAL ESTIMATION OF MEMBERSHIP FUNCTIONS

The problem of practical estimation of membership functions has not been systematically studied in the literature. Nevertheless, some ideas and methods have been suggested by several authors, independently.

a. Exemplification (Zadeh, Reference from III.3, 1972)

Let U be a universe of objects and A the name of a fuzzy set on U. μ_A can be estimated from partial information about it, such as the values that μ_A takes at a finite number of samples in U. "The definition of a fuzzy set by exemplification is an extension of the familiar linguistic notion of extensive definition." "The problem of estimating the membership function of a fuzzy set in U from the knowledge of its values over a finite set of points in U is the problem of abstraction which plays a central role in pattern recognition."

Example In order to build the membership function of A = "tall", we may ask a person whether a given height h is tall. To answer the person has to use one among several possible linguistic truth-values, e.g., *true, more or less true, borderline, more or less false, false*. The simplest method is then to translate these linguistic levels into numerical ones: 1, 0.75, 0.5, 0.25, 0, respectively. A discrete representation of the membership function is thus obtained by repeating the query for several heights.

b. Deformable Prototypes (Bremermann, 1976)

The idea behind this method is quite simple. Let P be a prototype that can be deformed by manipulating parameters $p_1, \ldots, p_n$. Given an object, one attempts to deform the prototype such that a maximal matching is obtained. The dissimilarity D between the object x and the prototype depends both on the minimal "distance" between them and the distortion "energy" of the deformation. Formally, we write

$$D(x) = \min_{p_1, \ldots, p_n} (m(x; p_1, \ldots, p_n) + w\delta(p_1, \ldots, p_n))$$

where m is a distance function between x and the prototype and δ is a distortion function weighted by w. A membership function μ_p can then be defined as

$$\mu_p(x) = 1 - (D(x)/\sup D).$$

c. Implicit Analytical Definition (Kochen and Badre, 1976)

The membership function is assumed continuous and differentiable and to have an S shape (i.e., we are concerned with fuzzy sets on $\mathbb{R}$). Consider, for instance, the adjective A = large. The marginal increase of a person's strength of belief that "x is A" is assumed proportional to the strength of

his belief that "x is A" and the strength of his belief that "x is not A."
That is,

$$\frac{d\mu_A}{dx}(x) = k\mu_A(x)(1 - \mu_A(x)),$$

whose solution is

$$\mu_A(x) = 1/(1 + e^{a-bx}).$$

The parameters a and b are estimated from statistical data. Thus, the
above method is more a justification of a shape than a quantitative
estimation procedure.

d. Use of Statistics

Membership functions can be estimated through polls, i.e., $\mu_A(x)$ is the
proportion of positive answers to the question, Does x belong to A? The
implicit assumption is that the probability of a positive answer from a
questioned person increases with the value $\mu_A(x)$; more specifically, the
probability of a positive answer is proportional to $\mu_A(x)$. This method was
used by Hersh and Caramazza (1976).

In the social sciences Nowakowska (1977) gave a measurement tool for
estimating the membership value of a person x in a social group A. Her
assumption is: if a subject x is asked about *his* membership in a fuzzy set
A, the probability of a positive response is an increasing function of the
value $\mu_A(x)$.

Lastly, another method may be considered: given a set of statistical data
in the form of a histogram, the induction from it of a possibility distribution is different from that of a probability distribution. In the first case the
histogram is normalized through an affine transformation that brings the
highest ordinate to 1; in the latter case the surface of the histogram is
brought to 1. When thus determining a possibility distribution, we postu-
late that from global precise knowledge about a population of events, we
can induce local imprecise knowledge about any element of this population. This latter assertion assumes that the population is homogeneous in
some sense. More specifically, recall the possibility–probability consistency
principle (II.5.B.c), which says that the possibility of an event is always
greater than or equal to its probability. Let h be a function from $\mathbb{R}$ to $\mathbb{R}^+$,
representing a smoothed histogram. The associated possibility and proba-
bility distributions should satisfy for any union D of disjoint intervals

$$\Pi(D) = \sup_{x \in D} h(x)/\sup h \geqslant \text{Prob}(D) = \int_D h(x)\,dx / \int_{-\infty}^{+\infty} h(x)\,dx \qquad (1)$$

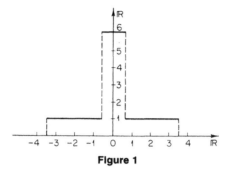

Figure 1

according to the consistency principle. The above inequality does not hold for any positive function h. However, functions such as

$$h(x) = \begin{cases} c \in \mathbb{R}^+ & \forall x \in I, \\ 0 & \text{otherwise}, \end{cases}$$

where I is an interval;

$$h(x) = \max(0, \alpha(1 - |x|/\lambda)) \qquad \text{(triangular function)},$$

and

$$h(x) = e^{-\alpha|x|} \qquad \text{for} \quad \alpha > 0$$

satisfy the above inequality. But it is false for the function pictured in Fig. 1, which may be regarded as "nonhomogeneous." Thus, it seems that only a histogram satisfying inequality (1) may be used to derive a possibility distribution consistent with the probability distribution issued from the histogram.

e. Relative Preferences Method (Saaty, 1974)

Let A be a fuzzy set on a discrete universe U. The membership values $\mu_A(x_i) = w_i$ for $x_i \in U$ are calculated from a set of data representing the relative membership values t_{ij} of an element x_i in A with respect to the membership of an element x_j in A. Saaty uses a scale divided into seventeen levels $\{\frac{1}{9}, \frac{1}{8}, \ldots, \frac{1}{2}, 1, 2, \ldots, 8, 9\}$; each level has a semantic interpretation: the larger t_{ij}, the greater the membership of x_i compared with that of x_j. The matrix T of the t_{ij}'s is such that $t_{ij} = 1/t_{ji}$. T is said to be consistent iff $\exists w = (w_1, \ldots, w_n)$ with $n = |U|$, such that $t_{ij} = w_i/w_j$ $\forall_i, \forall_j$. When T is consistent, T is transitive in the sense that

$$\forall i, j, k, \quad t_{ij} t_{jk} = t_{ik};$$

the rank of T is 1; the eigenvalues of T are 0, except one whose value is $n = |U|$; w is an eigenvector of T.

A way of evaluating the inconsistency of T is to calculate the difference between the greatest eigenvalue of T and the greatest ideal eigenvalue, i.e., n. The membership values w_i can be determined by finding the eigenvector of T such that $T \cdot w = nw$, where T is assumed as consistent as possible. This method seems appealing from a theoretical point of view when there is no prototype for the class A, but its practical applicability is limited by the size of $U \times U$ and by the difficulty of collecting consistent data. Moreover, the "arbitrariness" of the membership values is somewhat replaced by that of the t_{ij}.

f. Comparison of Subsets   (Fung and Fu, 1974)

Suppose A is a fuzzy set of U with a membership function μ_A; a fuzzy set $\tilde{A}$ on $\mathcal{P}(U)$ is induced from A, provided that U is finite, though the formula

$$\mu_{\tilde{A}}(\{x_1, \ldots, x_k\}) = \frac{1}{k} \sum_{i=1}^{k} \mu_A(x_i).$$

This definition has the intuitive meaning of an "average membership" of $\{x_1, \ldots, x_k\}$ in A.

A "preference" relation, denoted $\geqslant$ is defined in $\mathcal{P}(U)$ by

$$\forall S_1, S_2 \in \mathcal{P}(U), \quad S_1 \geqslant S_2 \quad \text{iff} \quad \mu_{\tilde{A}}(S_1) \geqslant \mu_{\tilde{A}}(S_2).$$

The interpretation of $S_1 \geqslant S_2$ is "S_1 matches A better than S_2." The data form a set of "preferences" between subsets of U; they can be translated, using the definition of μ_A, into a system of inequalities relating the membership values. These inequalities determine more or less strongly the $\mu_A(x_i)$. Other inductions of $\tilde{A}$ from A are possible. The applicability of the method is limited by the size of $\mathcal{P}(U)$. The "arbitrariness" now mainly lies in the induction process.

g. Filter Function   (MacVicar-Whelan, 1978)

MacVicar-Whelan introduced filter functions in order to identify the membership functions of fuzzy sets modeling adjectives such as *tall* in the framework of an experimental study of human height.

A *filter function* F is characterized by two parameters, the location NP of the neutral point ($F(NP) = \frac{1}{2}$) and the width $2w$ of the transition between nonmembership and membership. We are here interested in S-shaped fuzzy sets of $\mathbb{R}$.

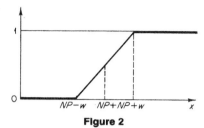

Figure 2

More specifically,

$$F(x; NP, w) = \begin{cases} 0 & \text{if } x \in (-\infty, NP - w], \\ (1/2w)(x - NP + w) & \text{if } x \in [NP - w, NP + w], \\ 1 & \text{if } x \in [NP + w, \pm\infty]. \end{cases}$$

F is pictured in Fig. 2. MacVicar-Whelan points out that a sophistication of the shape of the transition is useless because of the imprecision.

The concept of tallness is here related to a given population for which a normal probability distribution on the heights is known. Let $\bar{x}$ and σ be the parameters of this distribution. "A person is tall" is supposed to mean "this person has a large height," where *large* is modeled by the membership function μ such that

$$\mu(x) = F(x; \bar{x} + \alpha\sigma, \beta\sigma) \qquad \text{where} \quad x \text{ is a height.}$$

α and β have to be determined experimentally. And *small* is modeled in the same way by $1 - F(x; \bar{x} - \alpha\sigma, \beta\sigma)$.

h. Concluding Remarks

What is striking in the methods presented above is their lack of generality. For instance, it seems that there is no rule like maximum likelihood for probabilities to estimate possibilities. Anyway, it seems more important to become aware of how the human mind manipulates names of fuzzy sets than to figure out precisely numerical grades of membership since the perception process is itself fuzzy.

N.B.: Fuzzy measures must also be estimated; some hints on this problem are provided by Sugeno (Reference from II.5, 1977).

C. IDENTIFICATION OF FUZZY SET OPERATORS

Some experiments concerning the verification of the accurateness of fuzzy set operators have been reported in the literature.

In Hersh and Caramazza (1976) a group of people judge the size of black squares by means of 13 qualifiers such as "small," "large," "not small," "very large," "either small or large." Each person assigned a binary grade to each pair black square/qualifier. Membership functions were constructed as described in B.d. Complementation to 1 was shown to be a good model for negation, and max for either . . . or. In another context Oden (1977b) found that probabilistic operators ($\hat{+}$, $\cdot$) were more suitable for modeling disjunction and conjunction.

Zimmermann (1978) pointed out that the *and* operator could be either logical or compensatory. Empirically, he found that the min operator was a good model for logical *and*; on the contrary, the compensatory *and* expresses an aggregation of aspects which is not necessarily a conjunction—for instance, when we say that a car is attractive because it is fast "and" pretty. In this latter case the "and" may be translated by a product, an arithmetic mean, a geometric mean, etc. according to the situation, but rarely by min. The compensatory *and* is very common in decision-making. Other experiments were carried out by Rödder (1975).

The hedge "very" has also received particular attention in the literature. Zadeh (Reference from II.2, 1972) has conjectured that $\mu_{\text{very } A}(x) = [\mu_A(x)]^2$. Since then, Lakoff suggested that "very" operated also a translation, i.e., $\mu_{\text{very } A}(x) = [\mu_A(x - c)]^2$. Experimental verifications were carried out by Hersh and Caramazza (1976), who confirmed that the hedge modification involved some translation. Moreover, Kochen and Badre (1976) found that "very A" could be less fuzzy than A, which is consistent with squaring the membership function. In the context of his study of human height and the concept of tallness, MacVicar-Whelan (1978) empirically determined that the membership function of "very large" could be $F(x; \bar{x} + 2\alpha\sigma, \beta\sigma)$ (see B.g), i.e., a translation of $\alpha\sigma$. However, considering other studies on the hedge "very" in other contexts, he indicated that the shift could be more multiplicative than additive.

In natural language connectives and hedges are sometimes ambiguous and have no "universal" meaning. For instance, "and" may be as well logical as compensatory, the "or" may be "exclusive" or "inclusive," "very" may indicate an increase in precision (e.g., "very medium") or a change of category. Implicit categorization of the universe of discourse into different concepts has great influence on meaning. For example, consider a universe of heights roughly divided into "large heights" and "small heights"; then "not large" may be identical to "small," and "very large" is an increase in precision of "large." When the categorization is refined into "large," "medium," and "small," then "not large" will mean "small or medium." Lastly, if we add the categories "very small" and "very large," now "large" means "large and not very large," and "very large" is no longer more precise than "large." Note that the human mind can perceive only a small number of categories. Moreover, some aggregations of hedges, such as "not very large," may have an ambiguous structure: is it to be understood as "not (very large)" or "(not very) large"?

D. TOLERANCE ANALYSIS USING FUZZY SETS

Fuzzy sets not only model subjective categories, they may refer to the possibility of events, for instance, in the framework of tolerance analysis, as first suggested by Jain (Reference from II.2).

The tolerance interval of a measurement is the interval where it is possible that the actual value lies. More specifically, a flat fuzzy number (see II.2.B.e.η) can be viewed as a tolerance interval with no sharp boundary. A fuzzy set of $\mathbb{R}$ with several distinct maxima can model a set of imprecise measures of a given phenomenon. The membership value of a maximum may express the degree of relative reliability of the information that lies in it, while spreads model the imprecision and the fuzziness.

The use of fuzzy sets in tolerance analysis may throw some light on a well-known problem in measure processing: let $M_1, \ldots, M_n$ be n approximate measures obtained by means of several devices under the same operating conditions and which evaluate some quantifiable property of a given phenomenon. The M_i are assumed to be normalized fuzzy sets on $\mathbb{R}$. How are they to be processed, especially in the case where n is too small, so that probability theory cannot be applied? Let M be the result of the aggregation of the M_i. $M = \bigcap_i M_i$ is an optimistic one: it assumes invariance of the phenomenon, reliability of all measurement devices, and "closeness" of the M_i. When these assumptions are lacking, $\bigcap_i M_i$ is no longer normalized, and this aggregation is not very reliable; in that case $\bigcup_i M_i$ is a most valid aggregation. $(1/n) \cdot (\bigoplus_i M_i)$ may be a trade-off between the union and the intersection of the M_i in the sense that the fuzziness of this result is an average of the fuzziness of the M_i's.

Note that when $\bigcap_i M_i$ is not normalized, the gain in precision is counterbalanced by a loss in reliability, which makes the precision somewhat delusive; $\bigcup_i M_i$ is more reliable but less precise.

If the M_i are not equally reliable, two approaches may be considered. Let $r_i \in [0, 1]$ be the reliability of M_i. Two possible aggregation formulas are $M = \bigcup_i r_i M_i$ or $M = \bigcap_i M_i^{r_i}$ (see III.2.F.c) where the M_i are still normalized in both cases.

REFERENCES

Bremermann, H. (1976). Pattern recognition. *In* "Systems Theory in the Social Sciences" (H. Bossel, S. Klaszko, and N. Müller, eds.), pp. 116–159, Birkhaeuser, Basel. (Reference from IV.6, 1976b.)

Fung, L. W., and Fu, K. S. (1974). The kth optimal policy algorithm for decision-making in fuzzy environments. *In* "Identification and System Parameter Estimation" (P. Eykhoff, ed.), pp. 1052–1059. North-Holland Publ., Amsterdam. (Reference from IV.4, 1974a.)

Hersh, H. M., and Caramazza, A. (1976). A fuzzy-set approach to modifiers and vagueness in natural languages. *J. Exp. Psychol: Gen.* **105**, 254–276.

Kochen, M., and Badre, A. N. (1976). On the precision of adjectives which denote fuzzy sets. *J. Cybern.* **4**, No. 1, 49–59.

Lakoff, G. (1973). Hedges: A study in meaning criteria and the logic of fuzzy concepts. *J. Philos. Logic* **2**, 458–508. (Reference from IV.2.)

MacVicar-Whelan, P. J. (1977). Fuzzy and multivalued logic. *Int. Symp. Multivalued Logic, 7th, N.C.* pp. 98–102.

MacVicar-Whelan, P. J. (1978). Fuzzy sets, the concept of height and the hedge very. *IEEE Trans. Syst., Man Cybern.* **8**, 507–511.

Nowakowska, M. (1977). Fuzzy concepts in the social sciences. *Behav. Sci.* **22**, 107–115.

Oden, G. C. (1977a). Fuzziness in semantic memory: Choosing exemplars of subjective categories. *Mem. Cognit.* **5**, No. 2, 198–204.

Oden, G. C. (1977b). Integration of fuzzy logical information. *J. Exp. Psychol.: Hum. Percept. Perform.* **3**, No. 4, 565–575.

Rödder, W. (1975). "On 'and' and 'or' Connectives in Fuzzy Set Theory," Working Pap., 75/07. Tech. Univ. Aachen, Aachen. (Presented at EURO 1, Brussels, 1975.)

Saaty, T. L. (1974). Measuring the fuzziness of sets. *J. Cybern.* **4**, No. 4, 53–61.

Zimmermann, H. J. (1978). Results of empirical studies in fuzzy set theory. *In* "Applied General System Research" (G. J. Klir, ed.), pp. 303–312. Plenum, New York.

Chapter 2

FROM PROGRAMMING LANGUAGES TO NATURAL LANGUAGES

Natural languages are fuzzy in many respects. Traditional programming languages are not. The gap between them has been slightly reduced by the conception of some fuzzy programming languages. A brief survey of the corresponding works is provided in the first section. The second is devoted to the representation and interpretation of natural language sentences by means of fuzzy sets, according to Zadeh. Lastly, the application of natural language modeling to the representation of fuzzy dynamic systems is emphasized.

A. FUZZY PROGRAMMING LANGUAGES

Umano *et al.* (1978) have proposed the implementation for fuzzy-sets manipulation of a system that is a fuzzy version of the set-theoretic data structure (STDS) of Childs (NF 1968). The system, called FSTDS (fuzzy STDS) consists of a simple interpreter, a collection of fuzzy-set operations, and a data structure. The aim of the system is to make possible set-theoretic manipulations of type *n*, level *l* fuzzy sets, and *L*-fuzzy sets without paying attention to their representation in the computer. FSTDS is imbedded in FORTRAN, "because of its high portability." It has no control structure; however, owing to the connection between FSTDS and FORTRAN, it is possible to use FORTRAN control structures instead.

Fellinger (1974) has described a fuzzy system modeling language FSML which allows specification of individual nonfuzzy objects that may have fuzzy attributes. In order to visualize partial execution of conditional fuzzy instructions, the author uses a modified version of Petri nets (Holt, NF 1971), called E-nets (Nutt and Noe, NF 1973), which he extends to deal with fractional copies of tokens. This ability to reproduce tokens helps simulate the execution of fuzzy instructions in models of fuzzy systems.

A fuzzified version of PL1, called L.P.L. (which stands for linguistic-oriented programming language) has been developed by Adamo (1978a, b). Statements in the L.P.L. language are divided into basic statements (including inference and assignment statements) and control statements. Control statements are DO-END, IF-THEN-ELSE, PARALLEL, and DO-WHILE structures. The DO-END structure is similar to the one in PL1. The IF-THEN-ELSE structure is executed as described in III.3.B.c.γ. The DO-WHILE structure is an infinite set of nested conditional fuzzy structures whose execution requires the solution of a recursive fuzzy relational equation. The PARALLEL structure, which does not exist in PL1, is needed to realize a symmetrical execution of statements such as

> IF V = HIGH THEN perform statement 1
> ELSE IF V = LOW THEN perform statement 2
> ELSE do nothing

which, because it is fuzzy, does not provide the same results as

> IF V = LOW THEN perform statement 2
> ELSE IF V = HIGH THEN perform statement 1
> ELSE do nothing

Adamo (1978c) used L.P.L. for solving combinatorial and syntactic pattern recognition problems. The backtracking processes do not appear explicitly in the corresponding programs, but are "implicitly imbedded in the semantics."

Fuzziness has also been introduced in artificial intelligence languages in order to represent and manipulate fuzzy knowledge. An exhaustive survey of the various approaches to the representation and processing of fuzzy knowledge within the field of artificial intelligence was recently provided by Wahlster (1977). Kling (1973) was the first to deal with this problem and proposed a fuzzy version of PLANNER. The first fuzzy artificial intelligence programming language which was actually implemented is LeFaivre's FUZZY. "FUZZY acts as a many-valued programming language, in the sense that expressions can return both a value and a numeric modifier (called Z-value), which may be interpreted as a truth-value, degree of certainty, etc. . . . A fuzzy associative net is maintained by the system and 'procedure demons' may be defined for the control of fuzzy processes." Pattern-directed data access, a procedure invocation mechanism, and a backtrack control structure are also available. FUZZY is implemented in LISP (see Winston, NF 1977, for instance). FUZZY procedures may be used to define fuzzy algorithms (LeFaivre, 1974a). A description of another fuzzy artificial intelligence language is given by Wechsler (1976), who uses the question-answering system QA4, and proposes a model for medical diagnosis based on fuzzy procedural knowledge.

Other fuzzy programming languages are those of Mamdani (FSML, 1975) and Noguchi *et al.* (FLOU, 1976).

N.B.:   The paper by Chang and Ke (1978), concerned with the translation of "fuzzy queries," does not actually refer to fuzzy set theory; it actually deals with the interpretation of ambiguous questions in the framework of a data base: "fuzzy queries can be disambiguated by analyzing the queries against the information graph of a data base skeleton."

B. MODELING NATURAL LANGUAGES

Most of the sentences of a text in natural language contain fuzzy denotations. Moreover, "the numerous meaning representation, knowledge representation and query representation languages which have been described in the literature . . . are not oriented towards the representation of fuzzy propositions, that is propositions containing labels of fuzzy sets, and hence have no facilities for semantic—as opposed to syntactic—inference from fuzzy premises" (Zadeh, 1977a). Note that the fuzzy programming languages surveyed in Section A do not aim at modeling natural language. This section outlines Zadeh's approach to this problem.

a. The Concept of Meaning

By 1970, Zadeh (1971, 1972b) formalized the notion of "meaning" by equating it with a fuzzy subset on a universe of discourse generated by a kernel space.

A kernel space K can be any prescribed set of objects or constructs. In general, K is not sufficient to embed the meaning of any concept because some concepts may also involve n-tuples of elements of K and more generally a collection of fuzzy subsets on K. For instance, "much older than" is a label for a fuzzy set on K^2; in this fashion the term "very" may be equated with a subset of $\tilde{\mathcal{P}}(K) \times \tilde{\mathcal{P}}(K)$ since it is a function from $\tilde{\mathcal{P}}(K)$ to $\tilde{\mathcal{P}}(K)$. This motivates the following definition (Zadeh, 1971).

Let K be a kernel space and E a set that contains K and which is generated from K by a finite application of the operations of union, cartesian product, and collection of fuzzy sets. Then a *universe of discourse* U is a designated, not necessarily proper, subset of E.

Now let U be a universe of discourse and T a set of terms that play the roles of names of fuzzy sets on U. Let x be a term in T. The meaning of x, denoted $M(x)$, is the fuzzy subset on U whose membership function is $\mu_N(x, \cdot)$ where N is a naming relation on $T \times U$ (see III.3.A.i). $\mu_N(x, \cdot)$ can be viewed as a possibility distribution (Zadeh 1977a), i.e., it designates the objects (in U) that x possibly names. $M(x)$ may be specified in various ways, e.g., "by a table or by a formula or by an algorithm or by exemplification or in terms of other membership functions." (Zadeh, 1972b).

When the term must designate a precise object of U, the principle of "maximum meaningfulness" (Goguen, 1976) says that the "meaning" of the term is the object that has the maximum membership value in the fuzzy set named by the term.

b. Hedges

One of the basic problems in semantics is to evaluate the meaning of a composite term from knowledge of the meaning of each of its atomic subterms. We consider here the meaning of composite terms of the form $h \cdot x$ where h is a linguistic hedge such as "sort of," "very," . . . .The hedge h is viewed as a modifier of the meaning of x. Zadeh (1972a) defined some operators that may serve as a basis for modeling hedges:

normalization: $\mu_{\mathrm{norm}(A)}(u) = \mu_A(u)/(\sup \mu_A)$;
concentration: $\mu_{\mathrm{con}(A)}(u) = [\mu_A(u)]^2$;

dilation: $\mu_{\mathrm{dil}(A)}(u) = [\mu_A(u)]^{\frac{1}{2}}$;
contrast intensification:

$$\mu_{\mathrm{int}(A)}(u) = \begin{cases} 2\mu_A^2(u) & \text{for } \mu_A(u) \in [0, 0.5], \\ 1 - 2(1 - \mu_A(u))^2 & \text{otherwise.} \end{cases}$$

Examples of models of hedges are:

very $A = \mathrm{con}(A)$

more or less $A = \mathrm{dil}(A)$

plus$A = A^{1.25}$

slightly $A = \mathrm{int}[\,\mathrm{norm}(\mathrm{plus}\ A \text{ and not } (\mathrm{very}\ A))\,]$.

Thus, a small number of basic functions can produce a wide range of models of hedges. However, such an approach has some limits, which are discussed at length in Lakoff (1973). Significantly, hedges such as "very" are applied only to fuzzy concepts.

Remarks 1 Note that very A and more or less A can be viewed as fuzzy α-cuts (see II.2.A.e.γ), i.e., α-cuts with $\mu_{\tilde{\alpha}}(t) = t^2$ and $\sqrt{t}$, respectively ($t \in [0, 1]$).

2 A formal logic approach to the hedge "rather" was proposed by Kubinski (NF 1960). This approach strikingly contrasts with the one presented here.

c. Hierarchies of Fuzzy Concepts (Zadeh, 1971)

A fuzzy concept or simply a concept is a fuzzy set on the universe of discourse. Thus, if x is a term, then its meaning $M(x)$ is a concept. It is convenient to classify concepts according to their level (of abstraction) which is a rough measure of the complexity of characterization of a concept. More specifically, let K be the kernel space of the universe of discourse U. Then a term x and the corresponding concept $M(x)$ are at level 1 if $M(x)$ is a fuzzy subset on K^n for some finite n; x and $M(x)$ are at level 2 if $M(x)$ is a subset on $[\tilde{\mathscr{P}}(K)]^n$ for some finite n. More generally, x and $M(x)$ are at level l if $M(x)$ is a subset of $[\tilde{\mathscr{P}}^{l-1}(K)]^n$ for some finite n where $\tilde{\mathscr{P}}^{l-1}(K)$ stands for $\tilde{\mathscr{P}}(\cdots(\tilde{\mathscr{P}}(\tilde{\mathscr{P}}(K)))\cdots)$ with $l-1$ $\tilde{\mathscr{P}}$s in the expression. Note that when $n = 1$ and $M(x)$ is a concept at level l, then it is a level l fuzzy set on K (see II.2.C.a).

For example, K is a set of colored objects; then the concepts labeled "white," "yellow," or "green" are at level 1 because they can be represented as fuzzy sets on K; likewise the concepts labeled "redder than," "darker than" are at level 1 because they can be represented as fuzzy sets on K^2. On the other hand, the concept labeled "color" is essentially a collection of concepts such as $M(\text{white})$, $M(\text{yellow})$, ... and thus is a subset on $\tilde{\mathscr{P}}(K)$. "Color" is at level 2.

N.B.: Such hierarchies of concepts were studied by Goguen (1974) in the framework of category theory.

d. Complex Fuzzy Concepts

The storage capacity of computing systems is usually not sufficient to memorize explicitly all elements x of T, the set of terms, and their meanings $M(x)$. Complex concepts can then be defined by means of a grammar or more generally by an algorithm.

α. Generation of Terms by a Grammar (Zadeh, 1971)

In some restricted contexts, T is a formal language that can be generated by a formal grammar from a finite set of primitive terms. For instance, a term set of linguistic age values can be generated by a context free grammar from the primitive terms "young," "old," "very," "not," "and," "or." Each production is associated with a calculation rule of the membership function of the generated string: for example, the production $A \rightarrow$ very B is paired with $\mu_A(u) = [\mu_B(u)]^2$. Typical terms generated by such a grammar are "not very young," "not very young and not very old," "young and (old or not young)." However, some of these composite terms are intuitively hard to understand.

N.B.: Inagaki and Fukumura (Reference from III.3) use tree grammars (see III.3.A.f) to generate terms and their meanings.

An important example of this approach is the definition of a term set of linguistic truth values. The set of primitive terms is then {"true," "false," "very," "not," "and," "or"} where "true" is a fuzzy set on [0, 1], such that $\mu_{\text{true}}(1) = 1$ and "false" is the antonym of "true," i.e., $\mu_{\text{false}}(t) = \mu_{\text{true}}(1 - t)$ with $t \in [0, 1]$.

Remark Zadeh (Reference from II.2, 1975) calls a *linguistic variable* a complex $S = (T, U, G, M)$ where T is a term set, G a set of syntactic rules that generate T from a set of primitive terms, and M is a set of semantic rules that assign to each value x of S ($x \in T$) its meaning $M(x)$, which is a fuzzy subset on the universe of discourse U. A linguistic variable takes linguistic values that are names of fuzzy sets.

β. A Fuzzy Algorithmic Approach (Zadeh, 1976)

Complex concepts can be defined by means of fuzzy algorithms that have the structure of a *branching questionnaire*. These algorithms are said to be definitional (see III.3.B.e). The questions are fuzzy and are of the form, Is u A? where A is a name of a concept and u is an element of the universe of discourse. The answer to such a question may be fuzzy. If A is a primitive concept, the question is said to be atomic. A branching questionnaire is a representation in which the order of the constituent questions that are asked is determined by the answer to the previous questions. A branching questionnaire corresponds to a composite question which involves a complex concept. For instance, in "Is u big?" "big" means "long" and "wide" and "high." The answer will be deduced from the answers to the constituent questions, Is u long?, Is u wide?, Is u high?

More generally, a n-adique composite question Q, composed of n constituent questions $Q_1, \ldots, Q_n$, is characterized by its relational representation B whose tableau has the form

Q_1	$\cdots$	Q_j	$\cdots$	Q_n	Q
r_1^1	$\cdots$	r^j_1	$\cdots$	r_1^n	r_1
$\vdots$		$\vdots$		$\vdots$	$\vdots$
r_i^1	$\cdots$	r^j_i	$\cdots$	r_i^n	r_i
$\vdots$		$\vdots$		$\vdots$	$\vdots$
r_m^1	$\cdots$	r^j_m	$\cdots$	r_m^n	r_m

The universes of the constituent questions may be distinct. $\forall i, \forall j, r_i^j$ is a linguistic value, the name of a fuzzy set on U_j (universe of Q_j). r_i^j is a possible (authorized) answer to the question Q_j, which is either attributional (e.g., How old is Terry?) or classificational (e.g., Is Terry old?). r_i is a linguistic truth value (Q is classificational).

N.B.: 1. The tableau may not be complete, that is, certain combinations of the admissible answers to constituent questions may be missing from the tableau. This may imply that either the particular combination of answers cannot occur or the answer to Q corresponding to the missing entries is not known.

2. The components $S_1, \ldots, S_n$ of an n-ary linguistic variable S are said to be λ-*non-interactive* (Zadeh, Reference from II.2, 1975, λ means "linguistically") iff the assignment of specific linguistic values to S_{i_1}, $\ldots, S_{i_k}$ ($\forall k < n$) does not constrain the assignment of linguistic values to the linguistic variables of the complementary sequence $S_{j_1}, \ldots, S_{j_l}$ ($k + l = n$). When S_j takes its value in the set of possible answers for Q_j, for any j, the n variables S_j are λ-non-interactive if the tableau is complete. λ-non-interactivity and β-non-interactivity must not be confused (see II.3. A.b).

Several basic problems underlie the transformation of the relational representation of a complex concept into an efficient branching questionnaire. Of these, one is that of determining the conditional redundancies and/or restrictions that may be present in the relational representation. Another is that of determining the order in which the constituent questions must be asked in order to minimize the average cost of finding the answer to Q. This problem is considered at length in Zadeh (1976).

e. Linguistic Approximation

In the previous sections, we have been interested in the computation of the meaning of a composite term. We now consider the converse problem.

The linguistic approximation problem, i.e., find a term whose meaning is the same as or the closest possible to the meaning of an unlabeled fuzzy set, was first pointed out by Zadeh (1971).

Until now very few works have dealt with this problem. In order to solve it we may think of using a distance between fuzzy sets (see II.1.E.c), as Wenstøp (1976a), or Kacprzyk (Reference from V) did. However, when the term set is very large, a simple enumerative matching procedure requires too much computation time: the determination of the distance between two fuzzy sets involves all the elements of their discretized supports and thus may be very long.

To cope with this difficulty, Bonissone (1978) has recently proposed a pattern-recognition approach. The method proceeds in two main steps.

Four features are precalculated for all the elements of the term set (generated by a context-free grammar): the power (see II.1.D.a), a Shannon-like entropy (II.1.H), the first moment, and the skewness (third moment). The author has checked the weak correlation of these features. The first step consists in evaluating the four features of the unlabeled fuzzy set and to prescreen the term set in order to keep the closest terms in the sense of a quadratic weighted distance in the feature space. In the second step a Bhattacharyya (NF 1943) distance between the unlabeled fuzzy set and the meaning of each selected term is determined. The name of the closest labeled fuzzy set is then assigned to the unlabeled one.

f. Representation of Natural Language

To this day the most advanced work applying fuzzy set theory to the modeling of natural language is that of Zadeh (1977a, b), i.e., PRUF (an acronym for possibilistic relational universal fuzzy).

PRUF is a meaning representation language for natural languages. "Thus a proposition such as 'Richard is tall' translates in PRUF into a possibility distribution of the variable Height(Richard) which associates with each value of the variable a number in the interval $[0, 1]$ representing the possibility that Height(Richard) could assume the value in question. More generally a proposition, p, translates into a procedure P, which returns a possibility distribution, π^p, with p and π^p representing, respectively, the *meaning* of P and the *information* conveyed by p" (Zadeh, 1977a).

The theory underlying PRUF is that of approximate reasoning presented in III.1.E.

Some examples of translation into PRUF from Zadeh (1977a) are provided. In the translation of an expression e in a natural language into an expression E in PRUF, if w is a word in e, then its correspondent W in E is the name of a relation in D (the data base). E is a procedure whose form generally depends on the frame of the data base and, hence, is not unique.

Example 1 "Kent was walking slowly toward the door" translates, in PRUF, into:

$$\text{WALKING}\big[\text{Name} = \text{Kent}; \pi_{\text{speed}} = \text{SLOW}: \pi_{\text{time}} = \text{PAST};$$

$$\pi_{\text{direction}} = \text{TOWARD}(\text{Object} = \text{DOOR})\big].$$

Example 2 "Most men are tall." The frame of D comprises

$$\underline{\text{POPULATION}} \| \text{ Name} | \mu |$$

$$\underline{\text{MOST}} \| \rho | \mu |$$

where μ_i, in POPULATION, is the degree to which Name i is TALL and μ_j in MOST is the degree to which ρ_j is compatible with MOST. Then "Most men are tall" translates in PRUF into

$$\pi_{\text{Prop(TALL)}} = \text{MOST}$$

where

$$\text{Prop(TALL)} = \frac{\sum_i {}_\mu \text{POPULATION}[\text{Name} = \text{Name}_i]}{|\text{POPULATION}|} .$$

The numerator is the power (II.1.D.a) of the fuzzy set of tall men in POPULATION. |POPULATION| is the cardinality of POPULATION. π_a always stands for "possibillity distribution of the variable a."

PRUF can be used to translate propositions (declaration, assertions) and also questions. Very recently Zadeh (1978) outlined a possible extension of PRUF to the translation of imperatives (orders, commands). According to the "compliance criterion," the (nonfuzzy) response to an imperative must have the maximal membership value to the possibility distribution on the responses to this imperative. This principle is closely related to maximum meaningfulness principle (Goguen, 1976). The intended purpose of such an extension of PRUF is the execution of fuzzy instructions in Robotics.

For example, the command, "Please ask Mary to have a cup of coffee" is translated into

$$\big[\pi_{\text{strength}} = \text{MEDIUM}; \text{Issuer} = \text{Me}; \text{Addressee} = A; \text{Proposition:}$$

$$\text{REQUEST}\big[\pi_{\text{strength}} = \text{MEDIUM}; \text{Issuer} = A; \text{Addressee} = \text{Mary};$$

$$\text{Proposition: DRINK} \big[\text{Subject} = \text{Mary};$$

$$\text{Object} = \text{COFFEE}[\text{Vessel} = \text{CUP}]\big];$$

$$\pi_{\text{time}} = \text{CONTEXTUAL}\big]\big]$$

π_{strength} is the degree of imperativeness of the command. Note that the above example is a nested command. A denotes the person to whom the order is given.

Lastly, it is important to notice that a translation in PRUF is independent of the structure of the natural language considered.

A particular kind of imperative was modeled by Shaket (1976), who applied the maximum meaningfullness principle to the fuzzy designation of objects in a world of blocks. This world consists in rectangular parallelepipedal solid objects placed on a table. The system must understand such commands as: "Find a cube which is near a plate" or "Find the biggest blue plate." "In this system adjectives define a fuzzy set over a universe indicated in the noun. The truth value of an object in the noun group is found by multiplying and normalizing the values in the noun fuzzy set by those in the adjectives fuzzy set." Superlatives are modeled by a normalization and a concentration. For instance, if there are two objects $B1$ and $B2$ in the world, such that $\mu_{\text{long}}(B1) = 0.22$ and $\mu_{\text{long}}(B2) = 0.30.$, then $\mu_{\text{longest}}(B1) = (0.22/0.30)^2$ and $\mu_{\text{longest}}(B2) = 1$.

Note that such fuzzy instructions do not include actions because verbs other than "find" are not considered.

C. DEDUCTIVE VERBAL DYNAMIC SYSTEMS

Modeling natural language is crucial for the description of fuzzy systems. Recently, Wenstøp (1976a, b) proposed a verbally formulated simulation model for the representation of social phenomena. In traditional models causal relations must be precisely defined, even if the modeler has only a vague idea of their nature. To avoid the artificial step of translating vague ideas with inappropriate precision, the modeler should instead be allowed to formulate his models in natural language. The main point is that such verbal models may provide more significant information than artificially precise ones. The aim of the approach is the inference of the verbal model behavior from a linguistically described initial state. To make the verbal model deductive, it is necessary to:

specify a vocabulary and a grammar;
define a semantic model of the meaning of elementary terms of the vocabulary;
implement the syntactical–semantical model in a computer language (Wenstøp, 1976a, b, uses APL).

A verbal model consists in an ordered list of grammatically correct statements such that the (fuzzy) values of all independent variables that appear in any statement have been determined by previous statements in the list. Loops are allowed, which, according to Wenstøp, makes the behavior of the model hard to forecast a priori. Results are expressed in a verbal form owing to a linguistic approximation procedure (B.e).

The validation of verbal models is discussed in Wenstøp (1976b). First, one must be sure that the fuzzy sets used for the description of linguistic values are acceptable by normal intuition-based standards. Secondly, two modes of simulation exist for dynamic verbal models:

the values of the state variables, which are periodically recomputed, are fuzzy; when they are reentered as such in the model, the output usually gets fuzzier and fuzzier (see also III.2.F.a); this is the forecasting mode.

when the purpose is not prediction, but investigation of principal types of behavior, the tendency toward increased fuzziness must be removed; this can be achieved by restoring complete sharpness to the state variables at each iteration, owing to the maximum membership rule.

Verbal models may be useful in situations when part of the available information is not obtained by physical measurements or not quantifiable, especially for the description of dynamic systems where human behavior plays a prominent role.

Wenstøp (1976a, b) applied verbal models in a case study (an organization problem in a factory). A similar attempt can be found in Adamo (1977).

REFERENCES

Adamo, J. M. (1977). Towards introduction of fuzzy concepts in dynamic modeling. *Conf. Dyn. Modell. Control Natl. Econ., Vienna*, North-Holland, Amsterdam.

Adamo, J. M. (1978a). L.P.L. A fuzzy programming language: Syntactic aspects. M.I.A.G., Université Claude Bernard, Lyon, France. *Int. J. Fuzzy Sets Syst.* **3**, No. 2, 151–181, 1980.

Adamo, J. M. (1978b). Semantics for a fuzzy programming language. Part 1: Basic notions and logical expressions. Part 2: Elementary statements and control structure. M.I.A.G., Université Claude Bernard, Lyon, France. *Int. J. Fuzzy Sets Syst.* (to appear).

Adamo, J. M. (1978c). The L.P.L. language: Some applications to combinatorial and nondeterministic programming. M.I.A.G., Université Claude Bernard, Lyon, France.

Bonissone, P. P. (1978). "A Pattern-Recognition Approach to the Problem of Linguistic Approximation in Systems Analysis," Memo UCB/ERL M78/57. Electron. Res. Lab., Univ. of California, Berkeley.

Chang, S. K., and Ke, J. S. (1978). Data-base skeleton and its application to fuzzy query translation. *IEEE Trans. Software Eng.* **4**, No. 1, 31–44.

Fellinger, W. L. (1978). "Specification for a Fuzzy Systems Modelling Language." Ph.D. Thesis, Oregon State Univ., Corvallis.

Goguen, J. A. (1974). Concept representation in natural and artificial languages: Axioms, extension, and applications for fuzzy sets. *Int. J. Man-Mach. Stud.* **6**, 513–561. (Reference from II.4.)

Goguen, J. A. (1976). Robust programming languages and the principle of maximum meaningfulness. *Milwaukee Symp. Autom. Comput. Control* pp. 87–90.

Jouault, J. P., and Luan, P. M. (1975). "Application des Concepts Flous à la Programmation en Language Quasi-Naturel." Inst. Informat. Entreprise, C.N.A.M, Paris.

Kaufmann, A. (1975). "Introduction à la Théorie des Sous-Ensembles flous. Vol 2: Applications à la Linguistique, à la Logique et à la Sémantique." Masson, Paris. (Reference from I, 1975a.)

Kling, R. (1973). Fuzzy-PLANNER: Reasoning with inexact concepts in a procedural problem-solving language. (1st version.) *J. Cybern.* **3**, No. 4, 1–16. 2nd version, **4**, No. 2, 105–122 (1974).

Lakoff, G. (1973). Hedges: A study in meaning-criteria and the logic of fuzzy concepts. *J. Philos. Logic* **2**, 458–508. (Also in "Contemporary Research in Philosophical Logic and Linguistic Semantics" (D. Hockney, W. Harper, B. Freed, eds.), pp. 221–271. D. Reidel (1973), Dordrecht.)

LeFaivre, R. A. (1974a). "Fuzzy Problem-Solving," Ph.D. Thesis, Univ. of Wisconsin, Madison. (Also Tech. Rep. No. 37. Madison Acad. Comput. Cent., Madison, Wisconsin, 1974.)

LeFaivre, R. A. (1974b). The representation of fuzzy knowledge. *J. Cybern.* **4**, No. 2, 57–66.

Mamdani, E. H. (1975). "F.L.C.S.: A Control System for Fuzzy Logic." Queen Mary Coll., London.

Noguchi, K., Umano, M., Mizumoto, M., and Tanaka, K. (1976). "Implementation of Fuzzy Artificial Intelligence Language FLOU," Tech. Rep. Autom. Language IECE. Osaka.

Nowakowska, M. (1977). Fuzzy concepts in the social sciences. *Behav. Sci.* **22**, 107–115. (Reference from IV.1.)

Procyk, M. T. J. (1976). "Linguistic Representation of Fuzzy Variables," Fuzzy Logic Working Group. Queen Mary Coll., London.

Shaket, E. (1976). "Fuzzy Semantics for a Natural-Like Language Defined Over a World of Blocks," Memo No. 4, Artif. Intell. Comput. Sci. Dep., Univ. of California, Los Angeles.

Umano, M., Mizumoto, M., and Tanaka, K. (1978). F.S.T.D.S.-System: A fuzzy-set manipulation system. *Inf. Sci.* **14**, 115–159.

Wahlster, W. (1977). "Die Representation von vagen Wissen in natürlichsprachlichen Systemen der Künstlichen Intelligenz," Working Pap. No. 38. Inst. Informatik, Univ. Hamburg, Hamburg.

Wechsler, H. (1976). A fuzzy approach to medical diagnosis. *Int. J. Bio-Med. Comput.* **7**, 191–203.

Wenstøp, F. (1976a). Deductive verbal models of organizations. *Int. J. Man-Mach. Stud.* **8**, 293–311.

Wenstøp, F. (1976b). Fuzzy set simulation models in a systems dynamics perspective, *Kybernetes* **6**, 209–218.

Zadeh, L. A. (1971). Quantitative fuzzy semantics. *Inf. Sci.* **3**, 159–176.

Zadeh, L. A. (1972a). A fuzzy set-theoretic interpretation of linguistic hedges. *J. Cybern.* **2**, No. 3, 4–34. (Reference from II.2.)

Zadeh, L. A. (1972b). Fuzzy languages and their relation to human and machine intelligence. *Proc. Int. Conf. Man Comput. 1st, Bordeaux, 1970*, pp. 130–165. (Reference from III.3.) (M. Marois, ed.), S. Karger, Basel.

Zadeh, L. A. (1974). Linguistic cybernetics. *In* "Advances in Cybernetics and Systems" (J. Rose, ed.), Vol. 3, pp. 1607–1615. Gordon & Breach, London.

Zadeh, L. A. (1976). A fuzzy-algorithmic approach to the definition of complex or imprecise concepts. *Int. J. Man-Mach. Stud.* **8**, 249–291. (Also in "Systems Theory in the Social Sciences" (H. Bossel, S. Klaczko, and N. Müller, eds.), pp. 202–282. Birkhaeuser, Basel.)

Zadeh, L. A. (1977a). "PRUF–A Meaning Representation Language for Natural Languages," Memo UCB/ERL M77/61. Univ. Of California, Berkeley. [Also in *Int. J. Man-Mach. Stud.* **10**, No. 4, 395–460 (1978).]

Zadeh, L. A. (1977b). PRUF and its application to inference from fuzzy propositions. *Proc. IEEE Conf. Decision Control*, pp. 1359–1360.

Zadeh, L. A. (1978). "On Translation of Imperatives," Seminary 6-6-1978. Univ. of California, Berkeley.

Chapter 3

FUZZY SETS IN
DECISION-MAKING

Fuzzy decision-making is still in its early age; thus the reader must not expect to find here a new general theory. To date most of the works in this field either propose a philosophical background where already existing theories are reinterpreted or extend some specific problems to deal with fuzzy preference relations, fuzzy objective functions, fuzzy weightings, fuzzy votes, fuzzy utilities, fuzzy events, etc.

Section A is devoted to rank-ordering the elements of a set equipped with a given fuzzy binary preference relation. Fuzzy aggregation of criteria and aggregation of fuzzy criteria are considered in Section B. Section C is concerned with fuzzy group decision-making, and Section D with decision-making under fuzzy events and with fuzzy utilities.

A discussion of classical approaches in decision-making can be found in Luce and Raiffa (NF 1957).

A. FUZZY RANK-ORDERING

Let X be a finite set of possible objects (or actions) one of which must be chosen. It is difficult to define directly a linear preference ordering of the objects. Pairwise comparisons are more natural. Several ad hoc ranking methods are now surveyed. All of them deal with fuzzy relations.

i) Shimura (1973) has proposed an approach to the rank-ordering of objects from knowledge of numerical grades assigned to every object out of every pair in X. $f_y(x)$ denotes the (positive) "attractiveness" grade of x when the choice is limited to an element of $\{x, y\}$. These primitive evaluations can be reduced to relative preference grades $\mu(x, y)$ defined by (Shimura, 1973)

$$\forall x, y \in X, \quad \mu(x, y) = \frac{f_y(x)}{\max(f_y(x), f_x(y))} = \min\left(1, \frac{f_y(x)}{f_x(y)}\right).$$

$\mu(x, y) = 1$ as soon as x is at least as attractive as y. $\mu(x, x) = 1$. More generally, if $T = \{y_1, \ldots, y_m\} \subseteq X$, the relative preference grade of x over the elements of T is taken as

$$\mu(x, T) = \min_{i=1, m} \mu(x, y_i).$$

Intuitively, the most attractive element is $\hat{x}$ such that $\mu(\hat{x}, X) = \max_{x \in X} \mu(x, X)$. However, when $\max_{x \in X} \mu(x, X) \neq 1$, this result can be questioned. A sufficient condition for the existence of $\hat{x}$ such that $\mu(\hat{x}, X) = 1$ is (Shimura, 1973)

$$\forall x, y, z \in X, \quad \text{if} \quad f_y(x) > f_x(y) \quad \text{and} \quad f_z(y) > f_y(z),$$
$$\text{then} \quad f_z(x) > f_x(z). \tag{1}$$

When this condition holds, the most attractive object in $X - \{\hat{x}\}$ can be found in the same way. Repeated applications of this procedure yield a complete ranking of the objects.

N.B.:   The case when a template (or standard) object exists was also considered by Shimura (1973).

Another sufficient condition for the existence of a most attractive object $\hat{x}$ in X is (Shimura, 1973)

$$\forall x, y, z \in X, \quad \frac{f_z(x)}{f_x(z)} = \frac{f_y(x)}{f_x(y)} \frac{f_z(y)}{f_y(z)}. \tag{2}$$

This condition is more restrictive than (1).

   ii)   Saaty (1978) applied his method of determination of membership functions (see 1.B.e) to rank-ordering of objects. He assumes knowledge of the $\mu'(x, y)$ satisfying an "antisymmetry" property:

$$\mu'(x, y) \cdot \mu'(y, x) = 1.$$

$\mu'(x, y) > 1$ means x is preferred to y. Moreover, $\mu'(x, x) = 1$. μ' is said to be consistent iff $\forall x, y, z \in X$, $\mu'(x, z) = \mu'(x, y)\mu'(y, z)$, which is similar to condition (2).

   The objects are ranked according to weights $W(x)$, for all x in X, such that $\forall x, y \in X$, $\mu'(x, y) = W(x)/W(y)$. The existence of the $W(x)$ is guaranteed as soon as μ' is consistent. They are derived through a linear algebra method described in 1.B.e.

iii)   When μ is obtained in the form of a fuzzy preorder relation, a ranking of classes of noncomparable objects is always possible using the results of II.3.D.e, due to Orlovsky (1978).

iv)   Some hints for an alternative approach are now given. Assume μ is the membership function of a fuzzy relation and

$\mu(x, y) > 0.5$ means x is preferred to y,

$\mu(x, y) = 0.5$ means x and y have the same attractiveness.

Of course, when $\mu(x, y) > 0.5$, then $\mu(y,x) < 0.5$; and when $\mu(x, y) = 0.5$, then $\mu(y,x) = 0.5$. Moreover, $\mu(x,x) = 0.5$ for all x in X. An example of such a preference relation is a tournament relation that satisfies the equality

$$\forall x, y \in X, \quad \mu(x, y) + \mu(y,x) = 1.$$

Consider, for any object x, the fuzzy class dominated by x:

$$P_{\leqslant}(x) = \sum_{y \in X} \mu(x, y)/y,$$

i.e., the fuzzy set of objects to which x is preferred (see II.3.D.b). $P_{\leqslant}(x)$ expresses the global attractiveness of x in X. Using the transitive weak inclusion (II.1.E.c.α) denoted $\prec$, we state that the absolute attractiveness of the object x is greater than the attractiveness of y whenever $P_{\leqslant}(y)$ $\prec P_{\leqslant}(x)$ is true and $P_{\leqslant}(x) \prec P_{\leqslant}(y)$ is false. A consistency condition for μ is: $\forall x, y \in X, P_{\leqslant}(x) \prec P_{\leqslant}(y)$ and $P_{\leqslant}(y) \prec P_{\leqslant}(x)$ are not false at the same time. For assume this consistency condition does not hold for given x and y. Then:

$(P_{\leqslant}(x) \prec P_{\leqslant}(y)$ is false) is equivalent to $(\exists z \in X, \mu(x,z) > \frac{1}{2}$ and $\mu(y,z) \leqslant \frac{1}{2})$;

$(P_{\leqslant}(y) \prec P_{\leqslant}(x)$ is false) is equivalent to $(\exists t \in X, \mu(y,t) > \frac{1}{2}$ and $\mu(x,t) \leqslant \frac{1}{2})$.

These assertions mean: x is preferred to z; z is at least as attractive as y; y is preferred to t; and t is at least as attractive as x. Using transitivity, we conclude that x is preferred to x, which is "inconsistent."   Q.E.D.

The fuzzy dominated classes of a consistent μ form a (nonfuzzy) partially ordered set under $\prec$, and thus a ranking of disjoint subsets of noncomparable objects is possible.

N.B.:   A consistency index for fuzzy preference relations is described and discussed in Blin *et al.* (1973).

The presented approaches above may appear fragmentary in the sense that they have not yet been discussed in the framework of a general theory. Moreover, these methods are fuzzy only because they allow handling noncrisp preferences. At a further stage, one may think of dealing with linguistic preferences (i.e., $\mu(x, y)$ is a fuzzy number on $[0, 1]$.)

B. MULTICRITERIA DECISION-MAKING

Suppose now each object (or action) in X is assigned several numerical (or linguistic) evaluations. These evaluations refer either to local features of each object or to different global aspects (criteria) of the objects. These two pure situations may occur at the same time in real problems. In the first the partial evaluations refer to the same aspect. The problem of measuring the degree to which an object has the empirical property of its parts has been especially considered by Allen (1974). The semantics of the aggregation operators look similar in the two situations, but these operators will be discussed in the terminology of the second situation, i.e., criteria aggregation, for convenience. We study separately the cases when the evaluations are numerical (i.e., nonfuzzy) and linguistic (i.e., fuzzy).

a. Aggregation of Ordinary Criteria in the Framework of Fuzzy Set Theory

Let X be a set of n objects (or actions) $x_j, j = 1, n$, and $g_1, \ldots, g_m$ be m objective functions from X to $\mathbb{R}$ to be maximized. The set of "good" objects with respect to aspect i is the maximizing set G_i of g_i (see II.4.B.a). When the objectives are of equal importance, the fuzzy set D of optimal objects with respect to the m criteria may be defined as the intersection of all the maximizing sets G_i, i.e., $D = \bigcap_{i=1, m} G_i$ (Bellman and Zadeh, 1970). This aggregation is "pessimistic" in the sense that each object is assigned its worst evaluation. The corresponding "optimistic" aggregation is defined by $D = \bigcup_{i=1, m} G_i$ where each object is assigned its best evaluation. When the objectives are of unequal importance, let $r_i > 0$, $i = 1, m$, be m coefficients expressing the relative importance of each criterion; Yager (1977, 1978) has proposed the aggregation $D = \bigcap_{i=1, m} G_i^{r_i}$ (see also III.2.F.c). "The membership grade in all objectives having little importance ($r_i < 1$) becomes larger, and while those in objectives having more importance ($r_i > 1$) become smaller. This has the effect of making the membership function of the decision subset $D[\cdots]$ being more determined by the important objectives."

The above aggregation scheme assumes that the criteria cannot compensate each other. When this is no longer true, other schemes may be considered: product, arithmetic mean, geometric mean. Note that the last two are no longer associative. When these aggregations are weighted, we have for instance

$$\mu_D = \prod_{i=1}^{m} \left[\mu_{G_i} \right]^{r_i} \quad \text{or} \quad \mu_D = \sum_{i=1}^{m} w_i \mu_{G_i}$$

with $\sum_{i=1}^{m} w_i = 1$.

If the aggregation has to be insensitive to irregularities of the evaluations, we may think of using a "weighted median" such as Sugeno's integral (see II.5.A.b.α):

$$\mu_D(x_j) = \max_{k=1,m} \min\left(\mu_{G_{i_k}}(x_j), f(M_{i_k}) \right)$$

where $\mu_{G_{i_1}}(x_j) \leqslant \cdots \leqslant \mu_{G_{i_k}}(x_j) \leqslant \cdots \leqslant \mu_{G_{i_m}}(x_j)$, $M_{i_k} = \{i_k, \ldots, i_m\}$, and f is a fuzzy measure on the set of criteria; $f(M_{i_k})$ expresses the grade of importance of the subset of criteria M_{i_k}.

Lastly, Kaufmann (1975) has used a distance d between fuzzy sets (see II.1.E.c.β) to define D: more specifically, D minimizes the functional $\sum_{i=1}^{m} w_i d(D, G_i)$ where the w_i are weights.

N.B.: The weights may depend on the objects (or actions) considered.

Remark Roy (1975, 1976), has given a typology of criteria based on the existence or nonexistence of indifference or presumption of preference thresholds on the evaluations of objects. This typology can be interpreted in the framework of fuzzy set theory.

b. Fuzzy Aggregation and Fuzzy-Valued Criteria

α. Rating

Weights are usually subjectively assessed, sometimes linguistically. The $w_i(x_j)$ then take their values on a term set of linguistic values such as "very important," "more or less important," "not really important," etc. modeled by fuzzy numbers $\widetilde{w_i(x_j)}$ on $[0, 1]$ (or possibly $\mathbb{R}^+$). Moreover, in some situations the evaluation of the criteria are also fuzzy, i.e., $\mu_{G_i}(x_j)$ is a fuzzy number $\widetilde{\mu_{G_i}(x_j)}$ on $[0, 1]$. The linear aggregation scheme becomes

$$\widetilde{\mu_D(x_j)} = \left[\widetilde{w_1(x_j)} \odot \widetilde{\mu_{G_1}(x_j)} \right] \oplus \cdots \oplus \left[\widetilde{w_m(x_j)} \odot \widetilde{\mu_{G_m}(x_j)} \right].$$

Usually, in this formula the possible nonfuzzy values of the variables $w_i(x_j)$, fuzzily restricted by the $\widetilde{w_i(x_j)}$, are linked by the β-interactivity constraint $\sum_{i=1}^{m} w_i(x_j) = 1$ (see II.3.A.b). In the above formula the notation is then somewhat misused: because of the interactivity, $\oplus$ and $\odot$ are not exactly an extended sum and an extended product. The situation is similar to that of II.5.D.a.

Baas and Kwakernaak (1977) directly extended to fuzzy numbers the formula

$$\mu_D(x_j) = \sum_{i=1}^{m} w_i(x_j)\mu_{G_i}(x_j) \Big/ \sum_{i=1}^{m} w_i(x_j)$$

But, since the $2m$-ary operation

$$f(a_1, \ldots, a_m, b_1, \ldots, b_m) = \sum_{i=1}^{m} a_i b_i \Big/ \sum_{i=1}^{m} a_i$$

is neither increasing nor decreasing nor hybrid, the calculation of $\widetilde{\mu_D(x_j)}$ may be tricky (except for its 1-cut and its support, or when $m = 2$). However, in the first aggregation formula, it may seem natural to assume the β-interactivity constraint only holds for mean values of $\tilde{w}_i(x_j)$'s.

When the weights do not depend on the objects considered, it is not important to normalize them, and the calculation of $\widetilde{\mu_D(x_j)}$ becomes easy because of the noninteractivity. However, strictly speaking, D is no longer a type 2 fuzzy set since the support of $\widetilde{\mu_D(x_j)}$ is not generally included in [0, 1]. Fuzzy linear aggregation was also investigated in this case by Jain (1977).

Other fuzzy aggregation schemes are, for instance,

$\widetilde{\mu_D(x_j)} = \widetilde{\min}_{i=1,m} \widetilde{\mu_{G_i}(x_j)}$, i.e., $D = \sqcap_{i=1,m} G_i$ (see II.2.C.b);

$\widetilde{\mu_D(x_j)} = \widetilde{\min}_{i=1,m} [\widetilde{\mu_{G_i}(x_j)}]^{\tilde{r}_i}$; in this formula the fuzzy number $\widetilde{\mu_{G_i}(x_j)}$ is elevated to a fuzzy power $\tilde{r}_i$ (assumed to be a positive fuzzy number).

These two formulas generalize those of Bellman and Zadeh and of Yager.

β. Ranking

When the $\widetilde{\mu_D(x_j)}$ have been calculated by some aggregation method, it remains to rank the objects or actions x_j. The ranking of fuzzily rated objects is not obvious since no linear order exists among fuzzy numbers.

Jain (1977) has given a ranking procedure consisting of five steps:

(i)   find the support S of $\bigcup_j \widetilde{\mu_D(x_j)}$, $S \subset \mathbb{R} +$;
(ii) define the maximizing set M of S through the membership function $\mu_M(s) = [s/(\sup S)]^p$ where p is a parameter;
(iii) determine $M_j = [\widetilde{\mu_D(x_j)}] \cap M, j = 1, n$;
(iv) assign to each object x_j the membership value hgt(M_j);
(v)   rank the x_j according to hgt(M_j).

Baas and Kwakernaak (1977) used a preferability index μ_I whose value is, for the object x_j,

$$\mu_I(x_j) = \sup_{t_1,\ldots,t_k,\ldots,t_n} \min_{k=1,n} \mu_{d_k}(t_k)$$
$$\text{subject to} \quad t_j \geqslant t_k, \quad k = 1,n,$$

(3)

where μ_{d_k} is the membership function of the fuzzy number $\widetilde{\mu_D}(x_k)$. It generalizes formula (27) in II.2.B.g to n fuzzy numbers. The membership function μ_I gives only partial information on the preferability of the best action (there may be several x_j such that $\mu_I(x_j) = 1$). A fuzzy preferability value of the object x_j over the others is (Baas and Kwakernaak, 1977)

$$P_j = \widetilde{\mu_D}(x_j)$$

$$\ominus (m-1)^{-1} \left[\widetilde{\mu_D}(x_1) \oplus \cdots \widetilde{\mu_D}(x_{j-1}) \oplus \widetilde{\mu_D}(x_{j+1}) \oplus \cdots \oplus \widetilde{\mu_D}(x_n) \right]$$

(3) could be generalized using a fuzzy relation R that models for instance "much greater than" .... For any n, (3) becomes

$$\mu_I(x_j) = \sup_{t_1,\ldots,t_n} \min_{\substack{k=1,n \\ k \neq j}} \left(\mu_{d_j}(t_j), \mu_{d_k}(t_k), \mu_R(t_j,t_k) \right)$$

An alternative approach was recently suggested by Watson *et al.* (1979). Assume $m = 2$ for convenience. The ranking problem is viewed as one of implication: To what extent do the fuzzy ratings imply that object x_1 is better than object x_2 or conversely? This is formally translated by $X \to Y$ where X and Y are binary fuzzy relations on $\mathbb{R}$ such that

$$\mu_X(t_1,t_2) = \min(\mu_{d_1}(t_1), \mu_{d_2}(t_2)),$$

$$\mu_Y(t_1,t_2) = \begin{cases} 1 & \text{if } t_1 > t_2, \\ 0 & \text{otherwise.} \end{cases}$$

(A less rigid Y is possible.) The preference value of x_1 over x_2 is then

$$\mu_{I'}(x_1) = \inf_{t_1,t_2} \max(1 - \mu_X(t_1,t_2), \mu_Y(t_1,t_2)).$$

(4)

Note that the implication which is used is that introduced in III.1.B.b.α. $\mu_{I'}(x_1)$ corresponds to the less valid implication. $\mu_{I'}(x_2)$ is also calculated using

$$\mu_Y(t_1,t_2) = \begin{cases} 1 & \text{if } t_2 > t_1, \\ 0 & \text{otherwise.} \end{cases}$$

The best object x_j corresponds to the greatest $\mu_{I'}(x_j)$.

Note that formula (4) can also be written

$$\mu_{I'}(x_1) = 1 - \sup_{t_1, t_2} \min\big(\mu_{d_1}(t_1), \mu_{d_2}(t_2), \mu_{\bar{Y}}(t_1, t_2) \big)$$

$$= 1 - \mu_I(x_2), \text{ using formula (3).}$$

However, for $n > 2$

$$\mu_{I'}(x_j) = \inf_{t_1, \ldots, t_n} \max\left(1 - \min_{i=1,n} \mu_{d_i}(t_i), \min_{\substack{k=1,n \\ k \neq j}} \mu_Y(t_j, t_k) \right)$$

$$= 1 - \sup_{t_1, \ldots, t_n} \min\left(\min_{i=1,n} \mu_{d_i}(t_i), \max_{\substack{k=1,n \\ k \neq j}} \mu_{\bar{Y}}(t_j, t_k) \right)$$

It is easy to see that $1 - \mu_{I'}(x_j)$ can be defined by formula (3) where the supremum is now taken subject to the constraint $t_j \leqslant t_k$ for at least one $k \neq j$.

When $Y = R$ is a fuzzy relation, even for $n = 2$ the two approaches do not coincide any longer.

Another possibility for ranking fuzzy-rated alternatives is to calculate $\widetilde{\max}_{j=1,n} \widetilde{\mu_D}(x_j)$, i.e., looking for the fuzzy extremum of the fuzzifying function $\widetilde{\mu_D}$ from X to $[0, 1]$ on its domain X (see II.4.B.c.). An interesting index for the ranking is $\mu_{I''}(x_j) = \text{hgt}[\widetilde{\mu_D(x_j)} \cap \widetilde{\max}_{k=1,n} \widetilde{\mu_D(x_k)}]$.

Whatever method is chosen, the ranking can be questioned whenever a significant overlap between some $\widetilde{\mu_D}(x_j)$ exists. If this is the case, we may wish to define more precise partial evaluations and/or weights, when possible. Otherwise, the choice of an object will remain arbitrary. The main advantage of this approach is its making possible detection of such an indeterminacy.

N.B.: Fuzzy aggregation (using fuzzy weights) is obviously also worth considering for nonfuzzy-valued criteria.

c. Fuzzy Pareto-Optimal Set   (Zadeh, 1976)

The numerical aggregation of objective functions is not the only possible approach in multicriteria decision-making. Another is to define a preordering in the set X of objects as follows: x_j is preferred to x_k iff

$$\forall i = 1, m, \quad \mu_{G_i}(x_j) \geqslant \mu_{G_i}(x_k).$$

(The $\mu_{G_i}(x_j)$ are here assumed to be nonfuzzy.) To each object x_j can be associated the set $\Delta(x_j)$ of objects that dominate it, i.e., $\Delta(x_j) = \{x_k, x_k \text{ is preferred to } x_j\}$. Let C be the subset of objects that satisfy a prescribed constraint. Then $x_j \in C$ is said to be *undominated* iff $C \cap \Delta(x_j) = \{x_j\}$. The set of undominated objects in C contains the optimal solution, in the sense of Pareto, of the multiobjective decision problem. Zadeh (1976) has proposed a linguistic approach to this problem in order to reduce the size of the Pareto-optimal set by making use of fuzzy information regarding the trade-offs between objectives. These trade-offs are usually expressed in linguistic terms via fuzzy preference relations. Let ρ be the degree to which x_j is preferred to x_k. Then a partial linguistic characterization of ρ may be expressed, for $m = 2$, as (Zadeh 1976), "If ($\mu_{G_1}(x_j)$ is *much larger* than $\mu_{G_2}(x_k)$ and $\mu_{G_2}(x_j)$ is *approximately equal* to $\mu_{G_2}(x_k)$) or ($\mu_{G_1}(x_j)$ is *approximately equal* to $\mu_{G_1}(x_k)$ and $\mu_{G_2}(x_j)$ is *much larger* than $\mu_{G_2}(x_k)$), then ρ is *strong*." Here "much larger" and "approximately equal" are linguistic names of fuzzy binary relations in $[0, 1]^2$ and "strong" is a linguistic value of ρ. Such linguistic rules determine, once the evaluations of the objects are known, a type 2 fuzzy preference relation on $X \times X$, which may be used to define a fuzzy Pareto-optimal set, as suggested by Zadeh (1976).

C. AGGREGATION OF OPINIONS IN A SOCIAL GROUP. CONSENSUS

A very general class of decision-making problems is concerned with decisions made by a group. There are two main reasons why group decision models are attractive: first, they are easy to comment on and debate because of our intuition concerning social phenomena; secondly, according to Fung and Fu (1975), they are "means of reducing excessive subjectiveness due to idiosyncrasy of a single individual." We deal successively with the question of how best to aggregate individual choices into social preferences and with the formation of consensus.

a. Aggregation of Opinions in a Social Group

An axiomatic approach to rational group decision-making under uncertainty was presented by Fung and Fu (1975). Let X be a set of concurrent actions and m the number of individuals involved in the decision-making process. The preference pattern of every individual i is represented by an L-fuzzy set A_i on X. (An individual can formally be viewed as a criterion.) $\mu_{A_i}(x_j)$ denotes the degree of preference of action x_j by individual i. The authors give a set of axioms that a rational decision must satisfy:

    (i) L is an order topology induced by a linear order $\leqslant$ and is a connected topological space. The intervals in L are of the form $]a, b] = \{x \in L, a < x \leqslant b\}$ where $<$ denotes "$\leqslant$ but not $=$." L is said to be connected if L is not the union of two open disjoint nonempty sets in L. For instance, L cannot be a topological space on a discrete set, but $L = [0, 1]$ satisfies axiom i; however, L is not necessarily a bounded set. "The idea of using a topological structure instead of a numerical scale to describe psychological and social phenomena is not new (e.g., Lewin, NF 1936). . . . A feature of this approach is a generalization of fuzzy sets and decision theory to include the situation where the scale of memberships or risk functions is not necessarily established nor is a metric defined" (Fung and Fu 1975).

An aggregation $*$ is a binary operation on L; and an aggregate of two fuzzy sets A_1, A_2 is represented by $A_1 \circledast A_2$. The remaining axioms give the properties of a rational aggregation.

(ii) (law of independent components) There exists an operation $*$ on L such that $A_1 \circledast A_2 = \int_X \mu_{A_1}(x) * \mu_{A_2}(x)/x$, for all fuzzy sets A_1, A_2 on X, and $*$ is continuous in the sense of the topology of L.

(iii) (idempotency law) $\forall A_i \in \mathcal{P}_L(X)$, $A_i \circledast A_i = A_i$. This axiom asserts that if two individuals assign the same preference grade to an action, this grade is preserved in the aggregation of both opinions.

(iv) (commutativity) $\forall A_i, A_j \in \mathcal{P}_L(X)$, $A_i \circledast A_j = A_j \circledast A_i$. This axiom states that the aggregation must be symmetric.

(v) For $m \geqslant 3$, $A_1 \circledast A_2 \circledast \cdots \circledast A_m$ is inductively defined by $(A_1 \circledast A_2 \circledast \cdots \circledast A_{m-1}) \circledast A_m$.

(vi) (associativity) $\forall A_i, A_j, A_k \in \mathcal{P}_L(X)$, $A_i \circledast (A_j \circledast A_k) = (A_i \circledast A_j) \circledast A_k$.

"Although axioms (v) and (vi) are obviously acceptable in a set theoretic approach, their role in group-decision theory can be disputed" (Fung and Fu, 1975).

(vii) (nondecreasingness of $*$) $\forall x \in X$, $\forall B, C_1, C_2 \in \mathcal{P}_L(X)$ with $A_1 = B \circledast C_1$ and $A_2 = B \circledast C_2$ if $\mu_{C_1}(x) > \mu_{C_2}(x)$, then $\mu_{A_1}(x) \geqslant \mu_{A_2}(x)$. If an individual increases his preference grade of an action x, then the global preference grade of x in the aggregation cannot decrease.

The main result in Fung and Fu (1975), is the theorem: Let L and $*$ satisfy axioms (i)–(vii); then the only possible choices of $*$ are

(pessimistic aggregation) $\forall a, b \in L$, $a * b = \min(a, b)$;
(optimistic aggregation) $\forall a, b \in L$, $a * b = \max(a, b)$;
(mixed aggregation) $\exists \alpha \in L$ such that

$$\forall a \leqslant \alpha, \quad \forall b \leqslant \alpha, \quad a * b = \max(a, b),$$
$$\forall a \leqslant \alpha, \quad \forall b \geqslant \alpha, \quad a * b = \alpha,$$
$$\forall a \geqslant \alpha, \quad \forall b \geqslant \alpha, \quad a * b = \min(a, b).$$

Lastly, Fung and Fu (1975) proved that the only possible rational aggregation is the pessimistic (resp. optimistic) one when axiom (vii) is replaced by axiom (viii) (resp. (ix)):

(viii) There exists α in L and a lower limit 0 of L, such that, if $0 \leqslant x < \alpha$, then $0 * x = 0$.

(ix) There exists β in L and an upper limit 1 of L, such that, if $\beta < x \leqslant 1$, then $1 * x = 1$.

Axioms (i)–(vi) plus (viii) justify the *minimax principle*: the best action is x^* such that

$$\mu(x^*) = \sup_{x \in X} \min_{i=1, m} \mu_{A_i}(x).$$

These axioms provide a theoretical basis for some of the aggregation schemes described in Section B.a. However, to account for aggregation schemes of fuzzy-valued criteria (or preferences), some weaker structure seems to be needed for L, for instance, a topology induced in a partially ordered set (see Morita and Iida, 1975).

When the $\mu_{A_i}(x)$ are fuzzy numbers $\widetilde{\mu_{A_i}(x)}$, a linear aggregation was proposed by Nahmias (Reference from II.2). Suppose $w_1, \ldots, w_m$ are nonnegative weights, such that $\sum_{i=1}^{m} w_i = 1$, which reflect the relative importance of the opinion of each individual in the group decision. Nahmias claimed that the fuzzy grade $w_1 \widetilde{\mu_{A_1}(x)} \oplus \cdots \oplus w_m \widetilde{\mu_{Am}(x)}$ was a more reasonable description of the opinions of the group than a similar convex sum of random variables. This is because when the m individuals share the same opinion with regard to x, that is, $\widetilde{\mu_{A_i}(x)} = \tilde{a} \ \forall i = 1, m$, then the convex sum gives $\tilde{a}$ only when $\tilde{a}$ is a fuzzy number and not a random variable.

The aggregation of relative preferences in a group using fuzzy sets was considered by Blin and Whinston (1973) and Blin (1974). The opinions of the m individuals are assumed to be m linear orders over X. A social preference relation R is here a fuzzy relation obtained by aggregation of the individuals' preferences. Denoting by N_{ij} the number of individuals who prefer x_i to x_j, then possible definitions of R are

$$\mu_R(x_i, x_j) = N_{ij}/m$$

or

$$\mu_R(x_i, x_j) = \max(0, N_{ij} - N_{ji})/m.$$

Blin and Whinston noticed that the 1-cut of R, defined as above, is a nonfuzzy partial ordering. They obtained nonfuzzy linear orderings for R defined as a linear extension Λ of the 1-cut of R in the sense of Spilrajn's theorem (see II.3.D.c). Since several linear extensions may exist, the authors chose Λ^* maximizing

$$\sum_{(i,\,j)\in I(\Lambda)} \mu_R(x_i, x_j) \quad \text{where} \quad I(\Lambda) = \{(i, j), (x_i, x_j) \in \Lambda \text{ and } \mu_R(x_i, x_j) < 1\};$$

so in Λ^* the relative ranking of elements in a pair of initially non-comparable ones reflects the strength of preferences in the original fuzzy relation. An algorithm for computation of Λ^* can be found in Blin (1974).

b. Consensus

Ragade (1976, 1977) has modeled consensus formation in a group as a dynamical process by means of fuzzy automata. The description of how a consensus is reached aims at understanding the way decisions are made in a group. Fuzzy automata were already suggested as an interesting model for an individual's formulation of voting strategies in social choice theory by Hatten *et al.* (1975).

In a group each individual i is assumed to have a fuzzy profile of opinions $A_i(t)$ with regard to the n actions x_j at time t. Individual i perceives the opinion of j as

$$\Phi_{ij}(t) = T_{ij} * A_j(t)$$

where T_{ij} is the transformation matrix expressing that i does not perceive accurately j's opinions: $*$ is one of the four compositions:

$$\mu_{\Phi_{ij}(t)}(x_k) = \frac{1}{n} \sum_{l=1}^{n} t_{ij}^{kl} \mu_{A_j(t)}(x_l)$$

or

$$\mu_{\Phi_{ij}(t)}(x_k) = \max_l t_{ij}^{kl} \mu_{A_j(t)}(x_l)$$

or

$$\mu_{\Phi_{ij}(t)}(x_k) = \max_l \left(t_{ij}^{kl} \min_r \mu_{A_j(t)}(x_r) \right)$$

or

$$\mu_{\Phi_{ij}(t)}(x_k) = \frac{1}{n} \sum_{l=1}^{n} \left(t_{ij}^{kl} \min_r \mu_{A_j(t)}(x_r) \right).$$

The opinions of i, modified by the perceived opinions of other individuals, become at $t + 1$

$$A_i(t + 1) = A_i(t) \underset{j}{\perp} \Phi_{ij}(t)$$

where $\perp$ denotes for instance:

a max–min consensus: $A_i(t + 1) = A_i(t) \cup (\bigcap_{j \neq i} \Phi_{ij}(t))$; individual i "chooses to transform $A_i(t)$ by accepting any agreed improvement in the $\Phi_{ij}(t)$";

a min–max consensus: $A_i(t + 1) = A_i(t) \cap (\bigcup_{j \neq i} \Phi_{ij}(t))$; here individual i "chooses to transform $A_i(t)$ by rejecting any improvement in the $\Phi_{ij}(t)$."

Four other profile modification rules are given by Ragade (1977). A consensus is reached as soon as $A_i(t) = A_i(t + 1)$, $\forall i$.

N.B.:   Another model of how people perceive each other's behavior has been proposed by Vaina (Reference from V, 1978).

Let R denote the group-preference fuzzy relation when the consensus is reached, constructed for example according to Blin and Whinston. R is a *reciprocal* fuzzy relation, i.e.,

$$\forall x_j, x_k, \quad \mu_R(x_j, x_k) + \mu_R(x_k, x_j) = 1, \quad j \neq k,$$

$$\forall x_j, \quad \mu_R(x_j, x_j) = 0, \text{ by convention.}$$

$\mu_R(x_j, x_k) = 1$ means x_j is totally preferred to x_k; $\mu_R(x_j, x_k) = 0.5$ means x_j and x_k have equal preferences. Bezdek *et al.* (1977) have proposed scalar measures of consensus:

$$F(R) = \frac{2 \operatorname{tr}(R^2)}{n(n - 1)} \qquad \text{(average fuzziness)},$$

$$C(R) = \frac{2 \operatorname{tr}(RR^t)}{n(n - 1)} \qquad \text{(certainty)}.$$

$F(R)$ is supposed to express the average confusion exhibited by R and $C(R)$, the average "assertiveness." The following properties hold: $F(R) + C(R) = 1$; $F(R) \in [0, \frac{1}{2}]$; $C(R) \in [\frac{1}{2}, 1]$; $F(R) = \frac{1}{2}$ iff $C(R) = \frac{1}{2}$ iff $\mu_R(x_j, x_k) = 1/2$, $\forall j, \forall k, j \neq k$; $F(R) = 0$ iff $C(R) = 1$ iff $\mu_R(x_j, x_k) \in \{0, 1\}$ $\forall j, k$. (See Bezdek *et al.* (1977).) Other properties and discussions can be found in Bezdek *et al.* (1977).

D. DECISION-MAKING UNDER RANDOMNESS AND FUZZINESS

In this last section we survey some applications of fuzzy sets to more complex decision-making problems in which the choice of actions may depend not only on utility values but also on the states of nature or on possible or expected consequences of actions. States of nature, feasible actions, admissible consequences, utility values, available information, etc. may be fuzzy in practical situations. Fuzziness may also be introduced in statistical decision models where only probabilities of occurrence of events are known. Several more or less different attempts of this kind can be found in the literature. This field is still at an early stage of development.

a.   Choice of an Action According to the State of Nature

Let S be a set of q possible nonfuzzy states of nature s_k. s_k is assumed uncontrollable and describes the situation or environment in which a decision must be made. $X = \{x_1, \ldots, x_n\}$ is the set of actions that can be performed. Let u_{jk} be the nonfuzzy utility of performing x_j when the state of nature is s_k. When the state of nature is s_{k_0}, the best action is x_{j_0} such that

$$u_{j_0 k_0} = \max_j u_{jk_0}.$$

S is assumed finite here for convenience.

α.   Fuzzy State   (Jain, 1976)

A state of nature is fuzzy as soon as it is linguistically described or roughly preceived or approximately measured because of the complexity of the situation. The extension principle allows us to induce for each action x_j a fuzzy utility $\tilde{U}_j$ that reflects the lack of well-defined knowledge of the state:

$$\tilde{U}_j = \sum_{k=1}^{q} \mu_{\tilde{s}}(s_k)/u_{jk}$$

where $\tilde{s}$ is the fuzzy state, a fuzzy set on S. The fuzzy utilities $\tilde{U}_j$ can be ranked according to the methods described in B.b.β.

β.   Fuzzy State. Fuzzy Utilities

Let $\tilde{u}_{jk}$ be the fuzzy utility of performing x_j in the nonfuzzy state s_k. When only utilities are fuzzy, the problem is to rank those that correspond to the state of nature s_{k_0}, using methods of B.b.β. When both utilities and the state of nature are fuzzy, the extension principle allows assigning the membership value $\mu_{\tilde{s}}(s_k)$ to each fuzzy utility value $\tilde{u}_{jk}$. The utility of action x_j is now a level 2 fuzzy set (see II.2.C.a):

$$\tilde{\tilde{U}}_j = \sum_{k=1}^{q} \mu_{\tilde{s}}(s_k)/\tilde{u}_{jk}.$$

The problem of ranking level 2 fuzzy sets of $\mathbb{R}$ is still unsolved. A possible method for reducing $\tilde{\tilde{U}}_j$ to an ordinary fuzzy set is to consider $\tilde{u}_{jk}$ as deriving from a fuzzification kernel (see II.2.C.a). The reduced fuzzy utility $\tilde{U}_j$ will be

$$\tilde{U}_j = \bigcup_{k=1}^{q} \mu_{\tilde{s}}(s_k)\tilde{u}_{jk},$$

whose membership function is $\mu_{\tilde{U}_j}(z) = \max_k \mu_{\tilde{s}}(s_k)\mu_{\tilde{u}_{jk}}(z) \; \forall z$.

The reduced fuzzy utility $\tilde{U}_j$ can be ranked as in B.b.β. Another reduction method can be found in Jain (1976).

γ. *Probabilistic State. Fuzzy Utilities*

Assume that the state of nature is known only in probability, i.e., there is a probability distribution p over S such that $\sum_{k=1}^{q} p(s_k) = 1$. The decision problem when the utilities are nonfuzzy is classically solved by assigning to each action the nonfuzzy expected utility

$$U_j = \sum_{k=1}^{q} p(s_k)u_{jk};$$

and the expected best action is x_{j_0} such that

$$U_{j_0} = \max_j U_j.$$

When the utilities u_{jk} are fuzzy numbers $\tilde{u}_{jk}$, the same formula holds provided that we use an extended addition. The $\tilde{U}_j$ are now defined by

$$\tilde{U}_j = p(s_1)\tilde{u}_{j1} \oplus \cdots \oplus p(s_q)\tilde{u}_{jq}.$$

When the $\tilde{u}_{jk}$ are L-R type fuzzy numbers (see II.2.B.e), the $\tilde{U}_j$ are very easy to calculate and are ranked as in B.b.β. This situation was also studied by Jain (1978), who extended his approach to multicriteria (utilities) with fuzzy weights.

Sometimes, the probability of each state s_k is known only linguistically, i.e., $\widetilde{p(s_k)}$ (e.g., "very likely," "rather unprobable," etc.). This problem is considered by Watson *et al.* (1979). The membership function of $\tilde{U}_j$ is given by

$$\mu_{\tilde{U}_j}(z) = \sup_{\sum_{k=1}^{q} p_k v_k = z} \min_k \left(\mu_{\tilde{P}(s_k)}(p_k), \mu_{\tilde{U}_{jk}}(v_k) \right)$$

subject to $\sum_{k=1}^{q} p_k = 1$. The calculation of $\tilde{U}_j$ is not so easy as in the preceding case (for $q > 2$) because the linguistic probabilities are β-interactive (see II.5.D.a).

b. Statistical Decision-Making under Fuzzy Events

The main references of this section are Tanaka *et al.* (1976), Okuda *et al.* (1974, 1978), and Tanaka and Sommer (1977), who dealt with high-level decision-making. According to these authors, "much of the decision-making at the higher level might take place in a fuzzy environment, so that it is only necessary to decide roughly what action, what states, what parameters should be considered." Their formulation uses Zadeh's approach of probabilities of fuzzy events (see II.5.C.a).

Let us recall some definitions. Let S and S' be sets of states, $S = \{s_1, \ldots, s_q\}$, $S' = \{s_1', \ldots, s_r'\}$ with probabilities $p(s_k)$ and $p'(s_l')$, respectively. The probability of the fuzzy event A in S is

$$P(A) = \sum_{k=1}^{q} \mu_A(s_q)p(s_q).$$

Let $\rho(s_k, s_l')$ denote the joint probability of s_k and s_l'; then the joint probability of the fuzzy events A in S and B in S' is

$$P(A, B) = \sum_{k=1}^{q} \sum_{l=1}^{r} \mu_A(s_q)\mu_B(s_l')\rho(s_q, s_l').$$

The conditional probabilities $P(A \mid s_l')$ and $P(A \mid B)$ are defined by

$$P(A \mid s_l') = P(A, s_l')/p(s_l'), \qquad P(A \mid B) = P(A, B)/P'(B).$$

A decision problem with fuzzy events and fuzzy actions in the sense of Tanaka *et al.* (1976), is a 4-tuple $(S, \mathcal{X}, p, u)$ where $S = \{\tilde{s}_1, \ldots, \tilde{s}_r\}$ is a set of fuzzy states that are fuzzy events on $S = \{s_1, \ldots, s_q\}$ equipped with the probability distribution p; $\mathcal{X} = \{\tilde{x}_1, \ldots, \tilde{x}_n\}$ is the set of fuzzy actions; $u(\cdot, \cdot)$ is the utility function from $\mathcal{X} \times S$ to $\mathbb{R}$. S is assumed to be orthogonal, i.e.,

$$\sum_{l=1}^{r} \mu_{\tilde{s}_l}(s_k) = 1, \qquad \forall k = 1, q$$

(so that $\sum_{l=1}^{r} P(\tilde{s}_l) = 1$).

The expected utility of a fuzzy action $\tilde{x}_j$ is

$$U(\tilde{x}_j) = \sum_{l=1}^{r} u(\tilde{x}_j, \tilde{s}_l)P(\tilde{s}_l).$$

An optimal decision is a fuzzy action $\tilde{x}_0$ that maximizes $U(\tilde{x}_j)$. In the following the fuzzy state is assumed to be known through a message m_i belonging to $M = \{m_1, \ldots, m_t\}$, the set of possible messages. It is also supposed that a conditional probability $f(m_i \mid s_k)$ of receiving m_i, in state s_k, is known a priori. Using Bayes's formula to calculate the posterior probability $f(s_k \mid m_i)$ from $f(m_i \mid s_k)$ and $p(s_k)$, the expected utility $U(\tilde{x}_j \mid m_i)$ of the action $\tilde{x}_j$ when message m_i is received is

$$U(\tilde{x}_j \mid m_i) = \sum_{l=1}^{r} u(\tilde{x}_j, \tilde{s}_l)P(\tilde{s}_l \mid m_i).$$

The optimal decision $\tilde{x}(m_i)$ is the one that maximizes $U(\tilde{x}_j \mid m_i)$. The probabilistic information e is obtained through observation of the random message m whose probability distribution is f such that

$$f(m_i) = \sum_{k=1}^{q} p(s_k)f(m_i \mid s_k).$$

The expected utility of receiving the information e is

$$U(\tilde{x}(m)\mid m) = \sum_{i=1}^{t} U(\tilde{x}(m_i)\mid m_i)f(m_i).$$

The worth of the information e is defined by

$$V(e) = U(\tilde{x}(m)\mid m) - U(\tilde{x}_0).$$

When a message is characterisitic of a state (i.e., $f(m_i\mid s_k) = 0$ for $i \neq k$; $f(m_k\mid s_k) = 1$; $t = q$), the information is called probabilistic perfect information and denoted by e_∞.

Next, we consider fuzzy messages $\tilde{m}_1, \ldots, \tilde{m}_h$ that are fuzzy sets on M and satisfy the orthogonality condition

$$\sum_{d=1}^{h} \mu_{\tilde{m}_d}(m_i) = 1 \quad \forall i = 1, t.$$

Similarly, the expected utility of $\tilde{x}_j$ given $\tilde{m}_d$ is

$$U(\tilde{x}_j\mid \tilde{m}_d) = \sum_{l=1}^{r} u(\tilde{x}_j, \tilde{s}_l)P(\tilde{s}_l\mid \tilde{m}_d),$$

and the optimal decision $\tilde{x}(\tilde{m}_d)$ maximizes $U(\tilde{x}_j\mid \tilde{m}_d)$. The corresponding fuzzy information E has expected utility

$$U(\tilde{x}(\tilde{m})\mid \tilde{m}) = \sum_{d=1}^{h} U(\tilde{x}(\tilde{m}_d)\mid \tilde{m}_d)P(\tilde{m}_d)$$

where $\tilde{m}$ is a random fuzzy message. The worth of the fuzzy information E is $V(E) = U(\tilde{x}(\tilde{m})\mid \tilde{m}) - U(\tilde{x}_0)$. When a fuzzy message is characteristic of a fuzzy state (particularly, $r = h$), the corresponding information is called fuzzy perfect information and denoted E_∞. The following inequalities are proved in Tanaka *et al.* (1976):

$$V(E_\infty) \geqslant V(e_\infty) \geqslant V(e) \geqslant V(E).$$

These inequalities are consistent with intuition: $V(e) \geqslant V(E)$ is due to the fact the information E has fuzziness in addition to randomness. $V(e_\infty) \geqslant V(e)$ holds because e_∞ is better information than e; "$V(E_\infty) \geqslant V(e_\infty)$ is caused by the fact that our interest is not in S but in $\mathbb{S}$." (Tanaka *et al.* (1976)). The probabilistic entropies of $\mathbb{S}$ in the presence of fuzzy or nonfuzzy messages are also calculated by these authors.

N.B.: 1. Tanaka and Sommer (1977) proved that the probability of state s_k when two identical fuzzy messages $\tilde{m}_d$ are simultaneously received is $p(s_k\mid \tilde{m}_d, \tilde{m}_d) = p(s_k\mid \text{very } \tilde{m}_d)$, provided that $f(m_i\mid s_k) = 0$ for $i \neq k$, $f(m_k\mid s_k) = 1$ and $t = q$. Very $\tilde{m}_d$ is nothing but $(\tilde{m}_d)^2$. (See II.1.B.f.).

2. Fuzzy utilities would be worth considering in this framework.

c. Fuzzy Decision Analysis

In this section decisions are made according to an analysis of their consequences.

α. Opportunity Cost Calculation (Hägg, 1978)

Hägg (1978) recently suggested a means of extending decision analysis to take into account the possibility degrees of actions. Let $X = \{x_1, \ldots, x_n\}$ be the set of actions and $C = \{c_1, \ldots, c_m\}$ be the set of possible outcomes of the actions. $p(c_i | x_j)$ denotes the probability of outcome c_i when action x_j is performed. Traditionally, we have

$$\sum_{i=1}^{m} p(c_i | x_j) = 1, \qquad j = 1, n.$$

Suppose we now have a possibility distribution π over X. Hägg suggested the following interpretation of π: an external outcome c_0 may occur, which was unexpected, with a probability of occurrence $p'(c_0 | x_j) = 1 - \pi(x_j)$. The conditional probability distribution is modified to

$$p'(c_i | x_j) = p(c_i | x_j)\pi(x_j).$$

Given the payoff values $v_0, v_1, \ldots, v_m$ of the outcomes, then the expected opportunity costs of the actions are

$$\forall j = 1, n, \quad V_j = \sum_{i=0}^{m} p'(c_i | x_j)v_i.$$

v_0 may be difficult to estimate in real situations.

β. Fuzzy Behavioral Choice Model (Enta, 1976)

Classical choice models were criticized by Simon (NF 1967). The main reproach was the necessity, for the decision-maker, of assigning numerical payoffs and definite probabilities to outcomes. To avoid these difficulties, Simon proposed the following behavioral model.

Let X be a set of actions, C a set of outcomes, and S a set of states of nature. $\rho(x_j, s_k)$ denotes the set of possible outcomes of action x_j in state s_k. Assume there exists $C' \subset C$ containing the outcomes considered satisfactory. A good action x_j must satisfy the requirement

$$C_j = \bigcup_{k=1, q} \rho(x_j, s_k) \subseteq C'.$$

C_j is the set of possible consequences of x_j. The existence of good actions may be guaranteed by enlarging C' (lowering of the aspiration level) or by shrinking C_j (through gathering more information about $\rho(x_j, s_k)$).

Such a model and mechanisms have been "fuzzified" by Enta (1976). In his formulation C', s_k, $\rho(x_j, s_k)$ are fuzzy sets. The mechanisms that guarantee the existence of good actions may be straightforwardly extended by modifying the fuzziness of the different sets involved.

REFERENCES

Allen, A. D. (1974). Measuring the empirical properties of sets. *IEEE Trans. Syst., Man Cybern.* **4**, 66–73.

Baas, S. M., and Kwakernaak, H. (1977). Rating and ranking of multiple-aspect alternatives using fuzzy sets. *Automatica* **13**, 47–58.

Bellman, R. E., and Zadeh, L. A. (1970). Decision-making in a fuzzy environment. *Manage. Sci.* **17**, B141–B164. (Reference from III.4.)

Bezdek, J. C., Spillman, B., and Spillman, R. (1977). Fuzzy measures of preference and consensus in group decision-making. *Proc. IEEE Conf. Decision Control, New Orleans* pp. 1303–1308. [Revised version in *Int. J. Fuzzy Sets Syst.* **2**, 5–14. 1979, under the title: "Fuzzy relation spaces for group-decision theory: An application."]

Blin, J. M. (1974). Fuzzy relations in group decision theory. *J. Cybern.* **4**, No. 2, 17–22.

Blin, J. M., and Whinston, A. B. (1973). Fuzzy sets and social choice. *J. Cybern.* **3**, No. 4, 28–36.

Blin, J. M., Fu, K. S., Whinston, A. B., and Moberg, K. B. (1973). Pattern recognition in micro-economics. *J. Cybern.* **3**, No. 4, 17–27.

Dimitrov, V. (1976). Learning decision-making with a fuzzy automata. *In* "Computer-Oriented Learning Processes" (J. C. Simon, ed.), Noordhoff pp. 149–154.

Dorris, A. L., and Sadosky, T. L. (1973). A fuzzy set—theoretic approach to decision-making. *Nat. Meet. ORSA, 44th, San Diego, Calif.*

Dubois, D., and Prade, H. (1978). Comment on "Tolerance analysis using fuzzy sets" and "A procedure for multiple aspect decision-making." *Int. J. Syst. Sci.* **9**, No. 3, 357–360. (Reference from II.2.)

Enta, Y. (1976). "Fuzzy Choice Models." Dep. Bus. Adm., Hosei Univ., Tokyo.

Fung, L. W., and Fu, K. S. (1975). An axiomatic approach to rational decision-making in a fuzzy environment. *In* "Fuzzy Sets and Their Applications to Cognitive and Decision Processes" (L. A. Zadeh, K. S. Fu, K. Tanaka, and M. Shimura, eds.) pp. 227–256. Academic Press, New York.

Hägg, C. (1978). Possibility and cost in decision analysis. *Int. J. Fuzzy Sets Syst.* **1**, No. 2, 81–86.

Hatten, M. L., Whinston, A. B., and Fu, K. S. (1975). "Fuzzy Set and Automata Theory Applied to Economics," Reprint ser. No. 533, pp. 467–469. H. C. Krannert Grad. Sch., Purdue Univ., Lafayette, Indiana.

Jain, R. (1976). Decision-making in the presence of fuzzy variables. *IEEE Trans. Syst., Man Cybern.* **6**, No. 10, 698–703.

Jain, R. (1977). A procedure for multiple aspect decision making. *Int. J. Syst. Sci.* **8**, No. 1, 1–7.

Jain, R. (1978). Decision-making in the presence of fuzziness and uncertainty. *Proc. IEEE Conf. Decision Control, New Orleans* pp. 1318–1323.

Kaufmann, A. (1975). "Introduction à la Théorie des sous-Ensembles Flous. Vol. 3: Applications à la Classification et à la Reconnaissance des Formes, aux Automates et aux Systèmes, aux Choix de Critères." Masson, Paris. (Reference from I, 1975b.)

Morita, Y., and Iida, H. (1975). Measurement, information and human subjectivity described by an order relationship. *In* "Summary of Papers on Fuzzy Problems," No. 1, pp. 34–39. Working Group Fuzzy Syst., Tokyo.

Okuda, T., Tanaka, H., and Asai K. (1974). Decision-making and information in fuzzy events. *Bull. Univ. Osaka Prefect., Ser. A* **23**, No. 2, 193–202.

Okuda, T., Tanaka, H., and Asai, K. (1978). A formulation of fuzzy decision problems with fuzzy information, using probability measures of fuzzy events. *Inf. Control* **38**, No. 2, 135–147.

Orlovsky, S. A. (1978). Decision-making with a fuzzy preference relation. *Int. J. Fuzzy Sets Syst.* **1**, No. 3, 155–168. (Reference from II.3.)

Ragade, R. K. (1976). Fuzzy sets in communication systems and in consensus formation systems. *J. Cybern.* **6**, 21–38.

Ragade, R. K. (1977). Profile transformation algebra and group consensus formation through fuzzy sets. *In* "Fuzzy Automata and Decision Processes" (M. M. Gupta, G. N. Saridis, and B. R. Gaines, eds.), pp. 331–356. North-Holland Publ., Amsterdam.

Roy, B. (1974). La modélisation des préférences: Un aspect crucial de l'aide à la décision. METRA, **13**, 135–153.

Roy, B. (1975). Partial preference analysis and decision-aid. The fuzzy outranking relation concept. *I.I.A.S.A. Workshop* "Decision-Making with Multiple Conflicting Objectives" Vienna (Doc. SEMA, Montrouge, France).

Saaty, T. (1978). Exploring the interface between hierarchies, multiple objectives and fuzzy sets. *Int. J. Fuzzy Sets Syst.* **1**, No. 1, 57–68.

Shimura, M. (1973). Fuzzy sets concept in rank-ordering objects. *J. Math. Anal. Appl.* **43**, 717–733.

Tanaka, H., Okuda, T., and Asai, K. (1976). A formulation of fuzzy decision problems and its application to an investment problem. *Kybernetes* **5**, 25–30.

Tanaka, H., and Sommer, G. (1977). "On Posterior Probabilities Concerning a Fuzzy Information," Working Pap. No. 77/02. Inst. Wirtschaftswiss. R.W.T.H., Aachen, West Germany.

Watson, S. R., Weiss, J. J., and Donnell, M. (1979). Fuzzy decision analysis. *IEEE Trans. Syst., Man Cybern.* **9**, No. 1, 1–9.

Yager, R. R. (1977). Multiple objective decision-making using fuzzy sets. *Int. J. Man-Mach. Stud.* **9**, No. 4, 375–382.

Yager, R. R. (1978). Fuzzy decision-making including unequal objectives. *Int. J. Fuzzy Sets Syst.* **1**, No. 2, 87–95.

Yager, R. R., and Basson, D. (1975). Decision-making with fuzzy sets. *Decision Sci.* **6**, 590–600.

Zadeh, L. A. (1976). The linguistic approach and its application to decision analysis. *In* "Directions in Large-Scale Systems" (Y. C. Ho and S. K. Mitter, eds.) Plenum, New York, pp. 335–361.

Zeleny, M. (1976). The theory of displaced ideal. *In* "Multiple Criteria Decision-Making" (M. Zeleny, ed.), Lecture Notes in Economics and Mathematical Systems, Vol. 123, pp. 153–206. Springer-Verlag, Berlin and New York.

Chapter *4*

FUZZY CONTROL

There have been two kinds of applications of fuzzy sets in control theory. First, the Bellman and Zadeh (1970) approach to decision-making was used in optimal control problems in which the choice of performance criteria is both a matter of subjectivity and computational tractability. Secondly, Zadeh's linguistic approach to fuzzy systems has motivated many works dealing with the synthesis of fuzzy logic controllers for complex processes. This brief chapter successively surveys these two applications of fuzzy set theory.

A. FUZZY OPTIMAL CONTROL

Let us consider the discrete state equation of a linear time-invariant system: $s(t + 1) = As(t) + Bu(t)$, $t \in \mathbb{N}$, where $s(0)$ is an n-dimensional real initial state vector, $s(t)$ an n-dimensional real state vector, $u(t)$ an r-dimensional real control vector; A and B are $n \times n$ and $n \times r$ real matrices, respectively. The optimal control problem is to find a sequence of inputs (possibly of a fixed length) in order to reach a prescribed final state (i.e., the goal). There may exist constraints on the control sequence and on the intermediary states of the dynamic system.

a. Characterization of a Class of Fuzzy Optimal Control Problems
(Fung and Fu, 1974b)

Assume both the final time t_f and the control sequence $u(0), u(1)$, $\ldots, u(t_f - 1)$ are unknown. Let $V = U^* \times \mathbb{N}$ where U^* is the set of possible control sequences. Any element $v = (\theta, t_f) \in V$ represents a control sequence $= u(0), \ldots, u(t_f - 1)$ such that the process terminates at time t_f. Let $C(t), X(t), T$, and F be fuzzy constraints on the control value and on the state at time t, on the final time and on the final state, respectively. These can be viewed as fuzzy sets on V (using cylindrical extensions). $\mu_{C(t)}(v)$ is a membership value of $u(t)$ and depends inversely on the magnitude of $u(t)$. $\mu_{X(t)}(v)$ is a membership value of the state $s(t)$ that is reached at time t when the control sequence $\theta = u(0),, \ldots, u(t - 1)$ is applied. T is a fuzzily fixed final time. $\mu_T(v)$ decreases with t_f. F is a fuzzy target set. $\mu_F(v)$ is the membership value of $x(t_f)$ when $v = (u(0), \ldots, u(t_f - 1), t_f)$.

The fuzzy constraint sets defined above can be viewed as a collection of optimality criteria for v. The overall fuzzy goal set is obtained as a result of amalgamating the whole collection of criteria:

$$\mu_J(v) = \mu_T(v) * \mu_F(v) * \left(\mu_{C(0)}(v) * \mu_{C(1)}(v) * \cdots * \mu_{C(t_f - 1)}(v) \right)$$
$$* \left(\mu_{X(0)}(v) * \mu_{X(1)}(v) * \cdots * \mu_{X(t_f - 1)}(v) \right)$$

where $*$ denotes an aggregation operator such as min, max, product, etc. (see 3.B.a and 3.C.a).

b. Special Problems Using Particular Criteria

The fuzzy multistage decision-making problem stated above has been solved in the literature for special kinds of optimality criteria, namely when $*$ is min, product, and a linear convex sum.

α. Pessimistic Criterion

We suppose here $* = $ min. The optimality criterion becomes

$$\mu_J(u(0), \ldots, u(t_f), t_f)$$

$$= \min\left[\mu_T(t_f), \mu_F(s(t_f)), \min_{0 \leqslant t < t_f} \mu_{C(t)}(u(t)), \min_{0 \leqslant t < t_f} \mu_{X(t)}(s(t)) \right]. \quad (1)$$

An optimal control sequence $\hat{\theta}$ with terminal time $\hat{t}_f$ satisfies the condition

$$\mu_J(\hat{\theta}, \hat{t}_f) = \sup_{\substack{\theta \in U \\ t_f \in \mathbb{N}}} \mu_J(\theta, t_f).$$

Bellman and Zadeh (1970) dealt with the following subcase of fuzzy optimal control problems with pessimistic (see 3.B.a) criterion. The system under control is a finite deterministic automaton; there is no constraint on the state at time t, except for $t = t_f$, which is assumed given. Moreover, there is a constraint on the control sequence to be determined. The criterion is then

$$\mu_J(\theta) = \min\left[\mu_F(s(t_f)), \min_{0 \leqslant t < t_f} \mu_{C(t)}(u(t))\right].$$ (2)

To solve this problem a dynamic programming method was proposed by Bellman and Zadeh (1970). (See also Chang, 1969.) The method is carried out by applying the "optimality principle," which asserts that if $u(0)$, ..., $u(t_f - 1)$ is an optimal control sequence on $0, \ldots, t_f - 1$, then $\forall t < t_f - 1, u(t), \ldots, u(t_f - 1)$ is optimal on $\{t, \ldots, t_f - 1\}$. We have

$$\mu_J(\hat{\theta})$$

$$= \sup_{u(0), \ldots, u(t_f-2)} \min\left(\mu_{C(0)}(u(0)), \ldots, \mu_{C(t_f-2)}(u(t_f-2)), \mu'_{t_f-1}(s(t_f-1))\right)$$

where

$$\mu'_{t_f-1}(s(t_f-1))$$

$$= \sup_{u(t_f-1)} \min\left(\mu_{C(t_f-1)}(u(t_f-1)), \mu_F[\delta(s(t_f-1), u(t_f-1))]\right)$$

and $s(t+1) = \delta(s(t), u(t)), t \in \mathbb{N}$, is the state equation of the automaton. By iteration, we get the following equations which provide an optimal control sequence,

$$\mu'_{t_f-i}(s(t_f-i)) = \sup_{u(t_f-i)} \min\left(\mu_{C(t_f-i)}(u(t_f-i)), \mu'_{t_f-i+1}(s(t_f-i+1))\right)$$

$$s(t_f-i+1) = \delta(s(t_f-i), u(t_f-i)), \qquad i = 1, t_f,$$ (3)

with $\mu'_{t_f} = \mu_F$. Note that since the final state is known only fuzzily, the above equations must be iteratively solved for all possible final states. Bellman and Zadeh (1970) also solved the problem when t_f is not known and unconstrained and the final state must belong to a fixed ordinary set of states. Lastly Kacprzyk (1978) addressed the same problem, assuming a fuzzy constraint on the termination time.

Fung and Fu (1974b) solved the optimal control problem using (2) on a linear continuous unidimensional system. They also give a solution method for the same linear system with time-independent fuzzy constraints on the final time and the input sequence; the final state is assumed to reach a given time-dependent moving target $z(t)$, i.e., $s(t_f) = z(t_f)$. The corresponding optimality criterion is

$$\mu_J(\theta, t_f) = \min\left(\mu_T(t_f), \inf_{0 \leqslant t < t_f} \mu_C(u(t))\right).$$ (4)

The pessimistic criterion was also studied by Gluss (1973) for the optimal control of a single-input, single-output discrete system where inputs and outputs are valued on an infinite space. The final time is assumed known, and there are fuzzy constraints on the input and on the state at any time. The corresponding optimality criterion is

$$\mu_J(\theta) = \min(\mu_{X(0)}(s(0)), \ldots, \mu_{X(t_f)}(s(t_f)),$$
$$\mu_{C(0)}(u(0)), \ldots, \mu_{C(t_f-1)}(u(t_f-1))) \tag{5}$$

where $\mu_F = \mu_{X(t_f)}$. Gluss (1973) solved this fuzzy state regulation problem using a dynamic programming method.

Fung and Fu (1974a) have criticized the pessimistic criterion because it does not allow any trade-off between the membership values of the elementary criteria. This entails a "highly insensitive optimality criterion which virtually depends on the worst stage of the whole process" according to these authors.

β.   Other Optimality Criteria

Alternative optimality criteria that do not have the drawback mentioned in α have been studied in the literature. Fung and Fu (1974a) considered the optimal control of a finite deterministic automaton with a fuzzy goal expressed as a convex combination of the elementary criteria,

$$\mu_J(\theta) = \alpha\mu_F(s(t_f)) + \frac{1-\alpha}{t_f} \sum_{t=0}^{t_f-1} \mu_{C(t)}(u(t)).$$

They gave an algorithm for determination of the kth optimal policy (i.e., control sequence).

A linear convex optimality criterion was also studied by Gluss (1973). Moreover, he dealt with a product optimality criterion similar to (5) where product replaces min. He noticed that, when the membership functions of $C(t)$ and $X(t)$ are, for $t = 0, t_f - 1$,

$$\mu_{X(t)}(s(t)) = \exp\left[-a^2s(t)^2\right], \qquad \mu_{C(t)}(u(t)) = \exp\left[-u(t)^2\right],$$

and $\mu_F(s(t_f)) = \exp[-b^2s(t_f)^2]$, then we recover the usual quadratic optimality criterion

$$\sum_{t=0}^{t_f-1} \left(u(t)^2 + a^2s(t)^2\right) + b^2s(t_f)^2,$$

which expresses the necessity of keeping $s(t)$ "small" for all t, subject to the requirement that $u(t)$ is not too "large" for all t. Obviously, the quadratic criterion corresponds to a fuzzy objective.

The choice of an optimality criterion is however a matter of experience and seems very difficult to justify a priori.

Remark An extension of this approach to stochastic optimal control was carried out in a similar manner by Bellman and Zadeh (1970), Gluss (1973), and Jacobson (1976).

B. SYNTHESIS OF LINGUISTIC CONTROLLERS

Fuzzy controllers have been introduced by Mamdani (1974) and by Mamdani and Assilian (1975) for control of complex processes, such as industrial plants, especially when no precise model of the process exists and most of the a priori information is available only in qualitative form. It has been observed that a human operator is sometimes more efficient than an automatic controller in dealing with such systems. The intuitive control strategies used by trained operators may be viewed as fuzzy algorithms (Zadeh, Reference from III.3, 1973), which provide a possible method for handling qualitative informations in a rigorous way. This section gives a brief outline of this approach. More detailed surveys can be found in Mamdani (1977a) or in Tong (1977).

a. Structure of a Fuzzy Controller

The purpose of controllers is to compute values of action variables from observation of state variables of the process under control. The relation between state variables and action variables may be viewed as a set of logical rules. When this relation is only qualitatively known, fuzzy logical rules may be stated to implement an approximate strategy. An example of such a fuzzy rule is: if pressure error is *positive big* or *positive medium*, then if change in pressure error is *negative small*, then heat input change is *negative medium*, where "positive big" and "positive medium" are fuzzy sets on a discrete universe of pressure error values; similarly, "negative small" or "negative medium" are fuzzy sets, but not on the same universe.

Such rules are of the form:

$$\text{if } X \text{ is } A_i, \text{ then (if } Y \text{ is } B_i, \text{ then } Z \text{ is } C_i), \tag{6}$$

which is a conditional proposition (see III.1.E.d.δ). To translate this proposition, most authors used the min operator instead of a logical implication.[*] The conditional proposition is then equivalent to the fuzzy relation

$$\mu_{R_i}(X, Y, Z) = \min(\mu_{A_i}(X), \min(\mu_{B_i}(Y), \mu_{C_i}(Z)))$$

$$= \min(\mu_{A_i}(X), \mu_{B_i}(Y), \mu_{C_i}(Z)).$$

[*]Note that here, in propositions such as "X is A_i," X and A_i refer to the same universe.

When a set of n fuzzy rules is available, the resulting fuzzy relation R is the union of the n elementary fuzzy relations R_i, $i = 1, n$:

$$\mu_R(X, Y, Z) = \max_i \mu_{R_i}(X, Y, Z).$$

N.B.:   An intuitive justification of this method of translation and aggregation of the rules is the following: given the two consistent and nonredundant rules,

$$\text{if } X \text{ is } A, \text{ then } Y \text{ is } B,$$

$$\text{if } X \text{ is } \bar{A}, \text{ then } Y \text{ is unrestricted,}$$

we get

$$\mu_R(X, Y) = \max\left[\min(\mu_A(X), \mu_B(Y)), \min(1 - \mu_A(X), 1)\right]$$

$$= \max\left[1 - \mu_A(X), \min(\mu_A(X), \mu_B(Y))\right],$$

which is nothing but the implication $\overset{3}{\rightarrow}$ (see III.1.B.c).

If the state variables X and Y take fuzzy values A' and B', respectively, the fuzzy value C' of the action variable Z is obtained by applying the compositional rule of inference

$$C' = (A' \times B') \circ R$$

or

$$\mu_{C'}(Z) = \max_{X, Y} \min(\mu_{A'}(X), \mu_{B'}(Y), \mu_R(X, Y, Z)).$$

The A_i, B_i, and C_i are prescribed fuzzy sets on finite discretized universes which represent the possible ranges of measurement or action magnitudes.

In fuzzy controllers the inputs (e.g., X, Y) are usually precisely observed quantities, hence, not fuzzy; if, for instance, X_0 and Y_0 are the observed inputs, the compositional rule of inference reduces to

$$\mu_{C'}(Z) = \mu_R(X_0, Y_0, Z).$$

Furthermore, the output (e.g., Z) of the fuzzy controller, which must serve as input of the controlled process, thus has to assume nonfuzzy values. A decision procedure must be used to "unfuzzify" C', i.e., to obtain a nonfuzzy value "compatible" with C'. An obvious method is to choose the value that corresponds to the maximum of their membership function; when several values are possible, their average is chosen (mean of maxima method). Another obvious technique is to form an average based on the shape of the membership function.

N.B.:   1.   Of course, more complicated rules than (6) may be considered.

2.   Sets of linguistic conditional rules can be conveniently displayed in the form of decision tables. An example in Fig. 1 is from Kickert and

Change in error

	NB	NM	NS	NO	PS	PM	PB
PB				NM		NB	
PM							
PS	PS		NO	NM			
PO / NO	PM		PS	NO	NS		NM
NS				PM	NO	NS	
NM / NB		PB			PM		

Error

Figure 1. PB, positive big; PM, positive medium; PS, positive small; P0, positive zero; and analogously for negative.

Mamdani (1978) (the cells of the matrix contain the possible linguistic values of the outputs of the controller). Such decision tables are not necessarily complete. MacVicar-Whelan (1976), starting from a completed decision table (see Fig. 2a), first refined it as sketched in Fig. 2b, and suggested that "fuzzified" versions of Fig. 2b (see Fig. 2c) could be more realistic representations of the actual behavior of a human operator, whose

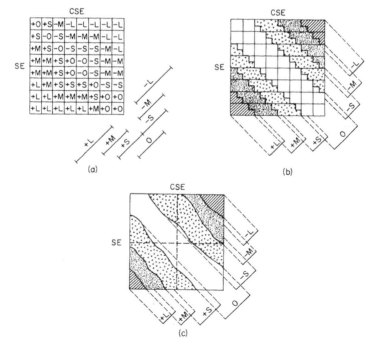

Figure 2. SE, speed error; CSE, change in speed error; L, large; M, medium; S, small.

strategy is modeled by the linguistic rules, in order to synthesize a fuzzy controller.

3.   A set of linguistic rules can also be expanded into a fuzzy flowchart. Actually, linguistic rules may be viewed as fuzzy instructions (see III.3.B.c. δ).

4.   Gaines (Reference from III.1, 1975) has made experiments to compare fuzzy and stochastic logics (see III.1.B.b.γ for the latter). It was concluded that no significant difference in control policy resulted from combining the fuzzy rules using any of the logics. "The robustness of the result to radical changes in the assumptions underlying the logical calculus used is an encouraging indication of the basic robustness of the technique."

5.   Willaeys and Malvache (1976) use a referential of fuzzy sets (see II.1.F.c.γ) in order to save computer memory storage.

b.   Determination of a Fuzzy Controller

The relation R is constructed by assuming three more or less arbitrary factors, whose choice depends on the experience of the designer. First, it is necessary to choose appropriate membership functions for the prescribed fuzzy sets. The second factor is the range of values in the various universes, i.e., the quantization level that can be widened or narrowed. The third factor is the set of rules itself. The spreads of the prescribed fuzzy sets and the quantization level must be fitted to the sensitivity of the process. Moreover, the number of prescribed fuzzy sets on a given universe must be sufficient so as to constitute a satisfactory covering for it. Hence, a good way of tuning the controller is to modify the set of control rules, i.e., add or delete rules or replace some C_i by other prescribed fuzzy sets.

They are three methodologies for the determination of a good set of rules:

(1)   A linguistic description of a control strategy used by a skilled operator will serve, provided that the speed of the process allows direct manual control. The rules obtained are of the form: if the error is *positive big* and the derivative of the error is *positive medium*, then set the control variable to *negative medium*. Anyway, the identification of the protocol used by the operator is not always easy.

(2)   When the speed of the process is too fast to be manually controlled, it is possible to analyze records of system responses to prototypes of input sequences. The rules obtained are then of the form: if the control variable is set to *positive medium* and the error is negative *medium* and the derivative of the error is *positive big* at time t, then at time $t + 1$ the error will be *negative small* and its derivative *positive medium*. Willaeys *et al.*

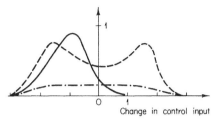

Change in control input

Figure 3.   ——one dominant rule; ——two contradictory rules; ——no satisfactory rule.

(1977b) compared these two approaches in a case study; because of the high speed of the process, they used an analog simulation to slow it in order to allow manual learning of control strategies. The best results were obtained with approach (1) because it is based on an effective learning procedure. However, the corresponding controller was less robust because perturbations were not simulated.

(3) A rule modification algorithm was proposed by Mamdani and Baaklini (1975) in order to automate the alteration of rules "by introducing a form of adaptive behaviour into the controller." This idea was also developed by Procyk (1976). See the next chapter for details.

The quality of the control rules used can be assessed by the shape of the membership functions of the calculated controller output fuzzy sets. The existence of a dominant control rule in a given state of the process is indicated by an output membership function presenting a single strong peak; a very low membership value of this maximum indicates that some rules are missing. When two distinct strong peaks exist, contradictory rules are present in the controller; see Fig. 3 (from King and Mamdani, 1977).

c. Performance and Results

The problem of stability of the controller was discussed by Mamdani (1976b, 1977a). He pointed out that "stability analysis relies on the availability of the mathematical model of the process." However, the main advantage of a fuzzy controller is that its synthesis does not require the existence of such a model. Hence the discussion on stability seems somewhat irrelevant for fuzzy controllers. They are implicitly assumed robust because they are based on human experience. "A confidence in the quality of control can always be obtained by running [the controller] in open-loop with the human operator present to make any changes in its structure to improve its performances." Nevertheless, Kickert and Mamdani (1978) have shown that, under certain restrictive assumptions, the fuzzy controller can be viewed as a multidimensional (multiple inputs, single output) multilevel relay. In this restricted framework a frequency domain stability analysis has been carried out by Kickert and Mamdani (1978) on an example for a system made up of a fuzzy controller and the modeled process. This analysis was possible because the fuzzy controller could be proved equivalent to a conventional nonlinear one. But generally a fuzzy controller cannot be described by an analytical function, so most of modern nonlinear system theory is not applicable.

Fuzzy controllers have been experimented with by many researchers who compared them with DDC algorithms or PID controllers, on highly nonlinear processes generally. The results obtained were good and sometimes better than those of classical methods. Numerous case studies can be found in the appended bibliography. Most of these deal with control of industrial processes such as warm water plants, heat exchanger systems, sinter plants, etc. The successful attempt of Pappis and Mamdani (1977) to apply fuzzy logic to the control of a traffic junction indicates that other problems can be investigated with this approach.

REFERENCES

Bellman, R. E., and Zadeh, L. A. (1970). Decision-making in a fuzzy environment. *Manage. Sci.* **17**, No. 4, B141–B154. (Reference from III.4.)

Braae, M., and Rutherford, D. A. (1978). Fuzzy relations in a control setting. *Kybernetes* **7**, 185–188.

Chang, S. S. L. (1969). Fuzzy dynamic programming and the decision-making process. *Proc. Princeton Conf. Inf. Sci. Syst., 3rd* pp. 200–203.

Chang, S. S. L. (1974). Control and estimation of fuzzy systems. *Proc. IEEE Conf. Decision Control*.

Chang, S. S. L. (1975). On risk and decision-making in a fuzzy environment. *In* "Fuzzy Sets and Their Application to Cognitive and Decision Processes" (L. A. Zadeh, K. S. Fu, K. Tanaka, and M. Shimura, eds.), pp. 219–226. Academic Press, New York.

Fung, L. W., and Fu, K. S. (1974a). The kth optimal policy algorithm for decision-making in fuzzy environments. *In* "Identification and System Parameter Estimation" (P. Eykhoff, ed.), pp. 1052–1059. North-Holland Publ., Amsterdam.

Fung, L. W., and Fu, K. S. (1974b). Characterization of a class of fuzzy optimal control problems. *Proc. Princeton Conf. Inf. Sci. Syst., 8th* (Reference also *In* "Fuzzy Automata and Decision Processes" (M. M. Gupta, G. N. Saridis, B. R. Gaines, eds.), 1977, pp. 209–219. North-Holland, Amsterdam.)

Gluss, B. (1973). Fuzzy multistage decision-making. *Int. J. Control* **17**, 177–192.

Jacobson, D. H. (1976). On fuzzy goals and maximizing decisions in stochastic optimal control. *J. Math. Anal. Appl.* **55**, 434–440.

Kacprzyk, J. (1978). Decision-making in a fuzzy environment with fuzzy termination time. *Int. J. Fuzzy Sets Syst.* **1**, No. 3, 169–180.

Kickert, W. J. M., and Mamdani, E. H. (1978). Analysis of a fuzzy logic controller. *Int. J. Fuzzy Sets Syst.* **1**, No. 1, 29–44.

Kickert, W. J. M., and Van Nauta Lemke, H. R. (1976). Application of a fuzzy controller in a warm water plant. *Automatica* **12**, 301–308.

King, J. P., and Mamdani, E. H. (1977). The application of fuzzy control systems to industrial processes. *Automatica* **13**, 235–242. (Reference also *In* "Fuzzy Automata and Decision Processes" (M. M. Gupta, G. N. Saridis, B. R. Gaines, eds.), 1977, pp. 321–330. North-Holland, Amsterdam.)

MacVicar-Whelan, P. J. (1976). Fuzzy sets for man-machine interaction. *Int. J. Man-Mach. Stud.* **8**, 687–697.

Mamdani, E. H. (1974). Application of fuzzy algorithms for control of simple dynamic plant. *Proc. Inst. Electr. Eng.* **121**, 1585–1588.

Mamdani, E. H. (1976a). Application of fuzzy logic to approximate reasoning using linguistic synthesis. *Proc. IEEE Int. Symp. Multiple-Valued Logic, 6th*, pp. 196–202.

Mamdani, E. H. (1976b). Advances in the linguistic synthesis of fuzzy controllers. *Int. J. Man-Mach. Stud.* **8**, 669–678.

Mamdani, E. H. (1977a). Application of fuzzy set theory to control systems: A survey. *In* "Fuzzy Automata and Decision Processes" (M. M. Gupta, G. N. Saridis, and B. R. Gaines, eds.), pp. 77–88. North-Holland Publ., Amsterdam.

Mamdani, E. H. (1977b). Application of fuzzy logic to approximate reasoning using linguistic systems. *IEEE Trans. Comput.* **26**, 1182–1191.

Mamdani, E. H., and Assilian, S. (1975). An experiment in linguistic synthesis with a fuzzy logic controller. *Int. J. Man-Mach. Stud.* **7**, 1–13.

Mamdani, E. H., and Baaklini, N. (1975). Prescriptive methods for deriving control policy in a fuzzy logic controller. *Electron. Lett.* **11**, 625–626.

Østergaard, J. J. (1977). Fuzzy logic control of a heat exchanger process. *In* "Fuzzy Automata and Decision Processes" (M. M. Gupta, G. N. Saridis, and B. R. Gaines, eds.), pp. 285–320. North-Holland Publ., Amsterdam.

Pappis, C. P., and Mamdani, E. H. (1977). A fuzzy logic controller for a traffic junction. *IEEE Trans Syst., Man Cybern.* **7**, No. 10, 707–717.

Procyk, M. T. J. (1976). "A Self-Organizing Controller for Single Input–Single Output Systems," Intern. Rep., Queen Mary Coll., London. (Reference from IV.5.)

Rutherford, D. A., and Bloore, G. C. (1976). The implementation of fuzzy algorithms for control. *Proc. IEEE* **64**, 572–573.

Rutherford, D. A., and Carter, G. A. (1976). A heuristic adaptive controller for a sinter plant. *Proc. IFAC Symp. Autom. Min., Miner. Met. Process., 2nd, Johannesburg.*

Sinha, N. K., and Wright, J. D. (1977). Application of fuzzy control to a heat exchanger system. *Proc. IEEE Conf. Decision Control, New Orleans* 1424–1428.

Tong, R. M. (1977). A control engineering review of fuzzy systems. *Automatica* **13**, 559–569.

Van Amerongen, J., Van Nauta Lemke, H. R., and Vander Veen, J. C. T. (1977). An autopilot for ships designed with fuzzy sets. *Conf. IFAC/IFIP Digital Comput. Appl. Process Control, 5th, LaHaye.*

Willaeys, D., and Malvache, N. (1976). Utilisation d'un référentiel de sous-ensembles flous, application à un algorithme flou. *Int. Conf. Syst. Sci., Wroclaw, Poland.* (Reference from II.1.)

Willaeys, D., Malvache, N., and Hammad, P. (1977a). Utilization of fuzzy sets for systems modelling and control. *Proc. IEEE Conf. Decision Control, New Orleans* pp. 1435–1439.

Willaeys, D., Mangin, P., and Malvache, N. (1977b). Utilisation des sous-ensembles flous pour la modélisation et la commande de systèmes; application à la régulation de vitesse d'un moteur soumis à de fortes perturbations. *Conf. IFAC/IFIP, Digital Comput. Appl. Process Control, 5th, LaHaye.*

Chapter **5**

FUZZY SETS IN
LEARNING SYSTEMS

"A learning system (or automaton) can be considered as a system (or automaton) which demonstrates an improvement of performance during its operation from a systematic modification of its structure or parameter values" (Fu, NF 1976). A very well-known model of a learning system is the variable structure stochastic automaton (see Varshavskii and Vorontsova, NF 1963; McLaren, NF 1966). In this model the evolution of transition probabilities or state probabilities reflect the information that the automaton has received from the input in such a way that the system performance can be improved during operation.

The same approach has been employed using a fuzzy automaton instead of a stochastic one, and more recently using a conditional fuzzy measure. This is the topic of Section A. A radically different learning process, for on-line improvement of fuzzy linguistic controllers, is presented in Section B. This chapter is just a short survey of the existing works.

A. LEARNING WITH AUTOMATA OR FUZZY CONDITIONAL MEASURES

A basic learning system is given in Fig. 1 (Wee and Fu, 1969). The unknown environment is assumed to be a system that on receiving the input u returns the output $y = f(u)$. The goal is to find $\hat{u}$ such that a given performance evaluator, which depends on y and u, is optimized. The learning system works as follows: first, a decision is made, i.e., a given u is chosen and sent to the environment, which ouputs $y = f(u)$. Secondly, the

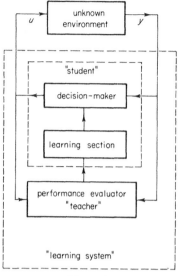

Figure 1

performance evaluator is used to compare the decision with the previous ones, from knowledge of u and y. Thirdly, the learning section is modified as a consequence and a new decision is made, which is supposed to be better than the previous ones. This procedure is iterated until convergence of the learning section, i.e., the goal has been attained. Here, the learning section consists in a fuzzy automaton or an inference model based on fuzzy integrals in the sense of Sugeno.

a. Fuzzy Automaton

Wee and Fu (1969) considered a fuzzy automaton with nonfuzzy inputs $i(t)$ and a time-dependent fuzzy transition relation $\delta(t)$. Let $\tilde{s}(t)$ be the fuzzy state of the automaton at time t, i.e., a fuzzy set on the finite set $S = \{s_1, \ldots, s_n\}$. The value i_l of $i(t)$ may depend on $y(t)$, the output of the unknown environment. The fuzzy state at time $t + 1$ is defined through a max–min composition:

$$\mu_{\tilde{s}(t+1)}(s_k) = \max_j \min\left(\mu_{\tilde{s}(t)}(s_j), \mu_{\delta(t)}(s_k, i_l, s_j)\right);$$

and alternatively a min–max composition:

$$\mu_{\tilde{s}(t+2)}(s_k) = \min_j \max\left(\mu_{\tilde{s}(t+1)}(s_j), \mu_{\delta(t+1)}(s_k, i_l', s_j)\right).$$

The learning behavior is reflected by having nonstationary fuzzy transition matrices with a convergence property. Wee and Fu (1969) have proposed the reinforcement algorithm

$$\mu_{\delta(t)}(s_k, i_l, s_j) = \mu_{\delta(t-1)}(s_k, i_l, s_k) \qquad \forall j \neq k,$$

$$\mu_{\delta(t)}(s_k, i_l, s_k) = \alpha_k \mu_{\delta(t-1)}(s_k, i_l, s_k) + (1 - \alpha_k)\lambda_k(t),$$

where $0 < \alpha_k < 1$ and $0 < \lambda_k(t) \leq 1$, $k = 1, n$. The α_k are constants that are related to the speed of learning. When the $\lambda_k(t)$ are known a priori, we are in the situation of a perfect teacher. Here, the $\lambda_k(t)$ depend on the performance evaluation, which serves thus as an unreliable teacher. Let $\lim_{t\to\infty} \lambda_k(t) = \hat{\lambda}_k$, then $\mu_{\delta(t)}(s_k, i_l, s_j) \to \hat{\lambda}_k$ when $t \to \infty$. Wee and Fu (1969) proved that $\mu_{\tilde{s}(t)}(s_k) \to \hat{\lambda}_k$ when $t \to \infty$. The convergence holds whether or not a priori information is available, i.e., the $\mu_{\tilde{s}(0)}(s_j)$ may be assigned any value in $[0, 1]$. Each state s_j of the fuzzy automaton corresponds to a possible input of the unknown environment. When $\tilde{s}(t)$ has been calculated, the decision (i.e., the choice of an s_j) is based on the maximum grade of membership:

$$\mu_{\tilde{s}(t)}(s_j) = \max_k \mu_{\tilde{s}(t)}(s_k).$$

However, a pure random choice is allowed if $\mu_{\tilde{s}(t)}(s_j)$ is below a given threshold.

Wee and Fu (1969) applied their learning model to pattern classification and control systems. Fu and Li (1969) used it for the determination of optimal strategies in games against a random environment and two-automaton zero-sum games. The fuzzy automaton was advantageously compared to the stochastic one. More recently, Saridis and Stephanou (1977) employed the same learning model in a coordination decision-making problem for the control of a prosthetic arm.

A slightly different learning model is that of Asai and Kitajima (1971a, b, 1972; Hirai et al., 1968). They considered a complete max–min fuzzy automaton, i.e., with an output map σ that is a time-varying fuzzy relation on $V \times S$ where V is the output universe. Their purpose was the optimization of a multimodal function. The domain of the function is divided into subdomains that correspond to the nonfuzzy states of the automaton; every subdomain is also divided into unit domains corresponding to the set of outputs of the automaton. Therefore, a global search can be executed by deciding the optimum output over the whole domain of the objective function, after a local search has been executed in order to find a candidate point in each subdomain. Global and local search are performed alternately. A reinforcement algorithm modifies the membership functions $\mu_{\delta(t)}(s_k, s_j)$ (the input is omitted) and $\mu_{\sigma(t)}(v_i, s_j)$ as follows:

$$\mu_{\delta(t+1)}(s_k, s_j) = \alpha\mu_{\delta(t)}(s_k, s_j) + 1 - \alpha \qquad \text{if} \quad I(t) > I_0,$$

$$\mu_{\sigma(t+1)}(v_i, s_j) = \alpha\mu_{\sigma(t)}(v_i, s_j) \qquad \text{if} \quad I(t) \leq I_0,$$

where $I(t)$ is the performance index at time t, which has to be for instance maximized, I_0 is a performance criterion which possibly depends on the previous results, and

$$\alpha = \min(0.99, \max(0.5, 1 - |(I(t) - I_0)/I_0|))$$

($\alpha < 1$ to ensure convergence). A success is obtained when $I(t) > I_0$ and the corresponding $\mu_{\delta(t+1)}(s_k, s_j)$ is increased; if $I(t) \leqslant I_0$ (failure), it is $\mu_{\sigma(t+1)}(v_i, s_j)$ that is decreased. In Kitajima and Asai (1974) time-varying subdomains are allowed. This method can be used for the adaptive control of dynamic systems; the performance index then evaluates the quality of control. (See also Jarvis, 1975.) An application of this approach to a nuclear engineering problem can be found in Serizawa (1973) and to structural identification of hierarchical systems in Tazaki and Amagasa (1977) (see also Chapter 8 for this application).

b. Conditional Fuzzy Measure

Sugeno and Terano (1977; Terano and Sugeno, 1977) recently proposed a learning model formulated using the concept of conditional fuzzy measure. It is similar to a Bayesian learning model in a stochastic environment. (See, e.g., Duda and Hart, NF 1973.)

Let X be a finite set and g_X a fuzzy measure on X. Let h be a function from X to $[0, 1]$. Assume $X = \{x_1, \ldots, x_n\}$ and $h(x_1) \leqslant \cdots \leqslant h(x_n)$. Then

$$\oint_X h(x) \circ g_X(\cdot) = \max_{i=1,n} \min(h(x_i), g_X(H_i))$$

where $H_i = \{x_i, x_{i+1}, \ldots, x_n\}$ (see II.5.A.b.α).

X is now viewed as a set of causes; let $Y = \{y_1, \ldots, y_m\}$ be a set of results. The problem is to estimate causes through a fuzzy information. Let g_Y be a fuzzy measure on Y; g_Y is assumed to be related to g_X through a conditional fuzzy measure $\sigma_Y(\cdot \,|\, x)$, i.e.,

$$g_Y(\cdot) = \oint_X \sigma_Y(\cdot \,|\, x) \circ g_X(\cdot).$$

g_X is viewed as an a priori weighting of causes by an estimator. $\sigma_Y(F \,|\, x)$, where $F \subseteq Y$, is the grade of fuzziness of the statement, "One of the elements of F results because of x." F is the information; in the deterministic case it is a singleton, but it may be a fuzzy set as well. $g_Y(\{y\})$ expresses the grade of fuzziness of the statement "y actually results," and "$\mu_F(y)$ represents the accuracy of the information objectively." We have

$$g_Y(F) = \oint_Y \mu_F(y) \circ g_Y(\cdot) = \oint_X \sigma_Y(F \,|\, x) \circ g_X(\cdot)$$

where $\sigma_Y(F|x) = \int_Y \mu_F(y) \circ \sigma_Y(\cdot\,|x)$ (see II.5.A.c). Since X is finite,

$$g_Y(F) = \max_{i=1,n} \min(\sigma_Y(F|x_i), g_X(\{x_i, x_{i+1}, \ldots x_n\}))$$

where the $\sigma_Y(F|x_i)$ are increasingly ordered.

After having new information F, the degree of confidence $g_Y(F)$ in F must be increased. This is done by modifying the fuzzy measure g_X through a reinforcement algorithm.

Let $\{x_1, \ldots, x_n\}$ be the set of x_i that are explicitly involved in the calculation of $g_Y(F)$. l is the smallest i such that $g_X(\{x_i, \ldots, x_n\})$ $\leqslant \sigma_Y(F|x_i)$. Following Sugeno and Terano (1977; Terano and Sugeno, 1977), $g_X(\cdot)$ and $\sigma_Y(\cdot\,|x)$ are assumed to be λ-fuzzy measures (see II.5.A.a. γ). The greater is $g_X(\{x_i\})$ $(i \geqslant l)$ and the smaller is $g_X(\{x_i\})$ $(i < l)$, the greater is $g_Y(F)$. Hence, $g_Y(F)$ is improved by the reinforcement rules:

$$g_X'(\{x_i\}) = \begin{cases} \alpha g_X(\{x_i\}) + (1 - \alpha)\sigma_Y(F|x_i), & i = l, n, \\ \alpha g_X(\{x_i\}), & i = 1, l-1, \end{cases}$$

with $0 < \alpha < 1$. Owing to the above expressions, $g_X'(\{x_i\})$, the new $g_X(\{x_i\})$, always remains smaller than $\sigma_Y(F|x_i)$ because it is useless to have it greater. α is related to the speed of convergence of $g_X(\{x_i\})$. The following properties are proved in Sugeno and Terano (1977):

the final values of the $g_X(\{x_j\})$ do not depend on a priori values, but are equal to $\sigma_Y(F|x_i)$ for x_i that makes $\sigma_Y(F|x_j)$ a maximum value and equal to zero for the other x_j when the same information F is repeatedly given;

when the same information F such that $\mu_F(y) = C$ $\forall y$ is repeatedly given, the $g_X(\{x_i\})$ converge to C $\forall i$.

Sugeno and Terano (1977; Terano and Sugeno, 1977) applied their learning model to the macroscopic search for a maximum of a multimodal function. The search domain is divided into blocks that correspond to the elements of Y. X is a set of criteria or types of clues through which one guesses whether a block contains the actual maximum. g_X expresses the grade of importance of subsets of criteria. The criteria may concern for instance the number of points examined in the previous searches or the average of the function obtained in previous searches. $\sigma_Y(\{y_j\}|x_i)$ evaluates the belief of finding an extremum in block y_j owing to the type of clue x_i. For instance, $\sigma_Y(\{y_j\}|x_i)$ may depend on the number of previously searched points in the block y_j. The available information F is given by

$$\mu_F(y_j) = \left(p_j - \min_k p_k\right) \Big/ \left(\max_k p_k - \min_k p_k\right)$$

where p_j is the maximum of the multimodal function found so far in block y_j. Note that F converges to the maximizing set of the function (see

II.4.B.a). g_X is subjectively initialized, and $\sigma_Y(\{y_j\} \mid x_i)$ is calculated at first from an initial random search. At each iteration a given number of new points are tested, the number of these points in block y_j is chosen in proportion to $g_Y(\{y_j\})$.

A current iteration works as follows: from the result of a search $\sigma_Y(\cdot \mid x_i)$ is calculated in each y_j and normalized (see II.5.A.a.γ); g_X is normalized; $g_Y(\{y_j\})$ is calculated from σ_Y and g_X; $g_Y(F)$ is then obtained, and the $g_X(\{x_i\})$ are corrected by the reinforcement rules. Then a new search is performed. This iteration is repeated until g_Y converges.

Sugeno and Terano (1977) have compared their model with a Bayesian learning model. Bayesian inference is now briefly reviewed. Let p_X be an a priori probability density on X and $\rho_Y(\cdot \mid x_i)$ a conditional probability density with respect to x_i. The conditional probability of a fuzzy event F is

$$\rho_Y(F \mid x_i) = \sum_{j=1}^{m} \mu_F(y_j)\rho_Y(y_j \mid x_i).$$

Learning is obtained through the Bayes formula, which yields the a posteriori probability density ρ_X on X after having the fuzzy information F:

$$\rho_X(x_i \mid F) = p_X(x_i)\rho_Y(F \mid x_i) \bigg/ \sum_{k=1}^{n} p_X(x_k)\rho_Y(F \mid x_k).$$

Note that when constant information $\mu_F(y_j) = C \; \forall y_j$ is given, we have $\rho_X(x_i \mid F) = p_X(x_i)$; that is, in Bayesian terms, obtaining constant information is the same as obtaining no information, i.e., learning nothing. However, the fuzzy model is able to distinguish between obtaining constant information and no information since under constant information the weighting g_X becomes uniform because the information is too fuzzy. Another difference between the Bayesian model and the one presented here is the possibility of controlling the speed of convergence by means of α. Lastly, Sugeno and Terano (1977) claimed that the fuzzy model could be expected to work "more effectively than a Bayesian learning model" under fuzzy information.

N.B.: Sugeno and Terano's learning model was used by Seif and Aguilar-Martin (1977) for classification of objects using a sensitive-skinned artificial hand.

B.   SELF-IMPROVEMENT OF FUZZY LINGUISTIC CONTROLLERS

The learning methods presented in A may be applied to adaptive nonfuzzy control of dynamic systems. In this section we are interested in

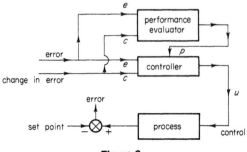

Figure 2

the design of adaptive fuzzy controllers, i.e., controllers defined by a set of fuzzy logical rules, as in 4.B.

Mamdani and Baaklini (Reference from IV.4) first suggested a prescriptive method for deriving the best control policy during run time in a fuzzy logic controller. The main idea is to automate the alteration of fuzzy control rules and thus obtain a self-regulating fuzzy controller. Such a controller can be useful when the system under control is subject to time-varying parameter changes and unknown disturbances.

A self-organizing controller for single-input, single-output systems has actually been implemented by Procyk (1977). The corresponding block diagram is shown in Fig. 2. The quality of control is periodically checked by the performance evaluator, which can modify the structure of the controller when the control is not satisfactory. This modification is supposed to improve the control strategy.

Controller and performance evaluator are both made up of a set of fuzzy inference rules relating $e(t), c(t)$ and $u(t)$, and $e(t), c(t)$ and $P(t)$, respectively, where $e(t)$ is the error at time t, $c(t) = e(t) - e(t-1)$, $u(t)$ is the control at time t, and $P(t)$ is the control modification at time t ($P(t)$ is possibly fuzzy). More specifically, $P(t)$ is the modification that should have altered the controller in order to improve its performance at time t. The rules of the performance evaluator implicitly define the band within which the process output is to be restricted. These rules determine the desired change in the controller to be made in order to keep the process output within the band. The rules of the controller are of the form

$$\text{If } e \text{ is } E_i, \text{ then, if } c \text{ is } C_i, \text{ then } u \text{ is } U_i.$$

Those of the performance evaluator are of form

$$\text{If } e \text{ is } E_j, \text{ then, if } c \text{ is } C_j \text{ then } p \text{ is } P_j,$$

where E_i, E_j, C_i, C_j, U_i, and P_j are prescribed fuzzy sets. The fuzzy output $P(t)$ of the performance evaluator is calculated from $e(t), c(t)$, and the set of fuzzy rules using max–min composition (see 4.B.a). $P(t)$ is used to modify the control strategy: consider a fuzzy rule i that contributed to the present bad performance; then U_i is modified to $U_i' = U_i \oplus P_i(t)$ where $\oplus$ denotes extended addition. Rule i in the controller is replaced by the rule:

If e is E_i, then, if c is C_i, then u is U_i'.

A detailed description of the implementation of a self-organizing controller can be found in Procyk (1977). It has been tested on first, second, and third order linear processes and nonlinear processes (Procyk, 1977).

REFERENCES

Asai, K., and Kitajima, S. (1971a). Learning control of multimodal systems by fuzzy automata. In "Pattern Recognition and Machine Learning" (K. S. Fu, ed.), pp. 195–203. Plenum Press, New York.

Asai, K., and Kitajima, S. (1971b). A method for optimizing control of multimodal systems using fuzzy automata. Inf. Sci. 3, 343–353.

Asai, K., and Kitajima, S. (1972). Optimizing control using fuzzy automata. Automatica 8, 101–104.

Fu, K. S., and Li, T. J. (1969). Formulation of learning automata and automata games. Inf. Sci. 1, 237–256.

Hirai, H., Asai, K., and Kitajima, S. (1968). Fuzzy automata and its application to learning control systems. Mem. Fac. Eng. Osaka City Univ. 10, 67–73.

Jarvis, R. A. (1975). Optimization strategies in adaptive control: A selective survey. IEEE Trans. Syst., Man Cybern. 5, No. 1, 83–94.

Kitajima, S., and Asai, K. (1974). A method of learning control varying search domain by fuzzy automata. In "Learning Systems and Intelligent Robots" (K. S. Fu and J. T. Tou, eds.), pp. 249–262. Plenum, New York.

Otsuki, S. (1970). A model for learning and recognizing machine. Inf. Process. 11, 664–671.

Procyk, M. T. J. (1977). "A Self-Organizing Controller for Single-Input Single-Output Systems," Res. Rep., No. 6. Fuzzy Logic Working Group, Queen Mary Coll., London.

Saridis, G. N., and Stephanou, H. E. (1977). Fuzzy decision-making in prosthetic devices. In "Fuzzy Automata and Decision Processes" (M. M. Gupta, G. N. Saridis, and B. R. Gaines, eds.), pp. 387–402. North-Holland Publ., Amsterdam.

Sasama, H. (1977). A learning model to distinguish the sex of a human name. In "Summary of Papers on General Fuzzy Problems," No. 3, pp. 19–24. Working Group Fuzzy Syst., Tokyo.

Seif, A., and Aguilar-Martin, J. (1977). Multi-group classification using fuzzy correlation LAAS-CNRS, Toulouse, France. [Also in Int. J. Fuzzy Sets Syst. 3, No. 2, 109–222, 1980.]

Serizawa, M. (1973). A search technique of control rod pattern for smoothing core power distributions by fuzzy automaton. J. Nucl. Sci. Technol. 10, No. 4, 195–201.

Sugeno, M., and Terano, T. (1977). A model of learning based on fuzzy information. Kybernetes 6, 157–166.

Tanaka, K. (1976). Learning in fuzzy machines (Execution of fuzzy programs). In "Computer-Oriented Learning Processes" (J. C. Simon, ed.), Noordhoff, pp. 109–148.

Tamura, S., and Tanaka, K. (1973). Learning of fuzzy formal language. *IEEE Trans. Syst., Man Cybern.* **3**, 98–102. (Reference from IV.8.)

Tazaki, E., and Amagasa, M. (1977). Heuristic structure synthesis in a class of systems using a fuzzy automaton. *Proc. IEEE Conf. Decision Control, New Orleans* pp. 1414–1418. (Reference from IV.8.)

Terano, T., and Sugeno, M. (1977). Macroscopic optimization using conditional fuzzy measures. *In* "Fuzzy Automata and Decision Processes" (M. M. Gupta, G. N. Saridis, and B. R. Gaines, eds.), pp. 197–208. North-Holland Publ., Amsterdam.

Wee, W. G. (1967). "On Generalizations of Adaptive Algorithm and Applications of the Fuzzy Sets Concept to Pattern Classification." Ph.D. Thesis, Purdue Univ., Lafayette, Indiana. (Reference from III.2.)

Wee, W. G., and Fu, K. S. (1969). A formulation of fuzzy automata and its application as a model of learning systems. *IEEE Trans. Syst., Sci. Cybern.* **5**, 215–223.

Wong, G. A., and Sheng, D. C. (1972). On the learning behaviour of fuzzy automata. *Int. Congr. Gen. Syst. Cybern., 2nd, Oxford.* [Also *In* "Advances in Cybernetics and Systems" Vol. 2 (J. Rose, ed.), 1975 pp. 885–896, Gordon & Breach, London.]

Chapter 6

PATTERN CLASSIFICATION WITH FUZZY SETS

It is well known that the concept of a fuzzy set first arose from the study of problems related to pattern classification (see Bellman *et al.*, 1966). This is not a posteriori surprising since the recognition of patterns is an important aspect of human perception, which is a fuzzy process in nature. Although a great amount of literature has been published dealing with fuzzy pattern classification, a unified theory is not available yet and a linguistic approach based on fuzzy sets is far from being completely developed. The topics of this chapter are clustered around three themes: pattern recognition, clustering methods, and information retrieval.

A. PATTERN RECOGNITION

Let Ω be a set of objects. A way of characterizing an object $p \in \Omega$ is to assign to it the values of a finite set of parameters considered relevant for the object. Each parameter is specific to a so-called *feature* of the object p. Thus, p can be associated to a mathematical object $x = M(p) = (m_1(p), \ldots, m_r(p))$ where m_i is the measurement procedure associated with feature i and $m_i(p)$ is the feature value. x is called a *pattern*. Note that there are usually many mathematical objects x that may be associated with p. The set of mathematical objects will be called pattern space. The above representation of an object does not take into account its structure. However, in some situations knowledge of this structure may be of great help in the recognition process. In this case the object is viewed as a formal structure that can be decomposed into *primitives*. These primitives may in turn be valued.

The purpose of pattern recognition is to assign a given object to a class of objects similar to it. According to Zadeh (1976), such a class is often a fuzzy set F—F is the label of the class. A recognition algorithm, when applied to an object p, yields the grade of membership $\mu_F(p)$ of p in a class F. For instance, when p can be modeled as a string of primitives that can be derived from a formal grammar, a recognition algorithm may consist in a parsing procedure. But the grade of membership of an object p in a class may also be the degree of its similarity to a typical object of the class, namely a prototype. When the explicit description of the recognition algorithm is known, this algorithm is said to be *transparent*; if such a description is not available, it is said to be *opaque* (Zadeh, 1976). Human perception usually uses opaque algorithms to recognize objects. The problem of pattern recognition is that of converting an opaque recognition algorithm R_{op} into a transparent one R_{tr}. Note that R_{op} acts on p and R_{tr} can act only on $M(p)$. The transformation of R_{op} into R_{tr} involves two steps:

(1) feature extraction: select a small set of measurement procedures m_i and/or a set of primitives in order to turn p into a mathematical object x (vector in a pattern space and/or formal structure);

(2) define a transparent algorithm R_{tr} that from $M(p)$ yields the grade of membership of p in a class F.

The first problem is generally the more difficult. However, we are mainly concerned here with step (2).

Fuzziness may be present at several levels in a pattern recognition problem: the pattern classes, the feature values, and even the transparent recognition algorithm may be fuzzy.

In the following, existing approaches involving fuzzy sets are surveyed; successively dealt with are semantic pattern recognition ($M(p)$ is a pattern vector in a feature space) and syntactic pattern recognition (p can be modeled as a string in a formal language).

a.   Semantic Pattern Recognition

One of the most intuitive ways of defining a fuzzy pattern class is to assign to each class a deformable prototype (Bremermann, 1976a, b; Albin, 1975). The grade of membership of a given object in the class depends on the deformation energy necessary to make the prototype close to the object and the remaining discrepancy between the object and the deformed prototype (see 1.B.b). Lee (1972) has given quantitative measures of the proximity of two n-sided polygons; however, he did not consider deformation energy. The proximity indices are based on angular and dimensional comparisons. Thus, for instance, triangles can be classified into "approximate right triangle," "approximate isosceles triangle," "ordinary triangle," etc. Siy and Chen (1974) have a similar approach in a handwritten numerical character recognition procedure. Each numeral is decomposed into primitives such as horizontal lines or portions of circles. The authors use proximity measures for the (semantic) identification of the primitives. However, the structural part of Siy and Chen's procedure (graph matching) is not fuzzy.

Kotoh and Hiramatsu (1973) propose a general approach for the representation of fuzzy pattern classes. A feature is viewed as a fuzzy partition of pattern space, i.e., each member of the fuzzy partition corresponds to a fuzzy value of this feature. For instance, if the possible fuzzy values of the feature "height" are "small," "medium," and "large," these values realize a fuzzy partition provided that the orthogonality condition

$$\mu_{small}(m(p)) + \mu_{medium}(m(p)) + \mu_{large}(m(p)) = 1 \qquad \forall p \in \Omega$$

holds ($m(p)$ denotes the height of p) (see II.1.B.b). A fuzzy pattern class is expressed by a logical expression of feature values that correspond to different features: for instance, the class of objects whose (height is "medium" or width is "narrow") and weight is "heavy." An algebra of fuzzy-valued features is then developed in Kotoh and Hiramatsu (1973). Operations such as refinement and unification of fuzzy-valued features, related to intersection and union of fuzzy sets, respectively, are introduced. Two pseudocomplementations of feature values are defined; these differ from the usual fuzzy set complementation: in the above example, "medium or large" and "medium and large" are the two pseudocomplements of "small." Note that in this approach each object is evaluated with respect to a fuzzy pattern class by means of a fuzzy logical expression (in the sense of III.1.A.a) that is specific to this pattern class. However, an opaque algorithm cannot always be reduced to the computation of a fuzzy logical expression.

More specifically, let F be a fuzzy pattern class defined by the fuzzy feature values $F_1, F_2, \ldots, F_r$ where F_i is a fuzzy value of feature i. An object p is thus characterized with respect to the class F by r membership values $\mu_{F_i}(m_i(p))$ denoted $\mu_i(p)$ for convenience. $\mu_F(p)$ is then constructed by aggregating the $\mu_i(p)$ in some manner. For instance, a "subjective" aggregation, when features are of unequal importance, could be the Sugeno (1973) integral. Other aggregation schemes are also possible, especially those presented in 3.B.a. It seems that the choice of an aggregation is very context-dependent.

Usually, there are several (fuzzy) pattern classes $F^1, \ldots, F^s$, and the recognition problem is to assign a given object p to a definite class. When the membership values $\mu_{F^j}(p)$ are available, p is assigned to the class k such that $\mu_{F^k}(p) = \max_j \mu_{F^j}(p)$ provided that $\mu_{F^k}(p)$ is sufficiently large. Otherwise, a new pattern class F^{s+1} may be created for p. Once more, a maximum meaningfulness principle has been applied.

Remark Perceptrons, introduced by Rosenblatt (NF 1961; see also Minsky and Papert, NF 1969), have been considered as decision machines in pattern recognition problems. The object p is accepted in a class F iff

$$\alpha_1 \mu_{F_1}(p) + \cdots + \alpha_r \mu_{F_r}(p) > \theta$$

where the F_i are crisp sets and the α_i and θ belong to $\mathbb{R}$. Kaufmann (1977, Reference from I) has recently considered "fuzzy perceptons" where the F_i are fuzzy sets; more general aggregations of the μ_{F_i} are possible.

Zadeh (1976) has suggested another approach to the pattern recognition problem. The features are linguistically valued, and the dependence between $\mu_F(p)$ and the feature fuzzy values $\tilde{m}_j(p)$ are expressed as an $(r + 1)$-ary fuzzy relation R_F on $X_1 \times \cdots \times X_r \times [0, 1]$ where X_j is the universe of $\tilde{m}_j(p)$. R_F is specific to the fuzzy pattern class F. R_F can be derived from a relational tableau (see 2.B.d.β) having n lines and $r + 1$ columns. Let ρ_i^j denote the current term of the tableau; each line i corresponds to the fuzzy rule: if $m_1(p)$ is ρ_i^1 and if ... and if $m_r(p)$ is ρ_i^r, then $\mu_F(p)$ is ρ_i^{r+1}, where ρ_i^j is a linguistic feature value for $j \leqslant r$ and ρ_i^{r+1} is a linguistic truth value. A first way of calculating $\mu_F(p)$ is to explicitly construct

$$R_F = \bigcup_{i=1,n} \bigcap_{j=1,r+1} \rho_i^j \, ;$$

then, knowing the linguistic feature values $\tilde{m}_i(p)$, $i = 1, r$, of an object p, $\mu_F(p)$ is obtained by max–min composition:

$$\mu_F(p) = \left[\tilde{m}_1(p) \times \cdots \times \tilde{m}_r(p) \right] \circ R_F.$$

Note that F is a type 2 fuzzy pattern class.

Another way of determining $\mu_F(p)$ is to build a branching questionnaire (see 2.B.d.β) by viewing each column j of the relational tableau as a set of possible answers to a question concerning feature j. Analogously, Chang and Pavlidis (1977) discussed certain theoretical aspects of fuzzy decision trees. A fuzzy decision tree is a tree such that each nonleaf node i has a k-tuple decision function f_i from Ω to $[0, 1]^k$ and k ordered sons. Each nonleaf son j of a node i corresponds to a question determined by the answer to the preceding question i. $f_i(p; j)$ valuates the branch from i to j.

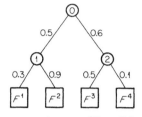

		min:	0.3	0.5	0.5	0.1
Figure 1.	$\mu_{F^i}(p)$:	product:	0.15	0.45	0.30	0.06

Each leaf corresponds to a pattern class. Each path of the decision tree from the root to a leaf l represents the decision $p \in F^l$. Each decision (leaf) is valued by the minimum (or the product) of the decision values $f_i(p; j)$ of the branches composing the path. The object is finally assigned to the pattern class that ends the best valued path. An example of a fuzzy decision tree is pictured in Fig. 1 with $k = 2$. The problem of optimizing a decision tree is to find the best path without computing the decision values of the others. Chang and Pavlidis (1977) use a branch–bound–backtrack method to optimize the fuzzy decision tree. Note that their approach differs from Zadeh's in which a decision tree (branching questionnaire) is characteristic of a pattern class and the leaves are the ρ_i^{r+1}. Moreover, in Chang and Pavlidis's model the truth values $f_i(p; j)$ are numerical and not linguistic.

Remark 1 *Recognition of binary discretized images using fuzzy logic.* Let $M(p)$ be a binary vector $(x_1, \ldots, x_r)$ that represents a discretized picture. To each pattern class F, Shimura (1975) associates four matrices $G_{00}, G_{01}, G_{10}, G_{11}$ where

$$\forall (k, l) \in \{0, 1\}^2, \quad G_{kl}(i, j) = \text{prob}\big[\, p \in F \,|\, x_i = k, x_j = l \,\big],$$

$$i = 1, r; \quad j = 1, r.$$

The values of the $G_{kl}(i, j)$ are learned through a reinforcement algorithm similar to those of 5.A.a. Max–min or min–max compositions between $M(p)$ and the $G_{kl}(i, j)$ are used to evaluate the compatibility of p with the pattern class F.

In Mukaidono (1977) the patterns are allowed to be noisy, i.e., each x_i belongs to [0, 1]. Let $I(p)$ be the original nonnoisy pattern and $M(p)$ be the noisy one. The noise, a vector N, is the absolute value of the difference $M(p) - I(p)$. It is easy to see that $M(p) = I(P) \triangle N(p)$ ($\triangle$ denotes the symmetrical difference (see II.1.B.f) associated with $(\tilde{\mathcal{P}}(\cdot), \cup, \cap, ^-)$ (see III.1.B.b.α)). Mukaidono (1977) studies the existence of a quantization threshold able to resolve the ambiguity (in the sense of III.1.A.b.α) caused by the noise.

2 A methodology for a speech understanding system using fuzzy set theory has been proposed by De Mori and Torasso (1976). The correspondence between spectrogram segments and lexicon words is described as a fuzzy relation. The first speech-understanding system using fuzzy sets seems to be the one by Brémont (Reference from V).

b. Syntactic Pattern Recognition

The idea behind syntactic pattern recognition is that certain pattern classes contain objects, such as geometric figures, that have an identifiable hierarchical structure that can be described by a formal grammar, called the *pattern grammar*. A basic set of pattern primitives is selected and forms the set of terminals of the grammar. The productions of the grammar are a list of allowable relations among the primitives. The pattern class is the set of strings generated by the pattern grammar. However, the concept of a formal grammar is often too rigid to be used for representation of real patterns, which are generally distorted and noisy, yet still retain much underlying structure. Stochastic techniques for describing such distorted and noisy patterns can be found in Fu (NF 1974).

Thomason (1973) has suggested that fuzzy languages could handle imprecise patterns when the indeterminacy is due to inherent vagueness. The fuzziness may lie in the definition of primitives or in the physical relations among them. Thus, the primitives become labels of fuzzy sets and the production rules of the grammar are weighted. The membership grade of a particular pattern in the class described by the grammar is calculated using max–min composition (see III.3.A.b), i.e., the grammar is fuzzy. The possibility of applying fuzzy grammars to the recognition of leukocytes and chromosomes is discussed in Lee (1973). Kickert and Koppelaar (1976) used an ordinary context-free grammar with a set of fuzzy primitives. A fuzzy set of strings compatible with the pattern to be recognized is generated. The method is applied to the recognition of handwritten capitals; the compatibility of the pattern with each of the 26 letters is calculated using the min operator. This approach is criticized by Stallings (1977) who compares it to a stochastic Bayesian one. Fractionally fuzzy grammars (see III.3.A.f) were used by DePalma and Yau (1975) for recognition of handwritten characters.

An important problem in syntactic pattern recognition is that of grammatical inference, i.e., given a set of structured patterns modeled by strings, find an automatic procedure that yields the production rules of a grammar capable of generating this set of patterns. A grammatical inference method may be useful for determining the grammar associated with the pattern class. Although such methods already exist for ordinary and stochastic grammars (see Fu and Booth, NF 1975), the inference of fuzzy grammars has not received much attention in the literature; see 8.C.

B. CLUSTERING ALGORITHMS

The primary objective of clustering techniques is to partition a given data set into so-called homogeneous clusters. The term *homogeneous* means that all points in the same group are close to each other and are not close to points in other groups. Clustering algorithms may be used to build pattern classes or to reduce the size of a set of data while retaining relevant information. In classical algorithms it is implicitly assumed that disjoint clusters exist in the set of data. However, the separation of clusters is a fuzzy notion, and the representation of clusters by fuzzy sets may seem more appropriate in certain situations. Whereas fuzzy pattern recognition has few practical applications yet, fuzzy set theory has given birth to several new interesting clustering techniques, which are described below. A survey of classical algorithms for pattern classification can be found in Duda and Hart (NF 1973).

a. **Detection of Unimodal Fuzzy Sets** (Gitman and Levine, 1970)

The method of unimodal fuzzy sets has been developed to overcome two drawbacks of usual clustering methods, namely their inability to handle large data sets (say 1000 points) and to detect clusters that exhibit complicated distributions in pattern space.

Let X be a finite set of vectors ($|X| = n$) in a metric space. Let d be the metric. For all $x^i \in X$, $i = 1, n$, denote by $\Gamma_{i\theta}$ the set $\{x \in X, d(x, x^i) \leqslant \theta\}$ where $\theta \in \mathbb{R}^+$. A fuzzy set A on X is constructed by assigning to each x^i the membership value $\mu_A(x^i) = |\Gamma_{i,\theta}|/n$. For a given θ, $\mu_A(x^i)$ is a measure of the concentration of points around x^i. The maxima of μ_A correspond to the "centers" of the clusters existing in X. The clustering procedure decomposes A into unimodal fuzzy sets (see II.1.F.a) and realizes the maximum separation among them. The procedure is divided in two main steps: first, local maxima are identified by a systematic search where both the order of the points according to their grade of membership and their order according to distance are used. The second step is the assignment of each point to a cluster. There are as many clusters as local maxima of μ_A. (For further details, see Gitman and Levine, 1970.) Note that the clusters obtained are not fuzzy sets.

b. Fuzzy Partition   (Ruspini's approach)

Ruspini (1969) has introduced the notion of a fuzzy partition to represent the clusters in a data set. A fuzzy partition is a family of fuzzy sets $F_1, \ldots, F_m$ on X such that

$$\forall x \in X, \quad \sum_{i=1}^{m} \mu_{F_i}(x) = 1.$$

"The advantage of a fuzzy set representation in cluster analysis is that stray points or points isolated between clusters as well as other types of uncertainties may be classified as such" (Ruspini, 1973b).

According to Ruspini (1973a), the problem of fuzzy clustering may be stated as follows. Given a finite data set X and a positive real-valued function δ (the distance or dissimilarity function), whose domain is X^2, such that

(1) $\forall x \in X, \delta(x, x) = 0,$
(2) $\forall x, y \in X, \delta(x, y) = \delta(y, x),$

find a fuzzy partition $F_1, \ldots, F_m$, where m is a priori known, such that close elements in X (in the sense of δ) will have similar classification (membership values) and dissimilar elements will have different classification. The classification of an element x is the vector $C(x) = [\mu_{F_1}(x) \cdots \mu_{F_m}(x)]$. One of the possible ways of satisfying the above requirement is to select the function $C(x)$ so as to minimize some suitably defined functional. Let us outline Ruspini's idea for constructing such a functional.

Let v be a function from $[0, 1]^m \times [0, 1]^m$ to $\mathbb{R}^+$ such that $v(a, a) = 0$ and $v(a, b) = v(b, a)$, and let f be a positive nondecreasing not identically zero real function of one real variable satisfying $f(0) = 0$, then the function C should be selected such that

$$\forall x, y \in X, \quad v(C(x), C(y)) = f(\delta(x, y)).$$

Generally, this equation has no solution. It is then relaxed into a minimization problem: find C minimizing

$$\sum_{x, y \in X} w(x)w(y)[v(C(x), C(y)) - f(\delta(x, y))]^2$$

where w is an appropriate weighting function. Usually, v is taken as a Euclidean distance. Various forms of f have been tried and discussed in Ruspini (1970, 1973a) where many experimental results are provided. A slightly different approach using association measures is described in Ruspini (1973b). The association measure between a point x and a fuzzy set F on X is taken as the inverse of a weighted average distance between x

and F. (The average distance between x and F is defined as

$$d(x, F) = \frac{1}{|F|} \sum_{i=1}^{|X|} \mu_F(x^i)\delta(x, x^i).$$

Ruspini's (1973b) idea is that the membership value of x in a fuzzy cluster F_j varies in proportion with the inverse of the average distance between x and F.

c. Fuzzy ISODATA

In some situations we are interested in finding not only a partition of a data set but also the most representative elements of the data set, i.e., the cluster *centers*. This is achieved by the ISODATA algorithm (Ball and Hall, NF 1967). This algorithm has been improved by allowing fuzzy clusters to be generated. First, the nonfuzzy version of ISODATA is reviewed.

Let X be a finite data set contained in a real vector space $V = \mathbb{R}^n$ and let d denote an arbitrary metric on V. Set diameters and set distances are defined by

$$\forall A \subset V, \quad \operatorname{diam} A = \sup_{x,\,y \in A} d(x, y),$$

$$\forall A \subset V, \quad \forall B \subset V, \quad d(A, B) = \inf_{\substack{x \in A \\ y \in B}} d(x, y).$$

d is assumed to be induced by a norm on V, i.e., a metric of the form $d(x, y) = \|x - y\|$ where $\| \; \|$ satisfies

$$\forall \alpha \in \mathbb{R}, \quad \forall u \in V, \quad \|\alpha u\| = |\alpha| \, \|u\|$$

and the triangle inequality. Let $\mathscr{F} = (F_1, \ldots, F_m)$ be a hard (i.e., nonfuzzy) partition of X. $\operatorname{conv}(F_i)$ denotes the convex hull of F_i in V (see II.3.C.e).

The subsets F_i of a nonfuzzy partition of X are said to be *compact well-separated* (CWS) clusters iff for all i, j, k with $j \neq k$, any pair (x, y) with x in F_i and y in $\operatorname{conv}(F_i)$ are closer together as measured by d than any pair (u, v) with u in F_j and v in $\operatorname{conv}(F_k)$ (Dunn, 1974a). This property can be quantified by the index

$$\beta(m, \mathscr{F}) = \left(\min_{1 \leqslant i \leqslant m} \min_{\substack{1 \leqslant j \leqslant m \\ j \neq i}} d(F_i, \operatorname{conv}(F_j)) \right) \Big/ \max_{1 \leqslant i \leqslant m} \operatorname{diam}(F_i).$$

According to Dunn (1974a), X can be partitioned into m CWS clusters relative to d iff

$$\bar{\beta}(m) = \max_{\mathscr{F}} \beta(m, \mathscr{F}) > 1$$

($\mathcal{F}$ belongs to the set of m-partitions of X). The problem of finding an $\mathcal{F}$ such that $\beta(m, \mathcal{F}) = \bar{\beta}(m)$ is very difficult. The above index is usually replaced by the simpler criterion

$$J(\mathcal{F}, v) = \sum_{i=1}^{m} \sum_{x \in F_i} d(x, v_i)^2$$

where v is an m-tuple of elements of conv(X) called the cluster centers and d is now supposed to be induced by an inner product:

$$d(x, y) = \left[(x - y)^t M (x - y) \right]^{\frac{1}{2}}.$$

M is called a sample covariance matrix. Usually, M is taken as the identity. $J(\mathcal{F}, v)$ can be interpreted as the average least square error of assimilating the elements of F_i to v_i, for all $i = 1, m$. The problem becomes: find $\mathcal{F}^*$ and v^*, for a given m, such that

$$J(\mathcal{F}^*, v^*) = \min_{\mathcal{F}} \inf_{v \in \text{conv}(X)} J(\mathcal{F}, v).$$

A local minimum of J is obtained by the following iterative method (ISODATA):

(1)   choose an $\mathcal{F} = F_1, \ldots, F_m$;
(2)   compute the centers v_i of the F_i;
(3)   construct a new partition $\hat{\mathcal{F}}$ according to the rule

$$x \in \hat{F}_i \qquad \text{iff} \qquad d(x, v_i) = \min_{1 \leq j \leq m} d(x, v_j);$$

(4)   if $\hat{\mathcal{F}} = \mathcal{F}$ stop; otherwise set $\mathcal{F} = \hat{\mathcal{F}}$ and go to step (2).

More details can be found in Dunn (1974a) where some limitations of the above algorithm are discussed. Every partition consisting in CWS clusters is necessarily a fixed point of ISODATA; however, there are examples where a fixed point of ISODATA is neither a global minimum of J nor a global maximum of $\beta(m, \mathcal{F})$. This is especially true for small values of $\bar{\beta}(m)$. ISODATA always yields some hard partition even when CWS clusters do not exist. Hence, when it is not known in advance that CWS clusters are actually present, "inferences drawn from ISODATA partitions can be very dangerous" (Dunn, 1974a).

To avoid this difficulty, Dunn (1974a) and Bezdek (1974a, b) have relaxed J to allow fuzzy partitions as global minima. More specifically, let $J_w(\mathcal{F}, v)$ be equal to

$$\sum_{i=1}^{m} \sum_{x \in X} \left[\mu_{F_i}(x) \right]^w d(x, v_i)^2, \qquad w \in \mathbb{R}^+$$

where

$$\mu_{F_i}(x) \in [0,1] \quad \text{and} \quad \sum_{i=1}^{m} \mu_{F_i}(x) = 1 \quad \forall x, \quad \forall i = 1, m.$$

Bezdek (1974a, b) and Dunn (1974a) have adapted the ISODATA algorithm to the minimization of $J_w(\mathcal{F}, v)$. Details can be found in the above references. In particular, it can be shown that $(\mathcal{F}, v)$ may be a local minimum of J_w, for $w \in]1, +\infty)$ and $v_i \notin X$, $i = 1, m$, only if

$$\mu_{F_i}(x) = \left\{ \sum_{j=1}^{m} \left[\frac{\|x - v_i\|^2}{\|x - v_j\|^2} \right]^{(w-1)^{-1}} \right\}^{-1}, \quad i = 1, m, \quad \forall x \in X.$$

$$v_i = \sum_{x \in X} [\mu_{F_i}(x)]^w x \Big/ \sum_{x \in X} [\mu_{F_i}(x)]^w, \quad i = 1, m.$$

The first formula replaces the nearest neighbor rule of step (3) of ISO-DATA. The iterative procedure is initialized either by an m-tuple of F_i or an n-tuple of the v_i. Dunn (1974a) proved that for $w = 1$, when the nearest neighbor rule is used, $\mathcal{F}$ is necessarily a hard partition.

The partition coefficient $\varphi_m(\mathcal{F})$ is defined by (Bezdek, 1974b)

$$\varphi_m(\mathcal{F}) = \frac{1}{|X|} \sum_{x \in X} \sum_{i=1}^{m} [\mu_{F_i}(x)]^2.$$

Note that $\forall x$,

$$1 = \left(\sum_{i=1}^{m} \mu_{F_i}(x) \right) \left(\sum_{j=1}^{m} \mu_{F_j}(x) \right) = \sum_{i=1}^{m} (\mu_{F_i}(x))^2 + \sum_{i \neq j} \mu_{F_i}(x)\mu_{F_j}(x).$$

Thus, when $\varphi_m(\mathcal{F}) = 1$, the F_i are pairwise disjoint and $\mathcal{F}$ is a hard partition. The minimum of φ_m is reached for $\mu_{F_i}(x) = m^{-1} \; \forall i \; \forall x$. The partition coefficient provides a quantitative measure of how "fuzzy" $\mathcal{F}$ is. The relation between the partition coefficient and CWS clusters has been studied by Bezdek (1974b) and Dunn (1974b). They proved that as $\bar{\beta}(m)$ increases the result of the fuzzy ISODATA algorithm becomes necessarily hard, and further the global minimum of J_w becomes arbitrarily close to the unique optimal CWS clustering of X corresponding to $\bar{\beta}(m)$. It is indicated in Bezdek (1974b) that $\varphi_m(\mathcal{F})$ may be used for testing the reliability of the solution of fuzzy ISODATA.

The influence of w on the result of the algorithm was discussed by Dunn (1974c). When w increases, the partition obtained, for a given X, becomes fuzzier and fuzzier. For $w = 2$, the result of the algorithm usually reflects the actual fuzziness of the clusters in X.

Gustafson and Kessel (1978) have recently generalized the fuzzy ISO-DATA algorithm to distances of the form $d(x, y) = (x - y)^t M(x - y)$ where M is no longer the identity.

Fuzzy ISODATA has been applied to medical taxonomy (Bezdek and Castelaz, 1977; Fordon and Fu, 1976), to Bayesian unsupervised learning (Bezdek and Dunn, 1975), and to feature selection for binary data (Bezdek, 1976b; Bezdek and Castelaz, 1977).

d. Graph-Theoretic Methods

The idea underlying the graph-theoretic approach to cluster analysis is to start from similarity values between patterns to build the clusters. The data are the entries of a fuzzy symmetrical relation R (or a distance matrix, in terms of dissimilarity). Usually, the methods described in the fuzzy-set literature yield nonfuzzy clusters. Several partitions are obtained together with their "degree of validity."

Flake and Turner (1968) determine a nonfuzzy partition made up of maximally coherent clusters. They use the coherence index

$$D(F) = \frac{2}{n(n - 1)} \sum_{x, y \in F} \mu_R(x, y)$$

where F is a nonfuzzy subset of X, the data set ($|X| = n$). Their algorithm is enumerative.

Tamura et al. (1971) start from a proximity relation (see II.3.C.d), compute its transitive closure, and construct the associated partition tree (see II.3.C.b). They obtain a nested sequence of nonfuzzy partitions. Dunn (1974d) indicated that this clustering method was related to the well-known single linkage approach (see, e.g., Duda and Hart, NF 1973).

Yeh and Bang (1975) define several kinds of clusters based on various notions of connectivity in a fuzzy symmetrical graph. For instance, a partition can be built from the λ-degree components of the fuzzy graph (see III.4.B.a). The authors notice that these methods are related to already-known techniques described in terms of distance rather than of similarity. However, the fuzzy graph approach is shown to be more powerful.

Recently, Bezdek and Harris (1978) have suggested that likeness relations included in the convex hull of the nonfuzzy equivalence relations in $X \times X$ (see II.3.C.e) could provide a basis for new clustering techniques. (See also Ruspini, 1977.)

e. Other New Methods

Instead of defining a fuzzy partition $\mathcal{F} = (F_1, \ldots, F_m)$ by the orthogonality condition $\sum_{i=1}^m \mu_{F_i}(x) = 1 \ \forall x \in X$, Zadeh (1976) has proposed the fuzzy affinity property in order to characterize fuzzy clusters induced by a fuzzy relation R. More specifically, $F_1, \ldots, F_m$ satisfy the fuzzy affinity property as soon as:

(1) both x, y, elements of X, have high grades of membership in some F_i iff (x, y) has a high grade of membership in R;

(2) if $x \in X$ has a high grade of membership in some F_i and $y \in X$ has a high grade of membership in some $F_j, j \neq i$, then (x, y) does not have a high grade of membership in R.

Note that the fuzzy affinity property implies some kind of transitivity for R. The set of pairs (x, y) having a "high" degree of membership in R can be found using a fuzzy α-cut R_{high} of R (see II.2.A.e.γ); "high" is here a fuzzy set on $[0, 1]$. "Basically the employment of fuzzy level sets for purposes of clustering may be viewed as an application of a form of contrast intensification."

Recently, Ruspini (1977) has dealt with a new approach to the cluster representation problem. A fuzzy partition is now viewed as a fuzzy set of fuzzy clusters. Classically, given a crisp equivalence relation R on $X \times X$ and denoting by $R(x)$ the set $\{y \in X, \mu_R(x, y) = 1\}$, a nonfuzzy subset C of X is said to be an R-cluster representation of X iff

$$\bigcup_{x \in C} R(x) = X. \tag{1}$$

If C contains no proper subset that is also an R-cluster representation of X, C is said to be a minimal representation of X. When R is fuzzy and X is finite, a fuzzy set C is said to be a fuzzy R-representation of X iff

$$\sum_{x \in X} \mu_R(x, y)\mu_C(x) \geq 1 \qquad \forall y \in X, \tag{2}$$

provided that $\sum_{x \in X} \mu_R(x, y) \geq 1 \; \forall y \in X$. The problem of finding a fuzzy minimal R-representation may be stated as: find a fuzzy R-representation C^* of X such that $H(C^*) = \inf H(C)$ where H is for instance the cardinality of C in the sense of II.1.D.a. In the conventional representation the set of clusters is $\{R(x), x \in C^*\}$—C^* is a set of cluster centers. In the fuzzy representation the set of clusters is a fuzzy set

$$\sum_{x \in X} \left[\mu_{C^*}(x) \Big/ \left(\sum_{y \in X} \mu_R(x, y)/y \right) \right],$$

using Zadeh's notation of fuzzy sets. The membership functions of the fuzzy clusters are $\mu_R(x, \cdot)$, their number is $|C^*|$, i.e., no longer an integer. $\mu_{C^*}(x)$ is the degree of eligibility of $\mu_R(x, \cdot)$ and is to be considered as the membership function of a fuzzy cluster.

N.B.:   Equation (2) generalizes (1) in the sense of the bold union $\cup$ (see II.1.B.e.). When C is not fuzzy, but R is, (1) can be extended into

$$\underset{x \in C}{\cup} R(x) = X$$

i.e., $\min(\sum_{x \in C} \mu_R(x, y), 1) = 1$; when C is fuzzy, this equation is obviously extended into (2).

Lopez de Mantaras (1978) deals with the case when the data set is not given at once, but the patterns arrive sequentially. The main features of his approach are:

(1)   relaxation of the orthogonality constraint that defines a fuzzy partition because the patterns are noisy; thus, too noisy patterns are allowed to have a very low degree of membership to each cluster;

(2)   it employs the concept of self-learning (see Lopez de Mantaras and Aguilar-Martin, NF 1978); the number of clusters is not known a priori.

C.   INFORMATION RETRIEVAL

An information retrieval system compares the specification of required items with the description of stored items and retrieves or lists all the items that match in some defined way that specification. An example of a fuzzy system describing an information retrieval process can be found in Negoita and Ralescu (1975, Chap. 4). We are concerned here with the clustering aspect of the problem.

Fuzzy approaches to information retrieval have been initiated rather early in the literature (Negoita, 1973a; Demant, 1971).

Let X be a set of documents. A fuzzy set on X is interpreted as a fuzzy cluster of documents. Let Y be a set of descriptors y_k, $k = 1, n$. A document $x \in X$ is described by the vector $(y_1(x), \ldots, y_n(x))$ where $\forall k, y_k(x) \in \{0, 1\}$. The probability that the descriptor y_k is present in any document of the cluster i is denoted p_{ik}. The membership function of the cluster i is μ_i such that (Negoita, 1973b)

$$\mu_i(x) = \sum_{k=1}^{n} p_{ik} y_k(x) \bigg/ \sum_{k=1}^{n} y_k(x).$$

A reasonable necessary condition for a clustering algorithm used for structuring the storage of documents is that every document should be assigned to at least one cluster. To take into account all the clusters, a document x is assigned to the cluster i as soon as

$$\mu_i(x) \geqslant \min_{j,k} \max_{x \in X} \min(\mu_j(x), \mu_k(x)).$$

Negoita (1973a) introduced the degree of relevance of a descriptor y_k to a document x as the truth value (belonging to $[0, 1]$) of the proposition "the document i has the descriptor y_k." Thus, there is a fuzzy relation R on $X \times Y$. It is supposed that $\forall x \in X, \exists y_k \in Y$ such that $\mu_R(x, y) > 0$. The fuzzy description of the document x is a fuzzy set $D(x)$ such that $\mu_{D(x)}(y) = \mu_R(x, y)$. A fuzzy relation ρ expressing the similarity between the documents can be induced on $X \times X$ (Negoita and Flondor, 1976) by

$$\mu_\rho(x, x') = \max_{y \in Y} \min(\mu_{D(x)}(y), \mu_{D(x')}(y)).$$

Clusters of similar documents can thus be considered.

Let g be a fuzzy measure on Y expressing the relative importance of the descriptors. A global evaluation $\delta(x)$ of a document x can be defined by means of Sugeno's integral:

$$\delta(x) = \oint_Y \mu_{D(x)}(y) \circ g(\cdot).$$

Such an evaluation may be helpful when searching for a document (Negoita and Flondor, 1976). For let q be a request whose fuzzy description is $D(q)$, the documents x that best match the request are such that a distance between $\delta(x)$ and $\delta(q)$ is minimum.

A linguistic approach to the representation and processing of fuzzy queries is described at length in Tahani (1977).

Remark   The problem of organizing the set of descriptors by means of fuzzy relations is considered by Reisinger (1974). The membership value of the link between two descriptors y and y' (i.e., the "association factor") is calculated from the numbers of documents in X that are partially characterized by both y and y', and by only one of them.

A similar approach is given by Radecki (1976) who uses the notion of fuzzy level set (II.1.C.).

REFERENCES

Adey, W. R. (1972). Organization of brain tissue: Is the brain a noisy processor? *Int. J. Neurosci.* **3**, 271–284.

Albin, M. (1975). "Fuzzy Sets and Their Application to Medical Diagnosis and Pattern Recognition." Ph.D. Thesis, Univ. of California, Berkeley.

Barnes, G. R. (1976). Fuzzy sets and cluster analysis. *Proc. Int. Joint Conf. Pattern Recognition 3rd.*

Bellacicco, A. (1976). Fuzzy classifications. *Synthese* **33**, 273–281.

Bellman, R. E., Kalaba, R., and Zadeh, L. A. (1966). Abstraction and pattern classification. *J. Math. Anal. Appl.* **13**, 1–7.

Bezdek, J. C. (1974a). Numerical taxonomy with fuzzy sets. *J. Math. Biol.* **1**, 57–71.

Bezdek, J. C. (1974b). Cluster validity with fuzzy sets. *J. Cybern.* **3**, No. 3, 58–73.

Bezdek, J. C. (1975). Mathematical models for systematics and taxonomy. *Proc. Annu. Int. Conf. Numer. Taxon., 8th*, pp. 143–164 (G. Estabrook, ed.), Freeman Co., San Francisco.

Bezdek, J. C. (1976a). A physical interpretation of fuzzy ISODATA. *IEEE Trans. Syst., Man Cybern.* **6**, 387–390.

Bezdek, J. C. (1976b). Feature selection for binary data: medical diagnosis with fuzzy sets. *AFIPS Natl. Comput. Conf. Expo., Conf. Proc.* (S. Winkler, ed.), pp. 1057–1068.

Bezdek, J. C., and Castelaz, P. F. (1977). Prototype classification and feature selection with fuzzy sets. *IEEE Trans. Syst., Man Cybern.* **7**, No. 2, 87–92.

Bezdek, J. C., and Dunn, J. C. (1975). Optimal fuzzy partitions: A heuristic for estimating the parameters in a mixture of normal distributions. *IEEE Trans. Comput.* **24**, 835–838.

Bezdek, J. C., and Harris, J. D. (1978). Fuzzy partitions and relations: An axiomatic basis for clustering. *Int. J. Fuzzy Sets Syst.* **1**, No. 2, 111–127. (Reference from II.3.)

Bremermann, H. (1976a). Pattern recognition by deformable prototypes. *In* "Structural Stability, the Theory of Catastrophes and Applications in the Sciences," Springer Notes in Mathematics, Vol. 25, pp. 15–57. Springer-Verlag, Berlin and New York.

Bremermann, H. (1976b). Pattern recognition. *In* "Systems Theory in the Social Sciences" (H. Bossel, S. Klaczko, and N. Müller, eds.), pp. 116–159. Birkhaeuser Verlag, Basel.

Chang, R. L. P., and Pavlidis, T. (1977). Fuzzy decision-tree algorithms. *IEEE Trans. Syst., Man Cybern.* **7**, No. 1, 28–35.

Conche, B. (1973). "Eléments d'une Méthode de Classification par Utilisation d'un Automate Flou." J.E.E.F.L.N., Univ. de Paris IX, Dauphine.

Conche, B. (1975). La classification dans le cas d'informations incomplètes ou non explicites. *In* "Séminaire, Contribution des Systèmes Flous à l'Automatique." Centr. Automat. Lille, Lille.

Demant, B. (1971). Fuzzy-Retrieval-Strukturen. *Angew. Inf., Appl. Inf.* **13**, 500–502.

De Mori, R., and Torasso, P. (1976). Lexical classification in a speech-understanding system using fuzzy relations. *Proc. IEEE Conf. Speech Process.*, pp. 565–568.

DePalma, G. F., and Yau, S. S. (1975). Fractionally fuzzy grammars with application to pattern recognition. *In* "Fuzzy Sets and Their Applications to Cognitive and Decision Processes" (L. A. Zadeh, K. S. Fu, K. Tanaka, and M. Shimura, eds.), pp. 329–351. Academic Press, New York. (Reference from III.3.)

Diday, E. (1972). Optimisation en classification automatique et reconnaissance des formes. *RAIRO—Oper. Res.* **3**, 61–96.

Diday, E. (1973). The dynamic clusters method and optimization in non hierarchical-clustering. *Conf. Optim. Tech., 5th*, pp. 241–254 (R. Conti, A. Ruberti, eds.), Springer Verlag, Berlin and New York.

Dunn, J. C. (1974a). A fuzzy relative of the ISODATA process and its use in detecting compact well-separated clusters. *J. Cybern.* **3**, No. 3, 32–57.

Dunn, J. C. (1974b). Well-separated clusters and optimal fuzzy partitions. *J. Cybern.* **4**, No. 1, 95–104.

Dunn, J. C. (1974c). Some recent investigations of a new fuzzy partitioning algorithm and its application to pattern classification problems. *J. Cybern.* **4**, No. 2, 1–15.

Dunn, J. C. (1974d). A graph theoretic analysis of pattern classification via Tamura's fuzzy relation. *IEEE Trans. Syst., Man Cybern.* **4**, 310–313.

Dunn, J. C. (1977). Indices of partition fuzziness and detection of clusters in large data sets. *In* "Fuzzy Automata and Decision Processes" (M. M. Gupta, G. N. Saridis, and B. R. Gaines, eds.), pp. 271–283. North-Holland Publ., Amsterdam.

Emsellem, B., and Rochfeld, A. (1975). Bases de données et recherche documentaire par voisinage. *METRA* **14**, No. 2, 321–334.

Flake, R. H., and Turner, B. L. (1968). Numerical classification for taxonomic problems. *J. Theor. Biol.* **20**, 260–270.

Fordon, W. A., and Fu, K. S. (1976). A non-parametric technique for the classification of hypertension. *Proc. IEEE Conf. Decision Control.*

Gitman, I., and Levine, M. D. (1970). An algorithm for detecting unimodal fuzzy sets and its application as a clustering technique. *IEEE Trans. Comput.* **19**, 583–593.

Gustafson, D. E., and Kessel, W. C. (1978). Fuzzy clustering with a fuzzy covariance matrix. Scientific Systems, Inc., Cambridge, Ma.

Kaufmann, A. (1975). "Introduction à la Théorie des Sous-Ensembles Flous. Vol. 3: Applications à la Classification et à la Reconnaissance des Formes, aux Automates et aux Systèmes, aux Choix des Critères." Masson, Paris. (Reference from I, 1975b.)

Kickert, W. J. M., and Koppelaar, H. (1976). Application of fuzzy set theory to syntactic pattern recognition of handwritten capitals. *IEEE Trans. Syst., Man Cybern.* **6**, No. 2, 148–151.

Koczy, L. T., and Hajnal, M. (1977). A new fuzzy calculus and its applications as a pattern recognition technique. *In* "Modern Trends in Cybernetics and Systems" (J. Rose and C. Bilciu, eds.), Vol. 2, pp. 103–118. Springer-Verlag, Berlin and New York.

Kotoh, K., and Hiramatsu, K. (1973). A representation of pattern classes using the fuzzy sets. *Syst.—Comput.—Controls* pp. 1–8.

Larsen, L. E., Ruspini, E. H., McNew, J. J., Walter, D. O., and Adey, W. R. (1972). A test of sleep staging systems in the unrestrained chimpanzee. *Brain Res.* **40**, 318–343.

Lee, E. T. (1972). Proximity measure for the classification of geometric figures. *J. Cybern.* **2**, No. 4, 43–59.

Lee, E. T. (1973). Application of fuzzy languages to pattern recognition. *Kybernetes*, **6**, 167–173.

Loginov, V. I. (1966). Probability treatment of Zadeh's membership functions and their use in pattern recognition. *Eng. Cybern.* pp. 68–69.

Lopez De Mantaras, R. (1978). "On Self-Learning Pattern Classification," Memo. Univ. of California, Berkeley.

Mukaidono, M. (1977). On some properties of a quantization in fuzzy logic. *Proc. IEEE Int. Symp. Multiple-Valued Logic*, pp. 103–106.

Negoita, C. V. (1973a). On the notion of relevance in information retrieval. *Kybernetes* **2**, 161–165.

Negoita, C. V. (1973b). On the application of the fuzzy sets separation theorem for automatic classification in information retrieval systems. *Inf. Sci.* **5**, 279–286.

Negoita, C. V., and Flondor, P. (1976). On fuzziness in information retrieval. *Int. J. Man-Mach. Stud.* **8**, 711–716.

Negoita, C. V., and Ralescu, D. A. (1975). "Applications of Fuzzy Sets to Systems Analysis," Chap. 7. Birkhaeuser Verlag, Basel. (Reference from I.)

Parrish, E. A., Jr., McDonald, W. E., Aylor, J. H., and Gritton, C. W. K. (1977). Electromagnetic interference source identification through fuzzy clustering. *Proc. IEEE Conf. Decision Control, New Orleans* pp. 1419–1423.

Radecki, T. (1976). Mathematical model of information retrieval system based on the concept of fuzzy thesaurus. *Inf. Process. & Manage.* **12**, 313–318.

Reisinger, L. (1974). On fuzzy thesauri. *Compstat, 1974, Proc. Symp. Comput. Stat.,* pp. 119–127 (G. Bruckman, F. Ferschl, L. Schmetterer, eds.), Physica-Verlag, Vienna.

Ruspini, E. H. (1969). A new approach to clustering. *Inf. Control* **15,** 22–32.

Ruspini, E. H. (1970). Numerical methods for fuzzy clustering. *Inf. Sci.* **2,** 319–350.

Ruspini, E. H. (1973a). New experimental results in fuzzy clustering. *Inf. Sci.* **6,** 273–284.

Ruspini, E. H. (1973b). A fast method for probabilistic and fuzzy cluster analysis using association measures. *Proc. Hawaii Int. Conf. Syst. Sci., 6th,* pp. 56–58.

Ruspini, E. H. (1977). A theory of fuzzy clustering. *Proc. IEEE Conf. Decision Control, New Orleans* pp. 1378–1383.

Saridis, G. N. (1974). Fuzzy notions in nonlinear system classification. *J. Cybern.* **4,** No. 2, 67–82. (Reference from III.2.)

Shimura, M. (1975). An approach to pattern recognition and associative memories using fuzzy logic. *In* "Fuzzy Sets and Their Applications to Cognitive and Decision Processes" (L. A. Zadeh, K. S. Fu, K. Tanaka, and M. Shimura, eds.), pp. 443–475. Academic Press, New York.

Siy, P., and Chen, C. S. (1974). Fuzzy logic for handwritten numerical character recognition. *IEEE Trans. Syst., Man Cybern.* **6,** 570–575.

Stallings, W. (1977). Fuzzy set theory versus Bayesian statistics. *IEEE Trans. Syst., Man Cybern.* **7,** 216–219.

Sugeno, M. (1973). Constructing fuzzy measure and grading similarity of patterns by fuzzy integrals. *Trans. S.I.C.E.* **9,** 361–370. (In Jpn., Engl. sum.)

Tahani, V. (1977). A conceptual framework for fuzzy query processing—A step towards very intelligent data systems. *Inf. Proces. & Manage.* **13,** 289–303.

Tamura, S., Higuchi, S., and Tanaka, K. (1971). Pattern classification based on fuzzy relations. *IEEE Trans. Syst., Man Cybern.* **1,** 61–66.

Thomason, M. G. (1973). Finite fuzzy automata, regular fuzzy languages and pattern recognition. *Pattern Recognition* **5,** 383–390.

Van Velthoven, G. D. (1975). Application of the fuzzy sets theory to criminal investigation. *Proc. Eur. Congr. Oper. Res., 1st, Brussels.*

Woodbury, M. A., and Clive, J. (1974). Clinical pure types as a fuzzy partition. *J. Cybern.* **4,** No. 3, 111–121.

Yeh, R. T., and Bang, S. Y. (1975). Fuzzy relations, fuzzy graphs and their applications to clustering analysis. *In* "Fuzzy Sets and Their Applications to Cognitive and Decision Processes" (L. A. Zadeh, K. S. Fu, K. Tanaka, and M. Shimura, eds.), pp. 125–149. Academic Press, New York. (Reference from III.4.)

Zadeh, L. A. (1976). "Fuzzy Sets and Their Application to Pattern Classification and Cluster Analysis," Memo UCB/ERL M-607. Univ. of California, Berkeley. [Also *In* "Classification and Clustering," Academic Press, 1977.]

Chapter *7*

FUZZY DIAGNOSIS

In this brief chapter the problem of the determination of the internal state of a system from a set of external observations is considered. States may be thought of as possible causes (e.g., diseases) and observations as effects (e.g., symptoms). In many practical situations the observations are fuzzy because they are partially qualitative; moreover, the relationship between causes and effects may be complex or ill known. The number of works dealing with fuzzy diagnosis is still rather small compared with decision-making and pattern classification using fuzzy sets. In Section A a fuzzy extension of the well-known Bayesian inference model is presented. Section B is devoted to the representation of causality by fuzzy relations.

A. DISCRIMINATION OF FUZZY STATES IN A PROBABILISTIC ENVIRONMENT

In most decision-making problems of large-scale systems, states are generally defined by fuzzy statements that roughly reflect a given situation. Asai *et al.* (1977) have formulated a method for discriminating such fuzzy states in probability space and have derived a diagnosis rule that minimizes the average probability of discrimination error. Let $S = (s_1, \ldots, s_n)$ be a set of nonfuzzy states; $p(s_i)$ denotes the a priori probability of being in state s_i. Let X be a set of possible observations. $p(x \mid s_i)$ is the probability of observing x when the state is s_i. Let F_1 and F_2 be two fuzzy states that

335

realize a fuzzy partition of S, i.e., $\forall s \in S$, $\mu_{F_1}(s) + \mu_{F_2}(s) = 1$. Only two fuzzy states are considered for convenience.

Generally, the a priori probabilities $p(s_i)$ are not known. They can be obtained from fuzzy prior information which is described by a fuzzy statement M. Only bounds on

$$P(M) = \sum_{i=1}^{n} \mu_M(s_i) p(s_i)$$

are available, say α_1 and α_2. The $p(s_i)$ are calculated using the principle of maximum entropy, i.e., they are solutions of the problem

$$\text{maximize} \quad -\sum_{i=1}^{n} p(s_i) \ln(s_i)$$
$$\text{subject to} \quad \alpha_1 \leqslant P(M) \leqslant \alpha_2.$$

The probability of the fuzzy state F_k when x is observed is

$$P(F_k \mid x) = \frac{P(x, F_k)}{P(x)} = \frac{\sum\limits_{i=1}^{n} \mu_{F_k}(s_i) p(x \mid s_i) p(s_i)}{P(x \mid F_1) P(F_1) + P(x \mid F_2) P(F_2)}$$

with

$$P(x \mid F_k) = \frac{P(x, F_k)}{P(F_k)} \quad \text{and} \quad P(F_k) = \sum_{l=1}^{n} \mu_{F_k}(s_i) p(s_i).$$

In these formulas $p(s_i)$, $p(x \mid s_i)$, $\mu_{F_1}(s_i)$, and $\mu_{F_2}(s_i)$ are assumed known for all $i = 1, n$.

The discrimination of a fuzzy state can be performed using the Bayes acceptance rule (extended to fuzzy states): F_1 is chosen iff $P(F_1 \mid s) > P(F_2 \mid s)$ and conversely. This rule corresponds to the minimization of the probability P_e of discrimination error. When the observation is made of a finite sequence of independent elementary observations $(x_1, \ldots, x_m) = x(m)$, Asai et al. (1977) give upper bounds for P_e. They have pointed out that when $m \to \infty$, P_e no longer converges to 0 in average value as in the nonfuzzy case. This fact is interpreted as follows: when discriminating fuzzy states, there is uncertainty in the meaning of the fuzzy states in addition to probabilistic uncertainty.

Lastly, the authors provide a rule for deciding when to stop the observations. Let H denote the entropy; we have

$$H(F_1, F_2 \mid x(m))$$
$$= -\left[P(F_1 \mid x(m)) \ln(P(F_1 \mid x(m))) + P(F_2 \mid x(m)) \ln(P(F_2 \mid x(m))) \right]$$

and

$$H(p\,|\,x(m)) = -\sum_{i=1}^{n} p(s_i\,|\,x(m))\ln(p(s_i\,|\,x(m))).$$

Asai *et al.* (1977) have proved that if

$$\max(P(F_1\,|\,x(m)), P(F_2\,|\,x(m))) \geq \max_i p(s_i\,|\,x(m)),$$

then $H(F_1, F_2\,|\,x(m)) < H(p\,|\,x(m))$, which means that the probabilistic uncertainty is greater than the uncertainty due to the fuzzy states. It is then worth getting new information x_{m+1} before discriminating F_1 and F_2.

B. REPRESENTATION OF CAUSALITY BY FUZZY RELATIONS

The use of fuzzy sets for medical diagnosis was first suggested by Zadeh (1969).

An approach to the modeling of medical knowledge by fuzzy relations is described by Sanchez (1977a). Let X be a set of symptoms, S a set of diagnoses, and $\mathcal{P}$ a set of patients. Two fuzzy relations are assumed to be given, namely Q on $\mathcal{P} \times X$ and T on $\mathcal{P} \times S$. Q expresses the fuzzy symptoms of the patients and R the fuzzy diagnoses given by a physician. In order to represent the medical knowledge inferred from Q and T, Sanchez proposes determining the greatest fuzzy relation R (in the sense of the usual fuzzy set inclusion) on $X \times S$ such that the proposition $R \overset{2}{\to} (Q \overset{2}{\to} T)$ is true (where $\overset{2}{\to}$ is Brouwerian implication, see III.1.B.c). This is equivalent to

$$\forall(x, s, p) \in X \times S \times \mathcal{P},$$

$$v\left(R \overset{2}{\to} \left(Q \overset{2}{\to} T\right)\right) = \mu_R(x,s)\,\alpha\,(\,\mu_Q(p,x)\,\alpha\,\mu_T(p,s)) = 1$$

where α is the operator introduced in II.1.G.a. Noting that for any propositions A, B, and C, we have

$$v\left(A \overset{2}{\to} B\right) = 1 \qquad \text{iff} \qquad v(A) \leqslant v(B)$$

and

$$v\left(A \overset{2}{\to} \left(B \overset{2}{\to} C\right)\right) = v\left((A \wedge B) \overset{2}{\to} C\right),$$

($\wedge$ denotes conjunction in the sense of III.1.B.b.α), we deduce

$$\forall(x, s, p) \in X \times S \times \mathcal{P}, \quad \min(\mu_R(x,s), \mu_Q(p,x)) \leqslant \mu_T(p,s),$$

i.e., $Q \circ R \subseteq T$. The greatest fuzzy relation R satisfying this inclusion is $R = Q^{-1} @ T$ (see II.3.E.a) which yields the medical knowledge associated with Q and T. Given a patient p having a fuzzy symptom $\tilde{x}$, a fuzzy set on X, the automated fuzzy diagnosis will be

$$\tilde{s} = \tilde{x} \circ (Q^{-1} @ T).$$

When the diagnosis is not satisfactory, the medical knowledge (R) can be improved by enlarging the set of patients diagnosed by a physician. Other formulations of medical diagnosis models can be found in Sanchez (1977b).

An alternative approach to diagnosis with fuzzy relations has been proposed by Tsukamoto and Terano (1977). They have illustrated their scheme of diagnosis on the detection of car troubles. S is now a set of possible failures, and X is still the set of symptoms. Let R be a fuzzy relation on $S \times X$ that models the causal link between failures and existence of symptoms; another fuzzy relation T on $S \times X$ reflects the causal link between failures and *observed* symptoms ($T \subseteq R$). Let $\tilde{s}$ be a fuzzy failure and $\tilde{x}$ a fuzzy symptom. The causality between failure and symptoms is expressed by the logical propositions

$$\forall x \in X, \quad P_1(x): \tilde{x}(x) \Rightarrow (\exists s \in S, (R(s,x) \wedge \tilde{s}(s)))$$
$$\forall s \in S, \quad \forall x \in X, \quad P_2(s,x): \tilde{s}(s) \Rightarrow \tilde{x}(x)$$

where $\tilde{x}(s), \tilde{s}(s), R(s,x)$ are predicates (such that

$$v(\tilde{x}(x)) = \mu_{\tilde{x}}(x), \qquad v(\tilde{s}(s)) = \mu_{\tilde{s}}(s), \qquad v(R(s,x)) = \mu_R(s,x));$$

$\Rightarrow$ denotes implication in the sense of III.1.B.b.β, $\wedge$ conjunction in the sense of III.1.B.b.α. $P_1(x)$ means that if a symptom x is observed, then at least one failure among those that cause s has occurred. Consequently, $v(P_1(x)) = 1$. $P_2(s,x)$ expresses that if a failure s occurs, then a symptom x is observed. The truth value of $P_2(s,x)$ is not necessarily 1, but greater than or equal to $\mu_T(s,x)$. The fuzzy propositions $P_1(x)$ and $P_2(s,x)$ translate into

$$\forall x \in X, \quad P_1(x): 0 \leqslant \mu_{\tilde{x}}(x) \leqslant \max_{s \in S} \min(\mu_{\tilde{s}}(s), \mu_R(s,x)),$$
$$\forall s \in S, \quad \forall x \in X, \quad P_2(s,x): 0 \leqslant \mu_{\tilde{s}}(s) \leqslant \min(1, 1 - \mu_T(s,x) + \mu_{\tilde{x}}(x)).$$

From knowledge of $\tilde{x}$, R, and T a fuzzy failure $\tilde{s}$ can be deduced by solving the system

$$\tilde{x} \subseteq \tilde{s} \circ R, \quad \forall s \in S, \quad \mu_{\tilde{s}}(s) \leqslant \min_{x \in X} (1, 1 - \mu_T(s,x) + \mu_{\tilde{x}}(x)).$$

In II.3.E.c a method for obtaining the solutions of the equation $\tilde{x} = \tilde{s}^* \circ R$ is described. These solutions, when they exist, are constructed from a set of

Φ-fuzzy sets $\{\Phi_i, i = 1, r\}$ by choosing the value of $\mu_{\tilde{s}*}(s)$ in $\mu_{\Phi_i}(s) = [\alpha_i(s), \beta(s)]$ provided that the same Φ_i is used to characterize $\tilde{s}*$. Noting that

$$\forall \tilde{s} \supseteq \tilde{s}*, \qquad \tilde{s} \circ R \supseteq \tilde{s}* \circ R = \tilde{x},$$

the exact intervals for choosing $\mu_{\tilde{s}}(s)$ such that $\tilde{x} \subseteq \tilde{s} \circ R$ are $[\alpha_i(s), 1]$. These intervals are then reduced by applying condition $P_2(s, x)$.

Remark   An artificial intelligence approach to diagnosis problems using fuzzy concepts has been outlined by Wechsler (1976). It is a medical expert system with the characteristics:

"the medical knowledge is represented procedurally." (i.e., contained in programs rather than in declarative structures);

it uses procedures which deal explicitly with statistically dependent symptoms through use of logical combination;

new information is added "via change or extension of procedure rather than through building a large data base to improve the statistical decision rules";

"inexact concepts (multi-valued) are allowed so as to deal with degrees of a symptom";

"the interpretation of inexactness is allowed to vary with context."

This approach is related to that developed by Shortliffe and Buchanan (NF 1975), i.e., the MYCIN system. The approximate reasoning used by MYCIN is based on measures of belief and disbelief (different from Shafer's (NF 1976) belief functions) rather than fuzzy set theory. Another related approach is that of Chilausky et al. (1976).

REFERENCES: PART IV, CHAPTER 7

Asai, K., Tanaka, H., and Okuda, T. (1977). On discrimination of fuzzy states in probability space. *Kybernetes* **6**, 185–192.

Chilausky, R., Jacobsen, B., and Michalski, R. S. (1976). An application of variable-valued logic to inductive learning of plant disease diagnostic rules. *Proc. IEEE Int. Symp. Multiple-Valued Logic, 6th*, pp. 233–240.

Sanchez, F. (1977a). Solutions in composite fuzzy relation equations: Application to medical diagnosis in Brouwerian logic. *In* "Fuzzy Automata and Decision Processes" (M. M. Gupta, G. N. Saridis, and B. R. Gaines, eds.), pp. 221–234. North-Holland Publ., Amsterdam. (Reference from II.3.)

Sanchez, E. (1977b). Inverses of fuzzy relations. Application to possibility distributions and medical diagnosis. *Proc. IEEE Conf. Decision Control, New Orleans* pp. 1384–1389. (Reference from II.3.)

Sanchez, E., and Sambuc, R. (1976). Relations floues. Fonctions Φ-floues. Application à l'aide au diagnostic en pathologie thyroidienne. *I.R.I.A. Med. Data Process. Symp., Toulouse*. (Reference from II.1.) Taylor & Francis.

Tsukamoto, Y., and Terano, T. (1977). Failure diagnosis by using fuzzy logic. *Proc. IEEE Conf. Decision Control, New Orleans* pp. 1390–1395.

Wechsler, H. (1976). A fuzzy approach to medical diagnosis. *Int. J. Bio-Med. Comput.* 7, 191–203. (Reference from IV.2.)

Zadeh, L. A. (1969). Biological application of the theory of fuzzy sets and systems. *In* "Biocybernetics of the Central Nervous System" (L. D. Proctor, ed.), pp. 199–212. Little, Brown, Boston, Massachusetts.

Chapter 8

FUZZY SETS IN THE IDENTIFICATION OF STRUCTURES

This chapter presents a few works that share the purpose of structural identification of systems, while based on fuzzy set theory. However, they are quite different in other respects. In the first section the construction of hierarchical models of organizations through fuzzy responses of a panel is considered. A learning method for the synthesis of a single-input, single-output system from knowledge of possible subsystems is then described. Lastly, first attempts at fuzzy grammar inference are reported.

A. FUZZY STRUCTURAL MODELING

Let S be a finite set of objects $s_i, i = 1, n$. These objects can be viewed as parts of a large system. The problem is to order S, i.e., to define how each part is related to others. The determination of the hierarchy underlying S is achieved by asking a panel of experts to supply the entries of an n by n relation matrix R, called the reachability matrix.

However, the process of collecting such data can become very long when n is a large number. To circumvent this difficulty, and also to avoid inconsistency in the data, Warfield (NF 1974a, b), has developed a computer-aided approach to the collection of binary entries of matrices describing the hierarchical structure of large systems. The main assumption is the transitivity of the relation obtained, which allows computation of entries from knowledge of others.

Warfield's method has been extended by Ragade (1976) to the case when the panel's answers are graded, i.e., R is a fuzzy reflexive, max–min transitive relation. $\mu_R(s_i, s_j)$ is the grade of dominance of s_i over s_j.

In a first phase S is partitioned in (nonfuzzy) clusters called subsystems, whose individual reachability matrix is known. In the second phase interconnection matrices between subsystems are filled.

First phase   The choice of an element in S, say s, is made. The panel must answer questions about the relations between s and the other objects. $S - \{s\}$ is then divided into four sets:

the lift set $L(s) = \{s_i \mid \mu_R(s, s_i) > 0\}$;
the feedback set $F(s) = \{s_i \in L(s) \mid \mu_R(s_i, s) > 0\}$;
the drop set $D(s) = \{s_i \mid \mu_R(s, s_i) = 0, \mu_R(s_i, s) > 0\}$;
the vacant set $V(s) = \{s_i \mid \mu_R(s, s_i) = \mu_R(s_i, s) = 0\}$.

The matrix is then arranged in block-triangular form:

		$L - F$	$F(s)$	s	$V(s)$	$D(s)$
$L(s)$ $\{$	$L - F$	R_{L-F}	0	0	0	0
	$F(s)$	$R_{F, L}$		K_F	0	0
	s	ρ_{L-F}	ρ_F	1	0	0
	$V(s)$	$R_{V, L-F}$	0	0	R_V	0
	$D(s)$	$R_{D, L}$		K_D	$R_{D, V}$	R_D

$\rho_{L-F}, \rho_F, K_F, K_D$ are supplied by the panel. Some blocks can be calculated by transitivity; we have, for instance,

$$\forall s_i \in F(s), \quad \forall s_j \in V(s), \quad \mu_R(s, s_j) = 0 \geqslant \min(\mu_R(s, s_i), \mu_R(s_i, s_j))$$

and

$$\mu_R(s_j, s) = 0 \geqslant \min(\mu_R(s_j, s_i), \mu_R(s_i, s));$$

hence, $R_{F, V} = 0 = R_{V, F}$. We also have .

$$\forall s_i \in D(s), \quad \forall s_j \in L(s), \quad \mu_R(s_i, s_j) \geqslant \min(\mu_R(s_i, s), \mu_R(s, s_j)) > 0.$$

Note that what is actually obtained are nonnull bounds on the $\mu_R(s_i, s_j)$ in $R_{D, L}, R_V, R_{D, V}$, and R_D. This process is iterated by choosing a new element in $L(s) - F(s)$ and partitioning this set as above, and so on, until drop sets, vacant sets, and nonfeedback parts of lift sets are singletons. $D(s)$ and $V(s)$ are similarly reduced.

Second phase Some lower off-diagonal blocks remain unknown, i.e., interconnection matrices. They are determined without extra information as follows. Let

$$R_{CC} = \begin{bmatrix} R_{AA} & 0 \\ R_{AB} & R_{BB} \end{bmatrix}$$

where R_{AB} is unknown. Using the transitivity of R_{CC}, i.e., $R_{CC} \circ R_{CC} = R_{CC}$, R_{AB} must satisfy

$$R_{AB} = (R_{AB} \circ R_{AA}) \cup (R_{AB} \circ R_{BB});$$

i.e., if R_{AA} is p by p, and R_{BB} q by q, and denoting by r_{ij} the entries of R_{CC},

$$\forall i > p, \quad \forall j \leqslant p, \quad r_{ij} = \max\left(\max_{k \leqslant p} \min(r_{ik}, r_{kj}), \max_{k > p} (r_{ik}, r_{kj}) \right)$$

$$= \max_k \min(r_{ik}, r_{kj}).$$

It is easy to see that by renaming the r_{ij}, the above equation can be formally written $z = z \circ T$ where z is a vector with pq components which are the r_{ij}, and T is a pq by pq matrix made up of R_{AA}, R_{BB}, and zeros. z is thus an eigenfuzzy set of T and can be calculated as in II.3.E.d. Another solution method is given in Ragade (1976).

The assumption of transitivity can be relaxed. Tazaki and Amagasa (1977a) define the semitransitivity of R by

$$\forall i, \quad \forall j, \quad \forall k, \quad \text{if } m_{ij} = \max_k \min(\mu_R(s_i, s_k), \mu_R(s_k, s_j)) \geqslant \theta$$

then $\mu_R(s_i, s_j) \geqslant m_{ij}$ where $\theta \in]0, 1]$ is a given threshold. The authors describe a procedure for constructing a semitransitive matrix (called a semireachability matrix) and deduce the structure of S.

When R is transitive and reflexive, a partial order on the elements of S is easily obtained as shown in II.3.D.c. When it is only semitransitive, a set of disjoint partial orders can still be obtained; this set depends on the value of θ (see Tazaki and Amagasa, 1977a).

B. HEURISTIC STRUCTURE SYNTHESIS

A system synthesis problem is now defined, and a heuristic fuzzy approach to this problem is proposed, following Tazaki and Amagasa (1977b). Let S be a single-input, single-output system made of n subsystems $S_1, \ldots, S_n$, as in Fig. 1.

Each subsystem $i \neq 1$ has an input x_i that is one of the outputs y_j, $j = 1, n - 1$, say $x_i = y_{j(i)}, j(i) < n$. The output y_i, $i = 1, n$, depends on the

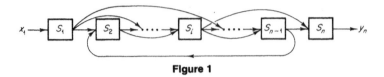

Figure 1

input x_i and a local decision d_i:

$$y_i = f_i(x_i, d_i).$$

No subsystem is connected to the input subsystem S_1; the output of S_n is not connected to any subsystem's input. n is assumed known and is the maximal number of connected subsystems in S. A performance criterion for S is selected of the form

$$\mathfrak{T}(x, y, d) = \sum_{i=1}^{n} \mathfrak{T}_i(x_i, y_i, d_i)$$

where $\mathfrak{T}_i$ is the objective function of the ith subsystem. The synthesis problem is stated as follows: find the admissible connections and the decisions that minimize $\mathfrak{T}$ under the constraints:

(1) $y_i = f_i(x_i, d_i)$, $i = 1, n$;
(2) $\exists j(i) < n, x_i = y_{j(i)}$, $i = 2, n$.

The heuristic algorithm solving this problem proceeds in four steps:

Step 1   Optimize each subsystem, for a nominal admissible input x_i, while relaxing (2), and determine the output and decision vectors in each subsystem.

Step 2   Calculate the matrix of a fuzzy relation R that expresses the discrepancy between the outputs and the inputs to which they are connected. Each term r_{ij} is of the form

$$r_{ij} = w_m \mu^1(x_i, y_j) + (1 - w_m) \mu^2(\mathfrak{T}_i, \mathfrak{T}_j), \qquad i = 1, n, \quad j = 1, n,$$

where (see Tazaki and Amagasa, 1977b) m indicates that $m - 1$ iterations have been run. $\mu^1(x_i, y_j)$ decreases with $|x_i - y_j|$ and is 1 for $x_i = y_j$. It evaluates the suitability of coupling the output of S_j with the input of S_i. $\mu^2(\mathfrak{T}_i, \mathfrak{T}_j)$ decreases with $\mathfrak{T}_i + \mathfrak{T}_j$ and valuates how S_i and S_j are optimized. $w_m \in [0, 1]$ is a weight calculated at each iteration by a reinforcement rule (see 5.A):

$$w_m = \alpha_m w_{m-1} + (1 - \alpha_m)\lambda \qquad \text{with} \quad \alpha_m = 1 - \frac{|\mathfrak{T}_m - \bar{\mathfrak{T}}_m|}{\mathfrak{T}_m} \qquad (\in\,]0, 1[)$$

$\mathfrak{I}_m$ is the value of $\mathfrak{I}$ at the mth iteration and $\bar{\mathfrak{I}}_m$ the average value of $\mathfrak{I}$ from the beginning of the run;

$$\lambda = 1 \quad \text{iff} \quad \mathfrak{I}_m < \bar{\mathfrak{I}}_m, \qquad \text{hence} \quad w_m > w_{m-1};$$

$$\lambda = 0 \quad \text{iff} \quad \mathfrak{I}_m \geq \bar{\mathfrak{I}}_m; \qquad \text{hence} \quad w_m < w_{m-1}.$$

And μ^2 is given a priority over μ^1.

Step 3 The set of admissible subsystems is a fuzzy set on S, say A_m at iteration m. A_0 is given a priori. The new admissible fuzzy set is

$$A_m = A_{m-1} \circ R.$$

The new structure of S is determined by two rules:

connect the input of S_i to the output of S_j such that

$$\mu_{A_m}(S_i) = \min(\mu_{A_{m-1}}(S_j), r_{ji});$$

when j is not unique, k is chosen such that

$$\max(\mu_{A_{m-1}}(S_k), r_{ki}) = \max_j \max(\mu_{A_{m-1}}(S_j), r_{ji}).$$

Conditions on R and A_0 are given in Tazaki and Amagasa (1977b) to make sure of the uniqueness of k.

Step 4 According to the structure determined in step 3, adjust the inputs of the connected subsystems; then return to step 1. The modification of the inputs of the connected subsystems is carried out through a reinforcement algorithm (see Tazaki and Amagasa, 1979). The inputs of the nonconnected subsystems remain unchanged. The algorithm stops when step 1 gives the same results as the preceding iteration.

Tazaki and Amagasa (1977b) claim that their method is more convenient than combinatorial enumerative or variational approaches from the point of view of computation time, and usually yields the optimal solution for small-sized systems.

Remark: A method for the structural decomposition of large dynamic systems is proposed by Dufour et al. (1976). It is based on the derivation of a partition-tree (see II.3.C.b.) from a similarity relation on a set of characteristic parameters. This relation is built from observations.

C. FUZZY GRAMMAR INFERENCE

The problem of grammatical inference is very important in syntactic pattern recognition. It consists in finding a formal grammar that generates a language containing a prescribed finite set of strings, and sometimes not containing another finite set of strings; however, the latter constraint is not considered in the following. Grammatical inference methods have been developed for ordinary regular and context free grammars, tree grammars, and unambiguous stochastic grammars. (See, e.g., Fu and Booth, NF 1975, for a survey.)

Inference methods for deriving fuzzy grammars would enhance the applicability of fuzzy language theory to syntactic pattern recognition problems. To date only two approaches exist in the literature of fuzzy sets, a reinforcement algorithm and an enumeration method.

a. Learning of Fuzzy Formal Language   (Tamura and Tanaka, 1973)

Let $G_1 = (V_N, V_T, P_1, S)$ be a fuzzy grammar where V_N is a set of nonterminals, V_T a set of terminals, P_1 a set of valued productions, and S is the starting symbol. V_N, V_T, and P_1 are chosen beforehand to cover a sufficient range. The set R of productions on which P_1 is a fuzzy set may contain improper production rules.

At time n the fuzzy grammar is $G_n = (V_N, V_T, P_n, S)$ where

$$P_n = \sum_R \mu_n(u \to v)/(u \to v), \qquad u, v \in V^* = (V_N \cup V_T^*).$$

For the purpose of learning, a finite set of strings $K_n = \{x_{n_i} | i = 1, N_n\}$ is given. Each string of K_n is parsed by G_n. To make the parsing possible, G_1 is assumed recursive. Let $Q(x_{n_i})$ be the subset of productions in R that can be used to generate x_{n_i}. The subset of productions that can be used to generate the strings in K_n is thus

$$Q(K_n) = \bigcup_{i=1, N_n} Q(x_{n_i}) \subseteq R.$$

The learning process consists in reinforcing the production in $Q(K_n)$ and weakening the others. More precisely, P_{n+1} is defined by

$$\mu_{n+1}(u \to v) = \alpha \mu_n(u \to v) + (1 - \alpha)\theta_n(u \to v), \qquad \alpha \in \,]0, 1[$$

where θ_n is the characteristic function of $Q(K_n)$.

N.B.: When G_n is ambiguous, one may wish to reinforce the productions of only one of the derivation chains that yield each x_{n_i}. The choice is made by an external supervisor. Denoting

$$\mathcal{L}_\lambda(G_n) = \{ x \in V^* | \, \mu_{G_n}(x) \geqslant \lambda \} \quad \forall \lambda \in [0, 1]$$

and by $\mathcal{L}(A)$ the nonfuzzy language generated by (V_N, B_T, A, S) with $A \subseteq R$, it can be proved that (see Tamura and Tanaka, 1973) if $\forall n \in \mathbb{N} -$

$\{0\}$, $K_n = K$, then $\forall \lambda \in]0, 1[$, $\exists N(\lambda)$ such that $\forall n \geqslant N(\lambda)$, $\mathcal{L}_\lambda(G_n) = \mathcal{L}(Q(K))$.

This result means that the only productions that remain valid are those in $Q(K)$, when K is always used as a training set. However, generally speaking, it is not clear what training set is adequate to intensify only a specified set of productions. Note that since θ_n is a characteristic function, the result is a nonfuzzy grammar, having $Q(K)$ as a production set.

b. A Combinatorial Approach (Lakshmivarahan and Rajasethupathy, 1978)

A slightly different kind of grammatical inference is now considered. Let $\Gamma = (V_N, V_T, R, S)$ be a nonfuzzy context-free grammar, and $\mathcal{L}(\Gamma)$ the nonfuzzy language generated by Γ. Let $E = \sum_i \mu_i / x_i$ be a given finite-support fuzzy set of V_T^*. The problem is to find a fuzzy set P of R, i.e., a fuzzy grammar $G = (V_N, V_T, P, S)$ such that $E \subseteq \mathcal{L}(G)$.

Let $k = |R|$. There are at most k different valuations for the productions, say $\rho_1, \rho_2, \ldots, \rho_k$. For any subset $X = \{P_{i_1}, \ldots, P_{i_r}\}$ of R, where P_{i_j} denotes the name of a production, let us define

$$C(X) = \min(\rho_{i_1}, \ldots, \rho_{i_r})$$

$$E_X = \left\{ x \in \mathcal{L}(\Gamma) \,|\, S \underset{X}{\Rightarrow} x \right\};$$

which is the set of strings derived by applying at least once each of the productions of X. Note that when Γ is not ambiguous, $\forall x_1, x_2 \in E_X$, $\mu_G(x_1) = \mu_G(x_2) = C(X)$.

$\mathcal{L}(\Gamma)$ is now decomposed into equivalence classes.

When Γ is not ambiguous, each string of $\mathcal{L}(\Gamma)$ belongs to only one set E_X since x can be derived only by one chain. The equivalence is thus defined by $\forall x, y \in \mathcal{L}(\Gamma)$, $x \sim y$ iff $X(x) = X(y)$ where $X(x)$ and $X(y)$ denote the set of productions necessary to derive x and y, respectively. The equivalence class of x is $E_{X(x)}$.

When Γ is ambiguous a string $x \in \mathcal{L}(\Gamma)$ may belong to several E_X. Hence $\sim$ is no longer an equivalence relation.

In the nonambiguous case the inference problem is solved as follows:

find all the subsets X^i of productions of Γ that give birth to complete derivation chains

$$S \to \alpha_1 \to \cdots \to \alpha_n \to x;$$

let M be the number of the X^i;

choose in each of the subsets E_{X^i} one string x^i whose length is $|X^i|$;

we thus obtain M strings to which are assigned weights $\mu_i, i = 1, M$ supposedly consistent, i.e., the system of nonlinear equations $\forall x^i$, $C(X^i) = \mu_i$ has a solution;

the valuations of the productions are the ρ_j, the solutions of the above system; generally, ρ_j is not unique.

N.B.:   1.   The sample set of strings E must have at most M elements to use the above algorithm. This algorithm indicates what kind of sample sets are worth considering.

2.   Some hints for dealing with the ambiguous case are proposed in Lakshmivarahan and Rajasethupathy (1978).

REFERENCES

Dufour, J., Gilles, G., and Foulard, C. (1976). Applications de la théorie des sous-ensembles flous à la partition des systèmes dynamiques complexes. *C. R. Acad. Sci.* Paris, Sér. A, **282**, 491–494.

Lakshmivarahan, S., and Rajasethupathy, K. S. (1978). Consideration for fuzzifying formal languages and synthesis of fuzzy grammars. *J. Cybern.* **8**, 83–100.

Ragade, R. K. (1976). Fuzzy interpretive structural modeling. *J. Cybern.* **6**, 189–211.

Tamura, S., and Tanaka, K. (1973). Learning of fuzzy formal language. *IEEE Trans. on Syst., Man Cybern.* **3**, 98–102.

Tazaki, E., and Amagasa, M. (1977a). Structural modeling in a class of systems using fuzzy sets theory. *Proc. IEEE Conf. Decision Control, New Orleans* pp. 1408–1413. [Also in *Int. J. Fuzzy Sets Syst.* **2**, No. 1, 87–104 (1979).]

Tazaki, E., and Amagasa, M. (1977b). Heuristic structure synthesis in a class of systems using a fuzzy automaton. *Proc. IEEE Conf. Decision Control, New Orleans* pp. 1414–1418. [Also in *IEEE Trans. Syst., Man Cybern.* **9**, No. 2, 73–79 (1979).]

Chapter 9

FUZZY GAMES

Like most system-oriented theories, game theory did not escape fuzzification, although the number of attempts is still rather small. Fuzzy games are intended to model conflict situations with imprecise information. Payoffs, strategies, coalitions, etc. may be fuzzy.

For instance, we can consider a two-person zero-sum game with a fuzzy payoff. Let S_1 and S_2 be the sets (assumed finite for simplicity) of the strategies of player I and of player II, respectively. Let $\tilde{P}(s_1, s_2)$, a fuzzy set on $\mathbb{R}$, denote the fuzzy payoff to player I when he chooses s_1 in S_1 and player II chooses s_2 in S_2. Player I wishes to maximize $\tilde{P}(s_1, s_2)$ and player II wishes to maximize $-\tilde{P}(s_1, s_2)$. Irrespective of what player II does, player I may secure for himself at least

$$\widetilde{\max_{s_1 \in S_1}} \ \widetilde{\min_{s_2 \in S_2}} \ \tilde{P}(s_1, s_2) = \tilde{v}_1.$$

Similarly, player II may secure for himself at least

$$\widetilde{\min_{s_2 \in S_2}} \ \widetilde{\max_{s_1 \in S_1}} \ \tilde{P}(s_1, s_2) = \tilde{v}_2.$$

$\tilde{P}(s_1, s_2)$, $\tilde{v}_1$, and $\tilde{v}_2$ may be viewed as possibility distributions on the actual value of the payoff. $\mathrm{hgt}(\tilde{v}_1 \cap \tilde{v}_2)$ values the possibility of the existence of a saddle point. Note that, for instance, the set of secure strategies s_1 of player I such that $\widetilde{\min}_{s_2 \in S_2} \tilde{P}(s_1, s_2) = \tilde{v}_1$ is a fuzzy set when the payoff function is fuzzy (see II.4.B.c).

In the remainder of this very short chapter a survey of the current literature is provided.

Orlovsky (1977) has considered the following two-person game. The sets of the possible (feasible) strategies of player I and player II are two fuzzy sets A_1 and A_2 on S_1 and S_2, respectively. But the two payoff functions P_1 (for player I) and P_2 (for player II) from $S_1 \times S_2$ to $\mathbb{R}$ are assumed nonfuzzy. Each player is here supposed to know the strategy chosen by the other player. Thus, player I maximizes his payoff $P_1(s_1, s_2)$ over his fuzzy strategy set A_1 for a given s_2. The fuzzy choice of player I is given by the membership function (see II.4.B.b)

$$\mu_I(s_1, s_2) = \begin{cases} \mu_{A_1}(s_1) & \text{if } s_1 \in \bigcup_{\lambda > 0} N(\lambda, s_2), \\ 0 & \text{otherwise,} \end{cases}$$

where

$$N(\lambda, s_2) = \left\{ s_1 \in S_1, P_1(s_1, s_2) = \sup_{s_1' \in D_\lambda^1} P_1(s_1', s_2) \right\},$$

$$D_\lambda^1 = \left\{ s_1, s_1 \in S_1, \mu_{A_1}(s_1) \geqslant \lambda \right\}.$$

The fuzzy choice of player II $\mu_{II}(s_1, s_2)$ is symmetrically defined. The *fuzzy equilibrium solution* is then introduced as a fuzzy set on $S_1 \times S_2$ whose membership function is

$$\mu_e(s_1, s_2) = \min(\mu_I(s_1, s_2), \mu_{II}(s_1, s_2)).$$

The fuzzy payoff $\tilde{P}_I$ of player I at the fuzzy equilibrium is (see II.4.B.b.β)

$$\mu_{\tilde{P}_I}(z) = \sup_{(s_1, s_2) \in P_1^{-1}(z)} \mu_e(s_1, s_2)$$

where $P_1^{-1}(z) = \{(s_1, s_2) \in S_1 \times S_2, P_1(s_1, s_2) = z\}$. $\tilde{P}_{II}$ is calculated similarly.

Ragade (1976) deals with two-person games where the preferences of the players are fuzzy. Let S_i be the strategy set (nonfuzzily restricted) of player i, $i = 1, 2$. Let $\mathcal{Q} = S_1 \times S_2$. For each player, M_i denotes a reflexive fuzzy preference relation on $\mathcal{Q} \times \mathcal{Q}$ ($\mu_{M_i}[(s_1, s_2), (s_1', s_2')] \in [0, 1]$, $i = 1, 2$). An element $(s_1, s_2) \in \mathcal{Q}$ is called an outcome. An outcome (s_1^*, s_2^*) is said to be λ_1-*rational* for player 1 if

$$\forall s_1 \in S_1, \quad \mu_{M_1}[(s_1, s_2^*), (s_1^*, s_2^*)] \geqslant \lambda_1.$$

The set of all λ_1-rational outcomes is denoted R_{1, λ_1}. R_{2, λ_2} is similarly defined. Since rational outcomes for both players correspond to equilibria, $R_{1, \lambda_1} \cap R_{2, \lambda_2}$ may be viewed as the set of outcomes in λ_1, λ_2-*equilibria*.

Butnariu (1978) questions the usual safest (worst-case) strategy rule of 2-person games. For a given player, the set of feasible strategies is fuzzy. It is obtained from the composition of a fuzzy relation on $S_1 \times S_2$ (expressing the preferences of the player conditioned by the other player's strategies) and the fuzzy set representing the estimation of the behavior of the other player.

Blaquière (1976), in the framework of n-person dynamic games with coalition, has introduced the notion of fuzzy optimality with respect to a set of players. The concepts of diplomacy and fuzzy diplomacy (in order to take into account the fact that "any subset of the set of players can try to improve its payoff by switching from one set of coalitions to another as time evolves") are also presented in the same reference.

Aubin (1974a, b, 1976) has introduced the concept of a *fuzzy core* in game theory. Let us consider an n-person game. Let N be the set of players and $\mathcal{Q}$ be a family of coalitions A (i.e., of subsets of N). $S(A)$ is the set of multistrategies of A. Let $\mathcal{P}_A = \{P_A^i\}_{i \in A}$ be the set of the real-valued loss functions of players i behaving as members of A. Let $\bar{s} \in S(N)$ be a multistrategy of the whole set of players. It is a *weak equilibrium* if

$$\exists \lambda \in M^n = \left\{ \lambda \in \mathbb{R}^{n*}, \forall i \in N, \lambda^i \geqslant 0 \text{ and } \sum_{i=1}^{n} \lambda^i = 1 \right\}$$

such that, for all coalitions A,

$$\sum_{i \in A} \lambda^i P_N^i(\bar{s}) \leqslant \inf_{s_A \in S(A)} \sum_{i \in A} \lambda^i P_A^i(s_A).$$

It is an *equilibrium* if $\exists \lambda \in \mathring{M}^n = \{\lambda \in M^n, \lambda^i > 0, \forall i \in N\}$. The *core* $C(\{S(A), \mathcal{P}_A\}_{A \in \mathcal{Q}})$ is the set of multistrategies $s \in S(N)$ that are not *rejected* by any coalition $A \in \mathcal{Q}$. By definition, a coalition A rejects $s \in S(N)$ if it can find $s_A \in S(A)$ yielding to each player i participating in A a loss $P_A^i(s_A)$ strictly smaller than the loss $P_N^i(s)$. Note that any equilibrium belongs to the core; the converse is false.

In order to "shrink" the core by allowing more coalitions to form and reject strategies Aubin (1976) embeds the set $\mathcal{Q}$ of coalitions A into the set $\mathcal{F}$ of *fuzzy coalitions* (fuzzy subsets of N). The game becomes "a fuzzy game." A fuzzy core can be defined (see Aubin, 1976). It is possible to show that "any equilibrium belongs to the fuzzy core and that the fuzzy core is contained in the set of weak equilibria" (see Aubin, 1976).

Another game situation involving coalitions and fuzzy sets in an economics context is considered in Féron (1976).

REFERENCES

Aubin, J. P. (1974a). Coeur et valeur des jeux flous à paiements latéraux. *C. R. Acad. Sci.* Paris, Sér. A, **279**, 891–894.

Aubin, J. P. (1974b). Coeur et équilibres des jeux flous sans paiements latéraux. *C. R. Acad. Sci.* Paris, Sér. A, **279**, 963–966.

Aubin, J. P. (1976). Fuzzy core and equilibrium of games defined in strategic form. *In* "Directions in Large-Scale Systems" (Y. C. Ho and S. K. Mitter, eds.), pp. 371–388. Plenum, New York.

Blaquière, A. (1976). Dynamic games with coalitions and diplomacies. *In* "Directions in Large-Scale Systems" (Y. C. Ho and S. K. Mitter, eds.), pp. 95–115. Plenum, New York.

Butnariu, D. (1978). Fuzzy games: A description of the concept. *Int. J. Fuzzy Sets Syst.* **1**, No. 3, 181–192.

Féron, R. (1976). Economie d'échanges aléatoires floue. *C. R. Acad. Sci.* Paris, Sér. A, **282**, 1379–1382. (Reference from V).

Nurmi, H. (1976). On fuzzy games, *Eur. Meet. Cybern. Syst. Res., 3rd, Vienna*.

Orlovsky, S. A. (1977). On programming with fuzzy constraint sets. *Kybernetes* **6**, 197–201. (Reference from II.4.)

Ragade, R. K. (1976). Fuzzy games in the analysis of options. *J. Cybern.* **6**, 213–221.

Chapter *10*

FUZZY SETS
AND CATASTROPHES

Zadeh's fuzzy set theory and catastrophe theory (Thom, NF 1973, NF 1974) appeared almost at the same time. However, they have been initially developed in very different frameworks. Each theory has encountered not only enthusiasm and approbation, but also criticism and even derision. The applications of both theories are concerned with system theory: approximate descriptions of complex processes for the former and models of discontinuous changes in the evolution of systems for the latter. It would be interesting to know whether both theories may be used concurrently. One may also be tempted to fuzzify catastrophe theory, but it is not clear that this would be fruitful. In fact, very few works dealing with these questions have been published to date. Thus, this chapter is somewhat different from the others in Part IV. No definition or result is provided. We just intend here to give some hints.

In order to have in mind the basic vocabulary, we begin with a brief description of one of the most widely used elementary catastrophes, the cusp (also called the Riemann–Hugoniot catastrophe). A potentiallike function

$$V(x; p, q) = \tfrac{1}{4} x^4 + \tfrac{1}{2} q x^2 + px$$

is supposed to be minimized as the system evolves, i.e.,

$$\frac{dx}{dt} = -\frac{\partial V}{\partial x} = -(x^3 + qx + p).$$

The set of equilibrium points $x^3 + qx + p = 0$ is the *manifold*. x is referred to as the state variable and p and q as the parameters (considered as

353

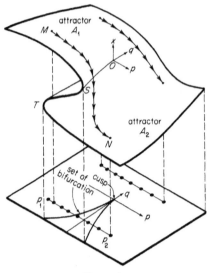

Figure 1

"slow" variables, while x is a "fast" one). The manifold and the set of bifurcation (projection of the folds of the manifold) $4q^3 + 27p^2 = 0$ are pictured in Fig. 1. There are two attractors (which correspond to minima of V) A_1 ($x > 0$, limited by the edge OS) and A_2 ($x < 0$, limited by the edge OT). If p goes from p_1 to p_2, the state trajectory follows the manifold smoothly until it reaches the edge OS of A_1, then there is a catastrophic jump (i.e., a discontinuity in the behavior of the system) to the lower attractor A_2 before continuing to N. But a path with fixed positive q avoids the fast jump. A presentation of the other types of elementary catastrophes (for other V) can be found in Thom (NF 1974).

In practical situations we never have to know explicitly what V is or what it represents; we need principally to know the type of the catastrophe (for instance, a cusp catastrophe), i.e., a qualitative description of the phenomenon to be modeled. The trajectories will then remain in a neighborhood of the theoretical manifold rather than on it. (See, e.g., Dixon, NF 1977.) Thus, we may be led to consider "fuzzy manifolds"—for instance, "fuzzy cusps" [as Kokawa et al. (1975, 1977) for modeling of the human concept-formational process: jump in the degree of confidence, in people's minds] with "fuzzy attractors." Besides, on ordinary manifolds boundaries between various attractors may be fuzzy.

Catastrophe theory may be very useful for modeling systems where humanistic components play an important role (for instance, the behavior of drivers in a traffic flow as modeled by Furutani (NF 1976a, b, NF 1977).

As pointed out by Flondor (1977), a classical notion of fuzzy set theory such as preference may be viewed as "a moment in the fight between different attractors." A different way of connecting catastrophe theory and fuzzy sets lies in the introduction of "catastrophic" membership functions (Zwick, *et al.*, 1978).

N.B.: Kitagawa (1975) has proposed using fuzzy topology as a basis for the introduction of fuzziness in catastrophe theory.

REFERENCES

Flondor, P. (1977). An example of a fuzzy system. *Kybernetes* **6**, 229.

Kitagawa, T. (1975). Fuzziness in informative logics. *In* "Fuzzy Sets and Their Applications to Cognitive and Decision Processes" (L. A. Zadeh, K. S. Fu, K. Tanaka, and M. Shimura, eds.), pp. 97–124. Academic Press, New York.

Kokawa, M., Nakamura, K., and Oda, M. (1975). Hint effect and a jump of logic in a decision process. *Syst.—Comput.—Controls* **6**, No. 3. (Reference from V, 1975b.)

Kokawa, M., Nakamura, K., and Oda, M. (1977). Fuzzy-theoretic and concept-formational approaches to inference and hint-effect experiments in human decision processes. *Proc. IEEE Conf. Decision Control, New Orleans* pp. 1330–1337. [Also in *Int. J. Fuzzy Sets Syst.* **2**, No. 1, 25–36 (1979).] (Reference from V, 1977b.)

Zwick, M., Schwartz, D. G., and Lendaris, G. G. (1978). Fuzziness and catastrophe. *Proc. Int. Conf. Cybern. Soc., Tokyo* pp. 1237–1241.

Part V

A SURVEY OF
POTENTIAL APPLICATIONS

The previous parts have developed at length the mathematics of fuzzy sets and presented various fuzzy approaches to system-oriented problems. Indeed, this is the main purpose of this book. Nevertheless, a review of the present fields of application is now provided in order to give examples of works where fuzzy concepts have been used.

Although some applications are actually based on real world data experiments, many others are not; and very often comparison with other techniques has not been made. Fuzzy set theory seems potentially promising; but, because of its novelty, the success of its applications is not completely established yet.

So far, fuzzy set theory seems to have been applied mainly to scientific areas where man is somewhat involved. However, there are some noticeable exceptions: the detection of hazards in switching circuits (Hughes and Kandel, Reference from III.1), functional approximation (Pavlidis and Chang, 1977), and quantum mechanics. Fuzzy logic for quantum mechanics is discussed by Almog (1978a, b) and Giles (1974), while Prugovečki (1973, 1974, 1975, 1976a, b, 1977) has introduced the notion of fuzzy events in the theory of measurement of observables. See also Ali and Doebner (1976, 1977).

Apart from these exceptions the applications concern the following fields: artificial intelligence and robotics, image processing, speech recognition, biological and medical sciences, control, applied operations research, economics and geography, sociology, psychology, linguistics, semiotics, and some more-restricted topics.

A. ARTIFICIAL INTELLIGENCE AND ROBOTICS

Since "Artificial Intelligence is the study of ideas which enable computers to do the things that make people seem intelligent" (Winston, NF 1977) and since, according to Zadeh (Reference from III.3, 1973), "The key elements in human thinking are not numbers but labels of fuzzy sets," the interaction between artificial intelligence and fuzzy set theory seems quite natural. Paradoxically, fuzzy artificial intelligence has not many proselytes yet. Some rationale and motivations may be found in Uhr (1975), Goguen (1975), and Hanakata (1974). Using fuzzy concepts, Goguen (1975) hopes to construct robust systems, i.e., systems "able to respond without program modification to slightly perturbed, or to somewhat inexactly specified situations." As a tool for modeling natural language (see IV.2.B), fuzzy set theory may be useful in man–machine communication. The problem of guiding a robot using fuzzy instructions has been considered by Gershman (1976) and Uragami et al. (1976). A system able to "understand" sentences that fuzzily designate objects has been developed by Shaket (1976). Rhodes and Klinger (1977a, b) have implemented an interactive flexible language (modeling hedges, as in fuzzy set theory—see IV.2.B.b) to modify graphic facial images. Schek (1977) has proposed an interactive robust system that is able to recognize slightly misspelled keywords, but the use of fuzzy concepts is limited to the idea of similarity. More theoretical is PRUF (Zadeh, 1977), which is a broad attempt to model semantic aspects of natural languages. Quite different, although related to robotics, is the work by Saridis and Stephanou (1977a, b; Stephanou and Saridis, 1976) where fuzzy automata and fuzzy grammars are employed for coordination and task organization in the hierarchical control of prosthetic devices.

N.B.: 1. For the application of fuzzy sets to computer science languages, see IV.2.A.

2. Becker (1973) seems to be the first to apply fuzzy sets to computer-aided design.

B. IMAGE PROCESSING AND SPEECH RECOGNITION

Works in image processing using fuzzy sets are rather scarce. A fuzzy relaxation approach to scene labeling can be found in Rosenfeld et al. (1976). The problem is to identify objects in a scene by using relationships among these objects to reduce or eliminate ambiguity. Nakagawa and Rosenfeld (1978) use local max and min operations for noise removal on gray-scale pictures. Jain and Nagel (1977) detect moving objects in a sequence of images by means of heuristic rules involving fuzzy texture indices on the level of gray of the pixels. However, the use of fuzzy set theory is only a marginal aspect of these works. Lastly, a computer aided system for art-oriented fuzzy image generation is described by Makarovitsch (1976, 1977).

The first attempt to apply fuzzy set theory to speech recognition seems to be that of Brémont (1975); see also Brémont and Lamotte (1974), Mas and Lamotte (1976). Speech understanding systems based on fuzzy relations were also proposed by De Mori and Torasso (1976) for lexical classification and by Coppo and Saitta (1976) for semantic analysis of sentences. Another problem (vowel and speaker recognition) is considered in Pal and Majumber (1977, 1978a, b), Pal *et al.* (1978).

C. BIOLOGICAL AND MEDICAL SCIENCES

The possibility of applying fuzzy set theory to biological and medical sciences was first discussed by Zadeh (Reference from IV.7). Fuzzy clustering algorithms have been used in the classification of EEG patterns (Adey, 1972; Larsen *et al.*, 1972), of ECG patterns (Albin, 1975), of hypertension (Fordon and Fu, 1976), of abdominal diseases (Bezdek and Castelaz, Reference from IV.6), of chromosomes (Lee, 1975), and of leukocytes (E. T. Lee, Reference from IV.6, 1973). Models of neurons based on fuzzy automata are described in Lee and Lee (1974) and Butnariu (1977). Sanchez (References from II.3, 1977a, c) has studied the representation of medical knowledge by means of fuzzy relations (see IV.7.B) for the purpose of automated diagnosis. An application to diagnosis in thyroid pathology can be found in Sanchez and Sambuc (Reference from II.1). Wechsler (Reference from IV.2) has described a medical expert system based on fuzzy concepts. Lastly, Malvache and Vidal (1974) have developed a fuzzy model of visual perception.

D. CONTROL

Applications of fuzzy sets to the linguistic control of mechanical systems are rather numerous. The reader is referred to the corresponding chapter (IV.4).

E. APPLIED OPERATIONS RESEARCH

Many case studies in operations research have been realized using fuzzy approaches. The distribution process of the customers of a given service within a whole system of service centers has been studied by Carlucci and

Donati (1977); a fuzzy mathematical model is proposed to predict this distribution. A methodology based on fuzzy sets for transportation network planning has been applied to the design of a bus network in a town (Dubois, 1977, 1978); more specifically, fuzzy sets are used in forecasting users' trips; users are assigned to paths calculated by means of a fuzzy Floyd's algorithm; the networks are evaluated through fuzzy criteria aggregation. A problem of optimal assignment of employees to work places, where data and constraints are verbally defined, has been studied by Kacprzyk (1976). A fuzzy logic controller of traffic in a single intersection of two one-way streets has been implemented by Pappis and Mamdani (1977). Prade (1977, 1979) deals with a real scheduling problem where the duration of the tasks and the availability of resources are incompletely specified—a fuzzy PERT method and other fuzzy tools are used. Sommer and Pollatschek (1976) have applied fuzzy linear programming to an air pollution regulation problem. Numerous practical production management problems with fuzzy features are described in Pun (1977). Ben Salem (1976) has developed a fuzzy multicriteria automatic decision-making procedure to determine the sequencing of operations accomplished by a machine tool. A. Jones (1974) models a computerized education system by specifying fuzzy relations among sets of media, objectives, and teaching modes. Other references can be found in the appended bibliography.

F. ECONOMICS AND GEOGRAPHY

Blin *et al.* (1973) and Hatten *et al.* (Reference from IV.3) have approached the problem of consumer choice in microeconomics using fuzzy relation or fuzzy automata. On a mathematical level S. S. L. Chang (Reference from III.2) has applied fuzzy set theory to economic modeling, economic forecasting, and economic policy. On a philosophical level rationales and discussions in favor of fuzzy approaches to economics and behavioral geography can be found in Ponsard (1975a) and Gale (1972), respectively. Ponsard (1977a, b) introduces Φ-fuzzy relations in central place theory in order to explain the hierarchical organization of an economic area. Deloche (1975) has proposed a taxonomic method based on fuzzy relations to determine boundaries of economic subregions. Fustier (1975) models the attractiveness of shopping centers, using the notion of fuzzy economical subzones.

G. SOCIOLOGY

Zadeh (1973) has suggested modeling human behavior (individual or group behavior) as a fuzzy system. It is assumed that basic system-theoretic concepts such as control, reinforcement, feedback, goal, constraints decisions, strategies, adaptation, and environment remain central to the discussion of human behavior. A similar attempt is made in Kaufmann (1977). Using verbal models, Wenstøp (Reference from IV.2, 1976a) is able to take into account human factors in the dynamic representation of organizations. Dimitrov (1977), Dimitrov and Cuntchev (1977) model the understanding of fuzzy imperatives by individuals or groups.

H. PSYCHOLOGY

Experimental verifications of the psycholinguistic reality of fuzzy sets and their operators are reported at length in Kochen (1975), Kochen and Badre (Reference from IV.1), Hersh and Caramazza (Reference from IV.1), Dreyfuss-Raimi et al. (1975), Oden (Reference from IV.1, 1977a, b), Oden and Anderson (1974), and MacVicar-Whelan (Reference from IV.1, 1978). An experimental study has been carried out by Kokawa et al. (1975a, b, 1977a, b). It deals mainly with memorizing, forgetting, and inference processes and with the effect of hints on subjective decisions.

I. LINGUISTICS

Flou sets were initiated by Gentilhomme (1968) in the framework of linguistics. Since then, other works have been published in fuzzy linguistics, as shown in the appended bibliography. For instance, Rieger (1974, 1976) has proposed a fuzzy-set approach to the textual analysis of eighteenth century German student lyric poetry.

Very different, although related, is the vast attempt, carried out by L. A. Zadeh, to model semantic aspects of natural language. His theory has been presented at length in III.1.E and IV.2.B. Lakoff's (1973) paper is a linguist's commentary on Zadeh's ideas about modeling hedges.

J. SEMIOTICS

Among the very few works that use fuzzy concepts in semiotics are those of Vaina. In her thesis (Vaina, 1976) a fuzzy reading of a short story by M. Eliade is presented: emergence and disappearance of themes and articulation of episodes are modeled using fuzzy sets. Vaina (1977) also outlines a semiotic approach to the problem of the coherence of a text, based on fuzzy topology. Also in Vaina (1978) a model of a relation of "with" between several people is given. The way people apprehend one another's behavior is viewed as a fuzzy multicriteria decision-making process, which induces a proximity relation between individuals. This model also involves concepts from modal logic.

Nowakowska (1976, 1978) describes a formal language of actions for dialogue purpose. The semantics of dialogues are modeled in the framework of fuzzy set theory.

K. OTHER TOPICS

Lastly, some particular topics are considered.

Damage assessment of structures   An original attempt that deserves mention is that of Blockley (1975, 1978), who analyzes human factors in the failure of mechanical structures. Subjectively assessed parameters make possible modification of the evaluation of the probability of failure.

Aid to creativity   "Aid to creativity" consists here in using a computer to generate, enumeratively, possibly unexpected solutions of a problem. Such solutions are constructed by assembling components picked out of different sets, each made of homologous elements. These sets are called "formational sets"; they may be fuzzy, and fuzzy relations can be defined between them in order to valuate the possibility of associating elements from different formational sets (see Cools and Peteau, 1974; Kaufmann *et al.*, 1973; Kaufmann, 1977).

Analysis of scientific literature   Some empirical analyses of the system of scientific literature by fuzzy sets are provided in Allen (1973) and Jones (1976).

REFERENCES

Numerical Calculus

Pavlidis, T., and Chang, R. L. P. (1977). Application of fuzzy sets in curve fitting. *Proc. IEEE Conf. Decision Control, New Orleans* pp. 1396–1400. [Also in *Int. J. Fuzzy Sets Syst.* **2**, No. 1, 67–74 (1979).]

Quantum Mechanics

Ali, S. T., and Doebner, H. (1976). On the equivalence of non-relativistic quantum mechanics based upon sharp and fuzzy measurements. *J. Math. Phys.* **17**, 1105–1111.

Ali, S. T., and Doebner, H. (1977). Systems of imprimitivity and representation of quantum mechanics on fuzzy phase spaces. *J. Math. Phys.* **18**, No. 2.

Almog, J. (1978a). Is quantum mechanics a fuzzy system? The vindication of fuzzy logic in microphysics (to appear). Hebrew University, Jerusalem.

Almog, J. (1978b). Semantic considerations on modal counterfactual logic with corollaries on axiomatizatility and recursiveness in fuzzy systems. Hebrew University, Jerusalem. [*Notre Dame J. Formal Logic* (to appear)].

Giles, R. (1974). A non classical logic for physics. *Stud. Logica* **33**, 397–415.

Prugovečki, E. (1973). A postulational framework for theories of simultaneous measurement of several observables. *Found. Phys.* **3**, 3–18.

Prugovečki, E. (1974). Fuzzy sets in the theory of measurement of incompatible observables. *Found. Phys.* **4**, 9–18.

Prugovečki, E. (1975). Measurement in quantum mechanics as a stochastic process on spaces of fuzzy events. *Found. Phys.* **5**, 557–571.

Prugovečki, E. (1976a). Probability measures on fuzzy events in phase space. *J. Math. Phys.* **17**, 517–523.

Prugovečki, E. (1976b). Localizability of relativistic particles in fuzzy phase space. *J. Phys. A: Math. Gen.* **9**, No. 11, 1851–1859.

Prugovečki, E. (1977). On fuzzy spin spaces. *J. Phys. A: Math. Gen.* **10**, No. 4.

Artificial Intelligence—Robotics

Becker, J. (1973). A structural design process: Philosophy and methodology. Ph.D. Thesis, University of California, Berkeley.

Gershman, A. (1976). "Fuzzy Set Methods for Understanding Vague Hints for Maze Running." M.S. Thesis, Comput. Sci. Dep., Univ. of California, Los Angeles.

Goguen, J. A. (1975). On fuzzy robot planning. *In* "Fuzzy Sets and Their Applications to Cognitive and Decision Processes" (L. A. Zadeh, K. S. Fu, K. Tanaka, and M. Shimura, eds.), pp. 429–447. Academic Press, New York.

Hanakata, K. (1974). A methodology for interactive systems. *In* "Learning Systems and Intelligent Robots" (K. S. Fu and J. T. Tou, eds.), pp. 317–324. Plenum, New York.

Makarovitsch, A. (1976). How to build fuzzy visual symbols. *Comput. Graph. & Arts* **1**, No. 1.

Makarovitsch, A. (1977). Visual fuzziness. *Comput. Graph. & Arts* **2**, 4–7.

Rhodes, M. L., and Klinger, A. (1977a). Modifying graphics images. *In* "Data Structures, Computer Graphics, and Pattern Recognition" (A. Klinger, K. S. Fu, and T. L. Kunii, eds.), pp. 385–411. Academic Press, New York.

Rhodes, M. L., and Klinger, A. (1977b). Conversational text input for modifying graphic facial images. *Int. J. Man-Mach. Stud.* **9**, 653–667.

Saridis, G. N., and Stephanou, H. E. (1977a). Fuzzy decision-making in prosthetic devices. *In* "Fuzzy Automata and Decision Processes" (M. M. Gupta, G. N. Saridis, and B. R. Gaines, eds. pp. 387–402. North-Holland Publ., Amsterdam. (Reference from IV.5.)

Saridis, G. N., and Stephanou, H. E. (1977b). A hierarchical approach to the control of a prosthetic arm. *IEEE Trans. Syst., Man Cybern.* **7**, No. 6, 407–420.

Schek, H. J. (1977). Tolerating fuzziness in keywords by similarity searches. *Kybernetes* **6**, 175–184.

Shaket, E. (1976). "Fuzzy Semantics for Natural-Like Language Defined over a World of Blocks," A.I. Memo No. 4. M.S. Thesis, Comput. Sci. Dep., Univ. of California, Los Angeles. (Reference from IV.2.)

Stephanou, H. E., and Saridis, G. N. (1976). A hierarchically intelligent method for the control of complex systems. *J. Cybern.* **6**, 249–261.

Uhr, L. (1975). Toward integrated cognitive systems, which *must* make fuzzy decisions about fuzzy problems. *In* "Fuzzy Sets and Their Applications to Cognitive and Decision Processes" (L. A. Zadeh, K. S. Fu, K. Tanaka, and M. Shimura, eds.), pp. 353–393. Academic Press, New York.

Uragami, M., Mizumoto, M., and Tanaka, K. (1976). Fuzzy robot controls. *J. Cybern.* **6**, 39–64.

Zadeh, L. A. (1977). "PRUF: A Meaning Representation Language for Natural Languages," Memo ERL M77/61. Univ. of California, Berkeley. (Reference from IV.2.)

Image Processing

Chang, S. K. (1971). Picture-processing grammar and its application. *Inf. Sci.* **3**, 121–148.

Jain, R., and Nagel, H. H. (1977). Analyzing a real-world scene sequence using fuzziness. *Proc. IEEE Conf. Decision Control, New Orleans* pp. 1367–1372.

Nakagawa, Y., and Rosenfeld, A. (1978). A note on the use of local min and max operations in digital picture processing. *IEEE Trans. Syst., Man Cybern.* **8**, No. 8, 632–635.

Rosenfeld, A., Hummel, R. A., and Zucker, S. W. (1976). Scene labeling by relaxation operations. *IEEE Trans. Syst., Man Cybern.* **6**, No. 6, 420–433.

Speech Recognition

Brémont, J. (1975). Contribution à la reconnaissance automatique de la parole par les sous-ensembles flous. Thèse. Université de Nancy, France.

Brémont, J., and Lamotte, M. (1974). Contribution à la reconnaissance automatique de la parole en temps réel par la considération des sous-ensembles flous. *C. R. Acad. Sci.* Paris, July 15th.

Coppo, M., and Saitta, L. (1976). Semantic support for a speech-understanding system, based on fuzzy relations. *Proc. Int. Conf. Cybern. Soc., Washington, D.C.* pp. 520–524.

De Mori, R., and Torasso, P. (1976). Lexical classification in a speech-understanding system using fuzzy relations. *Proc. IEEE Conf. Speech-Process.*, pp. 565–568. (Reference from IV.6.)

Mas, M. T., and Lamotte, M. (1976). Commande vocale de processus industriels. *Proc. Int. Cong. Cybern., 8th* 725–731. Namur, Belgium.

Pal, S. K., and Majumder, D. D. (1977). Fuzzy sets and decision-making approaches in vowel and speaker recognition. *IEEE Trans. Syst., Man Cybern.* **7**, 625–629.

Pal, S. K., and Majumder, D. D. (1978a). Effect of fuzzification and the plosive cognition system. *Int. J. Systems Sci.* **9**, 873–886.

Pal, S. K., and Majumder, D. D. (1978b). On automatic plosive identification using fuzziness in property sets. *IEEE Trans. Syst., Man Cybern.* **8**, 302–308.

Pal, S. K., Datta, A. K., and Majumder, D. D. (1978). Adaptive learning algorithms in classification of fuzzy patterns. An application to vowels in CNC context. *Int. J. Systems Sci.* **9**, 887–897.

Biological and Medical Sciences

Adey, W. R. (1972). Organization of brain tissue: Is the brain a noisy processor? *Int. J. Neurosci.* **3**, 271–284. (Reference from IV.6.)

Albin, M. (1975). "Fuzzy Sets and Their Application to Medical Diagnosis and Pattern Recognition." Ph.D. Thesis, Univ. of California, Berkeley. (Reference from IV.6.)

Butnariu, D. (1977). *L*-fuzzy automata. Description of a neural model. *In* "Modern Trends in Cybernetics and Systems" (J. Rose and C. Bilciu, eds.), Vol. 2, pp. 119–124. Springer-Verlag, Berlin and New York.

Fordon, W. A., and Fu, K. S. (1976). A non-parametric technique for the classification of hypertension. *Proc. IEEE Conf. Decision Control* (Reference from IV.6.)

Kalmanson, D., and Stegall, F. (1973). Recherche cardio-vasculaire et théorie des sous-ensembles flous. *Nouv. Presse Med.* **41**, 2757–2760.

Larsen, L. E., Ruspini, E. H., McNew, J. J., Walter, D. O., and Adey, W. R. (1972). A test of sleep staging systems in the unrestrained chimpanzee. *Brain Res.* **40**, 319–343. (Reference from IV.6.)

Lee, E. T. (1975). Shape-oriented chromosome classification. *IEEE Trans. Syst., Man Cybern.* **5**, 629–632.

Lee, S. C., and Lee, E. T. (1974). Fuzzy sets and neural networks. *J. Cybern.* **4**, No. 2, 83–103.

Malvache, M., and Vidal, P. (1974). "Application des Systèmes Flous à la Modélisation des Phénomènes de Prise de Décision et d'Appréhension des Informations Visuelles chez l'Homme." Final report ATP-CNRS, Université de Lille I.

Applied Operations Research

Ben Salem, M. (1976). "Sur l'Extension de la Théorie des Sous-Ensembles Flous à l'Automatique Industrielle." Ph.D. Thesis, Univ. de Paris VII.

Carlucci, D., and Donati, F. (1977). Fuzzy cluster of demand within a regional service system. *In* "Fuzzy Automata and Decision Processes" (M. M. Gupta, G. N. Saridis, and B. R. Gaines, eds.), pp. 379–385. North-Holland Publ., Amsterdam.

Dubois, D. (1977). Quelques outils méthodologiques pour la conception de réseaux de transports. Thèse Docteur-Ingénieur. Eco. Nat. Sup. Aéronaut. Espace, Toulouse.

Dubois, D. (1978). An application of fuzzy sets theory to bus transportation network modification. *Proc. Joint Automat. Control Conf., Philadelphia.*

Elliott, J. L. (1976). Fuzzy Kiviat graphs. *Proc. Eur. Comput. Congr. (EURO COMP. 76), London.*

Esogbue, A. O. (1977). On the application of fuzzy allocation theory to the modelling of cancer research appropriation process. *In* "Modern Trends in Cybernetics and Systems" (J. Rose and C. Bilciu, eds.), Vol. 2, pp. 183–193. Springer-Verlag, Berlin and New York.

Jones, A. (1974). Towards the right solution. *Int. J. Math. Educ. Sci. Technol.* **5**, 337–357.

Kacprzyk, J. (1976). Fuzzy set-theoretic approach to the optimal assignment of work places. *In* "Large Scale Systems Theory and Applications" (G. Guardabassi and A. Locatelli, eds.), pp. 123–131. Udine.

Kluska-Nawarecka, S., Mysona-Byrska, E., and Nawarecki, E. (1975). Algorithmes de commande pour une certaine classe de problèmes opérationnels construits avec utilisation de la simulation numérique. Rep. Institut de Fonderie, Cracovie, Poland.

Pappis, C. P., and Mamdani, E. H. (1977). A fuzzy logic controller for a traffic junction. *IEEE Trans. Syst., Man Cybern.* **7**, No. 10, 707–717. (Reference from IV.4.)

Prade, H. (1977). Exemple d'approche heuristique, interactive, floue, pour un problème d'ordonnancement. *Congr. AFCET Model. Maitrise Syst., Versailles, Editions Hommes Tech.* **2**, 347–355.

Prade, H. (1979). Using fuzzy set theory in a scheduling problem: A case study. *Int. J. Fuzzy Sets Syst.* **2**, No. 2, 153–165.

Pun, L. (1977). Use of fuzzy formalism in problems with various degrees of subjectivity. *In* "Fuzzy Automata and Decision Processes" (M. M. Gupta, G. N. Saridis, and B. R. Gaines, eds.), pp. 357–378. North-Holland Publ., Amsterdam.

Samoylenko, S. I. (1977). Application of fuzzy heuristic techniques to computer network design. *Proc. Int. Joint Conf. Artif. Intellig., 5th* p. 880, Cambridge, Massachusetts.

Sommer, G., and Pollatschek, M. A. (1976). "A Fuzzy Programming Approach to an Air-Pollution Regulation Problem," Working Paper No. 76-01. Inst. Wirtschaftswiss. R.W.T.H., Aachen, West Germany. (Reference from III.4.)

Stoica, M., and Scarlat, E. (1975). Fuzzy algorithm in economic systems. *Econ. Comput. Econ. Cybern. Stud. Res.* No. 3, 239–247.

Stoica, M., and Scarlat, E. (1977). Some fuzzy concepts in the management of production systems. *In* "Modern Trends in Cybernetics and Systems" (J. Rose and C. Bilciu, eds.), Vol. 2, pp. 175–181. Springer-Verlag, Berlin and New York.

Van Velthoven, G. (1977). Fuzzy models in personnel management. *In* "Modern Trends in Cybernetics and Systems" (J. Rose and C. Bilciu, eds.), Vol. 2, pp. 131–161. Springer-Verlag, Berlin and New York.

Economics and Geography

Blin, J. M., Fu, K. S., Whinston, A. B., and Moberg, K. B. (1973). Pattern recognition in micro-economics. *J. Cybern.* **3**, No. 4, 17–27. (Reference from IV.3.)

Deloche, R. (1975). "Théorie des Sous-Ensembles Flous et Classification en Analyse Économique Spatiale," Working Pap. IME, No. 11. Fac. Sci. Econ. Gestion, Dijon.

Féron, R. (1976). Economie d'échange aléatoire floue. *C. R. Acad. Sci.* Paris, Sér. A, **282**, 1379–1382.

Fustier, B. (1975). "L'Attraction des Points de Vente dans des Espaces Précis et Imprécis," Working Pap. IME, No. 10. Fac. Sci. Econ. Gestion, Dijon.

Gale, S. (1972). Inexactness, fuzzy sets, and the foundations of behavioral geography. *Geogr. Anal.* **4**, 337–349.

Gottinger, H. W. (1973). Competitive processes: Application to urban structures. *Cybernetica* **16**, No. 3, 177–197.

Massonie, J. (1976). L'utilisation des sous-ensembles flous en géographie. *Cahiers de Géographie de l'Université de Besançon*, France.

Ponsard, C. (1975a). L'imprécision et son traitement en analyse économique. *Rev. Econ. Polit.* No. 1, 17–37.

Ponsard, C. (1975b). Contribution à une théorie des espaces économiques imprécis. *Publ. Econ.* **8**, No. 2, 1–43. (Reference from II.1.)

Ponsard, C. (1976). "Esquisse de Simulation d'une Économie Régionale: L'Apport de la Théorie des Systèmes Flous," Working Pap. IME, No. 18. Fac. Sci. Econ. Gestion, Dijon.

Ponsard, C. (1977a). Hiérarchie des places centrales et graphes Φ-flous. *Environ. Plann. A* **9**, 1233–1252.

Ponsard, C. (1977b). "La Région en Analyse Spatiale," Working Pap. IME, No. 21. Fac. Sci. Econ. Gestion, Dijon.

Taranu, C. (1977). The economic efficiency—A fuzzy concept. *In* "Modern Trends in Cybernetics and Systems" (J. Rose and C. Bilciu, eds.), Vol. 2, pp. 163–173. Springer-Verlag, Berlin and New York.

Sociology

Dimitrov, V. (1977). Social choice and self-organization under fuzzy management. *Kybernetes* **6**, 153–156.

Dimitrov, V., and Cuntchev, O. (1977). Efficient governing of humanistic systems by fuzzy instructions. *In* "Modern Trends in Cybernetics and Systems" (J. Rose and C. Bilciu, eds.), Vol. 2, pp. 125–130. Springer-Verlag, Berlin & New York.

Itzinger, O. (1974). Aspects of axiomatization of behaviour: Towards an application of Rasch's measurement model to fuzzy logic. *COMPSTAT 74, Proc. Symp. Comput. Stat., Vienna* pp. 173–182. (G. Bruckman, F. Ferschl, and L. Schmetterer, eds.), Physics-Verlag.

Karsky, M., and Adamo, J. M. (1977). Application de la dynamique des systèmes et de la logique floue à la modélisation d'un problème de postes en raffinerie. *Cong. AFCET Modèl. & Maitrise Syst.* Versailles, Editions Hommes & Techniques.

Kaufmann, A. (1977). Théorie de l'opérateur humain. *In* "Introduction à la Theorie des Sous-Ensembles Flous. Vol. 4: Compléments et Nouvelles Applications." Masson, Paris. (Reference from I.)

Menges, G., and Skala, H. J. (1976). On the problem of vagueness in the social sciences. *In* "Information, Inference and Decision" (G. Menges, ed.), pp. 51–61. Reidel Publ., Dordrecht, Netherlands.

Zadeh, L. A. (1973). A system-theoretic view of behaviour modification. *In* "Beyond the Punitive Society" (H. Wheeler, ed.), pp. 160–169. Freeman, San Francisco, California.

Psychology

Dreyfuss-Raimi, G., Kochen, M., Robinson, J., and Badre, A. N. (1975). On the psycholinguistic reality of fuzzy sets: Effect of context and set. *In* "Functionalism" (R. E. Grossman, L. J. San., and T. J. Vance, eds.), pp. 135–149. Univ. of Chicago Press, Chicago, Illinois.

Kay, and Mc Daniel (1975). "Color Categories as Fuzzy Sets, " Working Pap. No. 44, Univ. of California, Berkeley.

Klabbers, J. H. G. (1975). General system theory and social systems: A methodology for the social sciences. *Ned. Tijdschr. Psychol.* **30**, 493–514.

Kochen, M. (1975). Applications of fuzzy sets in psychology. *In* "Fuzzy Sets and Their Applications to Cognitive and Decision Processes" (L. A. Zadeh, K. S. Fu, K. Tanaka, and M. Shimura, eds.), pp. 395–408. Academic Press, New York.

Kokawa, M., Nakamura, K., and Oda, M. (1975a). Experimental approach to fuzzy simulation of memorizing, forgetting, and inference process. *In* "Fuzzy Sets and Their Applications to Cognitive and Decision Processes" (L. A. Zadeh, K. S. Fu, K. Tanaka, and M. Shimura, eds.), pp. 409–428. Academic Press, New York.

Kokawa, M., Nakamura, K., and Oda, M. (1975b). Hint effect and a jump of logic in a decision process. *Syst.—Comput.—Control* **6**, No. 3, 18–24.

Kokawa, M., Oda, M., and Nakamura, K. (1977a). Fuzzy-theoretical dimensionality reduction method of multidimensional quantity. *In* "Fuzzy Automata and Decision Processes" (M. M. Gupta, G. N. Saridis, and B. R. Gaines, eds.) pp. 235–249. North-Holland Publ., Amsterdam.

Kokawa, M., Nakamura, K., and Oda, M. (1977b). Fuzzy-theoretic and concept-formational approaches to inference and hint effect experiments in human decision processes. *Proc. IEEE Conf. Decision Control, New Orleans* pp. 1330–1337. [Also in *Int. J. Fuzzy Sets Syst.* **2**, No. 1, 25–36 (1979).]

Oden, G. C., and Anderson, N. H. (1974). Integration of semantic constraints. *J. Verbal Learn. Verbal Behav.* **13**, 138–148.

Linguistics

Ballmer, T. T. (1976). "Fuzzy Punctuation or the Continuum of Grammaticality, " Memo ERL-M590. Univ. of California, Berkeley.

Gentilhomme, Y. (1968). Les ensembles flous en linguistique. *Cah. Linguist. Theor. Appl. (Bucharest)* **5**, 47–63. (Reference from II.1.)

Haroche, C. (1975). Grammaire, implicite et ambiguité (À propos des fondements de l'ambiguité inhérente au discours). *Found. Language* **13**, 215–236.

Joyce, J. (1976). Fuzzy sets and the study of linguistics. *Pac. Coast Philol.* **11**, 39–42.

Kaufmann, A. (1975). "Introduction à la Théorie des Sous-Ensembles Flous. Vol. 2: Applications à la Linguistique, à la Logique et à la Sémantique." Masson, Paris. (Reference from I, 1975a.)

Labov, W. (1973). The boundaries of words and their meanings. *In* "New Ways of Analyzing Variations in English" (C. J. N. Bailey and R. W. Shuy, eds.), Vol. 1, Georgetown Univ. Press, Washington, D.C.

Lakoff, G. (1973). Hedges: A study in meaning criteria and the logic of fuzzy concepts. *J. Philos. Logic* **2**, 458–508. (Reference from IV.2.)

Morgan, C., and Pelletier, F. (1977). Some notes concerning fuzzy logics. *Linguist. Philos.* **1**, No. 1.

Reddy, D. (1972). "Reference and Metaphor in Human Language." Ph.D. Thesis, Univ. of Chicago, Chicago, Illinois.

Rieger, B. (1974). Eine "tolerante" Lexikonstruktur. Zur Abbildung natürlich-sprachlicher Bedeutung auf unscharfe—Mengen in Toleranzraumen. *Z. Literatur Ling.* **16**, 31–47.

Rieger, B. (1976). Fuzzy structural semantics. On a generative model of vague natural language meaning. *Eur. Meet. Cybern. Syst. Res., 3rd, Vienna.*

Semiotics

Nowakowska, M. (1976). Towards a formal theory of dialogues. *Semiotica* **17**, 291–313.

Nowakowska, M. (1978). Formal theory of group actions and its applications. *Philosophica* **21**, 3–32.

Vaina-Pusca, L. (1976). "Lecture Logico-Mathématique de la Narration—Modèle Sémiotique." Inst. Rech. Ethnol. Dialectales C.C.E.S., Univ. Paris IV, Paris.

Vaina, L. (1977). "Fuzzy Sets in the Semiotic of Text." Unpublished manuscript. M.I.T. Cambridge, Ma.

Vaina, L. (1978). Semiotics of "with." *Versus Quad. Studi Semiot.* **19**.

Damage Assessment of Structures

Blockley, D. I. (1975). Predicting the likelihood of structural accidents. *Proc. Inst. Civ. Eng.* **59**, Part 2, 659–668.

Blockley, D. I. (1978). Analysis of subjective assessments of structural failures. *Int. J. Man-Mach. Stud.* **10**, 185–195.

Aid to Creativity

Cools, M., and Peteau, M. (1974). Un programme de stimulation inventive: STIM 5. *RAIRO—Oper. Res.* **8**, No. 3, 5–19.

Kaufmann, A. (1977). Procédures de stimulation inventive. *In* "Introduction à la Théorie des Sous-Ensembles Flous. Vol. 4: Compléments et Nouvelles Applications." Masson, Paris. (Reference from I.)

Kaufmann, A., Cools, M., and Dubois, T. (1973). "Stimulation Inventive dans un Dialogue Homme-Machine Utilisant la Méthode des Morphologies et la Théorie des Sous-Ensembles Flous," IMAGO Discuss. Pap. No. 6, Univ. Cathol. de Louvain, Louvain, Belgium.

Analysis of Scientific Literature

Allen, A. D. (1973). A method for evaluating technical journals on the basis of published comments through fuzzy implications: A survey of the major IEEE Transactions. *IEEE Trans. Syst., Man Cybern.* 3, 422–425.

Jones, W. T. (1976). A fuzzy set characterization of interaction in scientific research. *J. Am. Soc. Inf. Sci.*, Sept./Oct., 307–310.

CONCLUSION

Obviously, a basic concept in fuzzy set theory is the idea of a set without sharp boundaries. Another very important concept is that of fuzzy correspondences (represented by fuzzy relations). The sup–min composition allows building images of fuzzy sets through fuzzy correspondences. When the correspondences are just ordinary functions, sup–min composition reduces to the extension principle. Moreover, it should be noted that there is often no canonical way to extend classical concepts into fuzzy ones. For instance, operators other than "min" have been pointed out in this book and require further investigations.

Some applications of fuzzy set theory, such as switching logic or clustering analysis for example, turn only the first idea to account. They may appear more multivalent than fuzzy, in the sense that only grades of membership (rather than fuzzy sets as a whole) are manipulated. Similarly, multivalent logics only underlie fuzzy set theories without providing a sufficient framework for approximate reasoning.

Fuzzy sets allow information to be approximately summarized in a humanlike fashion, or modeling ill-known data. Fuzzy-set theory provides the right tool for the manipulation of this information, i.e., for approximate reasoning or for a generalized tolerance analysis. From this point of view the specification of a fuzzy system consists in a linguistic description of its behavior and/or assignment of fuzzy parameters to an ordinary mathematical model. A fuzzy system has a "possibilistic" interpretation, whereas a stochastic system has a probabilistic one. Fuzziness may lie in the system itself or in its model. It is mainly a matter of human perception.

A great amount of work has been already accomplished. However, the ability to apply fuzzy concepts to practical problems requires a somewhat deeper understanding of the specificity of Zadeh's theory.

REFERENCES IN THE
NONFUZZY LITERATURE

Arbib, M. A., and Manes, E. G. (1975). "Arrows, Structures, and Functors: The Categorical Imperative." Academic Press, New York.

Arbib, M. A., and Zeiger, H. P. (1969). On the relevance of abstract algebra to control theory. *Automatica* **5**, 589–606.

Aumann, R. J. (1965). Integrals of set-valued functions. *J. Math. Anal. Appl.* **12**, 1–12.

Ball, G., and Hall, D. (1967). A clustering technique for summarizing multivariate data. *Behav. Sci.* **12**, 153–155.

Bhattacharyya, A. (1943). On a measure of divergence between two statistical populations defined by their probability distributions. *Bull. Calcutta Math. Soc.* **35**, 99–109.

Birkhoff, G. (1948). "Lattice Theory," Colloq. Publ., Vol. 25. Am. Math. Soc., New York.

Black, M. (1937). Vagueness: An exercise in logical analysis. *Philos. Sci.* **4**, 427–455.

Bobrow, L. S., and Arbib, M. A. (1974). Machines in a category. *In* "Discrete Mathematics," pp. 644–707. Saunders, Philadelphia, Pennsylvania.

Brainard, W. S. (1969). Tree generating regular system. *Inf. Control* **14**, 217–231.

Childs, D. L. (1968). Feasibility of a set-theoretic data structure—A general structure based on a reconstituted definition of relation. *Inf. Process.* **68**, 420–430.

Chomsky, N., and Schützenberger, M. P. (1970). The algebraic theory of context-free languages. *In* "Computer Programming and Formal Systems" (P. Braffort and D. Hirschberg, eds.), pp. 118–161. North-Holland Publ., Amsterdam.

Choquet, G. (1953). Theory of capacities. *Ann. Inst. Fourier* **5**, 131–295.

Chou, S. M., and Fu, K. S. (1975). "Transition Networks for Pattern Recognition," TR-EE 75-39. Purdue Univ., Lafayette, Indiana.

Cohen, L. J. (1973). A note on inductive logic. *J. Philos.* **70**, No. 2, 27–40.

Dempster, A. P. (1967). Upper and lower probabilities induced by multivalued mappings. *Ann. Math. Statist.* **38**, 325–339.

Dixon, D. J. (1977). Catastrophe theory applied to ecological systems. *Simulation* **29**, No. 1, 1–15.

Duda, R., and Hart, P. (1973). "Pattern Classification and Scene Analysis." Wiley (Interscience), New York.

Floyd, R. W. (1962). Shortest paths. *Commun. ACM* **5**, 345.

Fu, K. S. (1974). "Syntactic Methods in Pattern Recognition." Academic Press, New York.

Fu, K. S. (1976). Learning with stochastic automata and stochastic languages. *In* "Computer Oriented Learning Process" (J. C. Simon, ed.), pp. 69–107. Noordhoff, Leyden.

Fu, K. S., and Booth, T. L. (1975). Grammatical inference: Introduction and survey. *IEEE Trans. Syst., Man Cybern.* **5**, Part 1, 95–111; Part 2, 409–423.

Fu, K. S., and Huang, T. (1972). Stochastic grammars and languages. *Int. J. Comput. Inf. Sci.* **1**, No. 2, 135–170.

Furutani, N. (1976a). A new approach to traffic behaviour. I. Modelling of "following-defence" behaviour. *Int. J. Man-Mach. Stud.* **8**, 597–615.

Furutani, N. (1976b). A new approach to traffic behaviour. II. Individual car and traffic flow. *Int. J. Man-Mach. Stud.* **8**, 731–742.

Furutani, N. (1977). A new approach to traffic behaviour. III. Steering behaviour and the butterfly catastrophe. *Int. J. Man-Mach. Stud.* **9**, 233–254.

Halkin, H. (1964). Topological aspects of optimal control of dynamical polysystems. *Contrib. Differential Equations* **3**, 377–385.

Hansen, E. (1965). Interval arithmetic in matrix computation. Part 1. *SIAM (Soc. Ind. Appl. Math.) J. Numer. Anal.* **2**, 308–320.

Hansen, E. (1969). "Topics in Interval Analysis." Oxford Univ. Press, London and New York.

Hansen, E., and Smith, R. (1967). Interval arithmetic in matrix computation. Part 2. *SIAM (Soc. Ind. Appl. Math.) J. Numer. Anal.* **4**, 1–9.

Holt, A. W. (1971). Introduction to occurrence systems. *In* "Associative Information Techniques" (E. L. Jacks, ed.), pp. 175–203. Am. Elsevier, New York.

Hopcroft, J. E., and Ullmann, J. D. (1969). "Formal Languages and Their Relation to Automata." Addison-Wesley, Reading, Massachusetts.

Hughes, G. E., and Cresswell, M. J. (1972). "An Introduction to Modal Logic." Methuen, London.

Kubinski, T. (1960). An attempt to bring logic nearer to colloquial language. *Stud. Logica* **10**, 61–72.

Lewin, K. (1936). "Principles of Topological Psychology." McGraw-Hill, New York.

Li, T. J., and Fu, K. S. (1969). On stochastic automata and languages. *Inf. Sci.* **1**, 403–419.

Ling, C.-H. (1965). Representation of associative functions. *Publ. Math. Debrecen* **12**, 189–212.

Lopez De Mantaras, R., and Aguilar-Martin, J. (1978). Classification par auto-apprentissage à l'aide de filtres numériques adaptatifs. *Congr. AFCET/IRIA Reconnaissance Formes Trait. Images.*

Luce, R. D., and Raiffa, H. (1957). "Games and Decisions," Wiley, New York.

McCluskey, E. J. (1965). "Introduction to the Theory of Switching Circuits." McGraw-Hill, New York.

McLaren, R. W. (1966). A stochastic automaton model for synthesis of learning systems. *IEEE Trans. Syst., Sci. Cybern.* **2**, 109.

Minsky, M., and Papert, S. (1969). "Perceptrons." MIT Press, Cambridge, Massachusetts.

Moore, R. E. (1966). "Interval Analysis." Prentice-Hall, Englewood Cliffs, New Jersey.

Murray, F. B., ed. (1972). "Critical Features of Piaget's Theory of the Development of Thought." Univ. of Delaware Press, Newark.

Nutt, G. J., and Noe, J. D. (1973). Macro *E*-nets for representation of parallel systems. *IEEE Trans. Comput.* **22**, No. 8, 718–729.

Prim, R. C. (1957). Shortest connection matrix network and some generalizations. *Bell Syst. Tech. J.* **36**, 1389–1401.

Rabin, M. O. (1963). Probabilistic automata. *Inf. Control* **6**, 230–245.

Rescher, N. (1969). "Many-Valued Logic." McGraw-Hill, New York.

Robinson, J. A. (1965). A machine-oriented logic based on the resolution principle. *J. Assoc. Comput. Mach.* **12**, No. 1, 23–41.

Rosenblatt, F. (1961). "Principles of Neurodynamics: Perception and the Theory of Brain Mechanisms." Spartan Books, Washington, D.C.

Roy, B. (1969–1970). "Algèbre Moderne et Théorie des Graphes," Vols. 1 and 2. Dunod, Paris.

Schweizer, B. and Sklar, A. (1963). Associative functions and abstract semi-groups. *Publ. Math. Debrecen* **10**, 69–81.

Shackle, G. L. S. (1961). Decision, Order, and Time in Human Affairs. Cambridge University Press, London and New York, pp. 67–103.

Shafer, G. (1976). "Mathematical Theory of Evidence." Princeton Univ. Press, Princeton, New Jersey.

Shortliffe, E. H., and Buchanan, B. G. (1975). A model of inexact reasoning in medicine. *Math. Biosci.* **23**, 351–379.

Simon, H. A. (1967). A behavioral model of rational choice. *In* "Organizational Decision-Making" (M. Alexis and C. Z. Wilson, eds.), pp. 174–184. Prentice-Hall, Englewood Cliffs, New Jersey.

Thom, R. (1973). "Stabilité Structurelle et Morphogenèse." Benjamin, New York.

Thom, R. (1974). "Modèles Mathématiques de la Morphogenèse," 10.18 Paperback No. 887, Union Générales d'Editions, Paris.

Varshavskii, V. I., and Vorontsova, I. P. (1963). On the behaviour of stochastic automata with variable structure. *Autom. Remote Control (USSR)* **24**, 327.

Warfield, J. N. (1974a). Developing subsystem matrices in structural modeling. *IEEE Trans. Syst., Man Cybern.* **4**, No. 1, 74–80.

Warfield, J. N. (1974b). Developing interconnection matrices in structural modeling. *IEEE Trans. Syst., Man Cybern.* **4**, No. 1, 81–87.

Winston, P. H. (1977). "Artificial Intelligence." Addison-Wesley, Reading, Massachusetts.

Woods, W. A. (1970). Transition network grammars for natural language analysis. *Commun. ACM* **13**, 591–602.

Zadeh, L. A. (1969). The concepts of system, aggregate, and state in system theory. *In* "System Theory" (L. A. Zadeh and E. Polak, eds.), Inter-University Electronic Series, Vol. 8, pp. 3–42. McGraw-Hill, New York.

LIST OF THE MOST COMMONLY USED SYMBOLS

GENERAL MATHEMATICAL SYMBOLS

$=$	equal to
$\equiv$	equal to (by definition)
$<$	less than; $>$, greater than
$\leqslant$	less than or equal to; $\geqslant$, greater than or equal to
$\forall$	For all
$\exists$	there exists at least one
$\exists!$	there exists one and only one
$\in$	belongs to
$\{x, \dots\}$	set of elements x
iff	if and only if
$\mathcal{P}(X)$	set of subsets of X
Y^X	set of functions from X to Y
$\mathbb{N}$	set of natural integers
$\mathbb{R}$	set of real numbers
$[a, b]$	closed real interval
$[a, b[$	real interval closed in a, open in b
$[a, +\infty)$	set of real numbers greater than or equal to a
e^x	exponential of x

ln	Napierian logarithm
$\lvert a \rvert$	absolute value of the number a
A^t	transpose of a matrix A
$\mathrm{tr}(A)$	sum of diagonal terms of a matrix A
$\sum_{i=1}^n$	sum of n numbers indexed by i; $\sum_i$, sum of numbers indexed by i
$\prod_{i=1}^n$	product of n numbers indexed by i
$\int_a^b$	integral over an interval $[a,b]$
$0, 1$	least and greatest elements of a lattice (respectively)
$\varnothing$	empty set

LOGICAL SYMBOLS

$v(P)$	truth value of proposition P
$P \wedge Q$	conjunction of P and Q, $v(P \wedge Q) = \min(v(P), v(Q))$
$P \vee Q$	disjunction of P and Q, $v(P \vee Q) = \max(v(P), v(Q))$
$\neg P$	negation of P, $v(\neg P) = 1 - v(P)$
$P \wedge Q$	conjunction of P and Q, $v(P \wedge Q) = \max(0, v(P) + v(Q) - 1)$
$P \vee Q$	disjunction of P and Q, $v(P \vee Q) = \min(1, v(P) + v(Q))$
$\mapsto$	any implication
$\rightarrow$	implication, $v(P \rightarrow Q) = \max(1 - v(P), v(Q))$
$\Rightarrow$	implication, $v(P \Rightarrow Q) = \min(1, 1 - v(P) + v(Q))$
$\twoheadrightarrow$	implication, $v(P \twoheadrightarrow Q) = 1 - v(P) + v(P) \cdot v(Q)$
α	$a \, \alpha \, b = 1$ iff $a \leqslant b$; $a \, \alpha \, b = b$ iff $b < a$

FUZZY SETS SYMBOLS

μ_A	membership function of a fuzzy set A on a universe U
$\int_U \mu_A(u)/u$	Zadeh's notation of a fuzzy set A on a universe U
$\sum \mu_A(u)/u$	Zadeh's notation of a fuzzy set A on a discrete universe U
$\cap$	intersection of fuzzy sets

$\cup$	union of fuzzy sets
$\bar{A}$	complement of the fuzzy set A
$\subseteq$	inclusion of fuzzy sets
$\prec$	weak inclusion of fuzzy sets
$\cap$	bold intersection of fuzzy sets
$\cup$	bold union of fuzzy sets
$\hat{+}$	probabilistic sum of fuzzy sets
$\cdot$	product of fuzzy sets
$\sqcap$	intersection of type two fuzzy sets, $\mu_{A\sqcap B}(x) = \min(\mu_A(x), \mu_B(x))$
$\sqcup$	union of type two fuzzy sets, $\mu_{A\sqcup B}(x) = \max(\mu_A(x), \mu_B(x))$
$\overset{\sqcap}{A}$	complement of a type two fuzzy set, $\mu_{\overset{\sqcap}{A}}(x) = 1 \ominus \mu_A(x)$
$\sqsubseteq$	inclusion of type two fuzzy sets
$|A|$	cardinality of a fuzzy set A
$\mathrm{supp}\,A$	support of a fuzzy set A
$\mathrm{hgt}(A)$	height of a fuzzy set A
$\tilde{\mathscr{P}}(X)$	set of fuzzy sets on X
$\tilde{\mathscr{P}}_n(X)$	set of type n fuzzy sets on X
$\tilde{\mathscr{P}}^l(X)$	set of level l fuzzy sets on X
$\mathscr{P}_L(X)$	set of L-fuzzy sets on X
$A \times B$	cartesian product of the fuzzy sets A and B

FUZZY RELATION SYMBOLS

$\mathrm{dom}(R)$	domain of the fuzzy relation R
$\mathrm{ran}(R)$	range of the fuzzy relation R
$c(R)$	cylindrical extension of R
$\mathrm{proj}[R; U]$	projection of R on the universe U
$R \circ Q$	sup–min composition of the fuzzy relations R and Q
$R \overline{\circ} Q$	inf–max composition of the fuzzy relations R and Q
$R @ Q$	sup–α composition of fuzzy relations R and Q
$\hat{R}$	transitive closure of a fuzzy relation R

EXTENDED OPERATIONS ON FUZZY SETS ON ℝ

$\oplus$	addition
$\ominus$	subtraction
$\odot$	multiplication
$\odot\!\!\!\cdot$	division
$\widetilde{\max}$	maximum
$\widetilde{\min}$	minimum

OTHER SYMBOLS

$\int\!\!\!\!\!\!-$	Sugeno's integral
Π	possibility measure
P	probability measure
π	possibility distribution

AUTHOR INDEX

Numbers in italics refer to the pages on which the complete references are listed.

SUBJECT INDEX

Printed and bound by CPI Group (UK) Ltd, Croydon, CR0 4YY

03/10/2024

01040414-0007